AF361983

Aerial Robotics for Inspection and Maintenance

Aerial Robotics for Inspection and Maintenance

Editors

Alejandro Suarez
Jonathan Cacace
Matko Orsag

MDPI • Basel • Beijing • Wuhan • Barcelona • Belgrade • Manchester • Tokyo • Cluj • Tianjin

Editors
Alejandro Suarez
University of Seville
Spain

Jonathan Cacace
University of Naples Federico
II
Italy

Matko Orsag
University of Zagreb
Croatia

Editorial Office
MDPI
St. Alban-Anlage 66
4052 Basel, Switzerland

This is a reprint of articles from the Special Issue published online in the open access journal *Applied Sciences* (ISSN 2076-3417) (available at: https://www.mdpi.com/journal/applsci/special_issues/Aerial_Robotics_for_Inspection_and_Maintenance).

For citation purposes, cite each article independently as indicated on the article page online and as indicated below:

LastName, A.A.; LastName, B.B.; LastName, C.C. Article Title. *Journal Name* **Year**, *Volume Number*, Page Range.

ISBN 978-3-0365-4637-7 (Hbk)
ISBN 978-3-0365-4638-4 (PDF)

Cover image courtesy of IEEE Access

Contents

About the Editors

Alejandro Suarez

Alejandro Suarez (Dr.) received his degree in telecommunication engineering and M.Sc. degree in automation and robotics from the University of Seville, Spain, in 2012 and 2014, respectively, and his Ph.D. degree in robotics in 2019. Since 2012, he has been with the Robotics, Vision and Control Group at the University of Seville, working on several European projects, including FP7 EC-SAFEMOBIL, H2020 AEROARMS, and H2020 HYFLIERS. For three months he stayed at the Robotics and Mechatronics Institute, DLR, Oberpfaffenhofen. He is currently working on the AERIAL-CORE H2020 Project, the ERC Advanced GRIFFIN Project, and other projects with companies involving robotic manipulation. He is the author of 35 articles in international conferences and journals related to aerial robotics. His research interests include aerial robotic manipulation, humanoid robots, and the development of anthropomorphic, compliant, and lightweight robotic arms (LiCAS).

Jonathan Cacace

Jonathan Cacace (Dr.) was born in Naples, Italy, on December 13, 1987. He received his master's degree (magna cum laude) in computer science from the University of Naples Federico II in 2012 and his Ph.D. degree in robotics in 2016 from the same institution. Currently, he is working as an Assistant Professor at the PRISMA Lab (Projects of Robotics for Industry and Services, Mechatronics and Automation Laboratory) at the University of Naples Federico II, where he is involved in several research projects in the fields of human–robot interaction in Industry 4.0, the autonomous control of UAVs for inspection and maintenance, and robotic manipulation. He serves as an Associate Editor for the IEEE ICRA conference and IEEE's Robotics and Autonomous Letters.

Matko Orsag

Matko Orsag (Dr.) is an Associate Professor at the University of Zagreb Faculty of Electrical Engineering and Computing. He is and has been involved in various projects in the fields of robotics, control, and automation. In 2011/2012 he worked as a visiting researcher at Drexel University, Philadelphia, USA as a recipient of the Fulbright exchange grant. He has coauthored over 60 scientific and professional journal and conference papers, a book chapter, and a monography in aerial robotics. He has helped organized multiple international conferences, serves as an editor of several journals, and is currently serving as the National Coordinator of European Robotics Week as well as the IEEE RAS Croatian Section Chair. In 2019 he received a Croatian Academy of Engineering young scientist award, "Vera Johanides".

Editorial

Aerial Robotics for Inspection and Maintenance: Special Issue Editorial

Alejandro Suarez [1,*], Jonathan Cacace [2] and Matko Orsag [3]

[1] GRVC Robotics Laboratory, University of Seville, Camino de los Descubrimientos s/n, 41092 Seville, Spain
[2] Department of Electrical Engineering and Information Technology, University of Naples Federico II,
     Via Claudio 21, 80125 Naples, Italy; jonathan.cacace@unina.it
[3] Faculty of Electrical Engineering and Computing, University of Zagreb, Unska 3, 10000 Zagreb, Croatia;
     matko.orsag@fer.hr
* Correspondence: asuarezfm@us.es

Citation: Suarez, A.; Cacace, J.; Orsag, M. Aerial Robotics for Inspection and Maintenance: Special Issue Editorial. *Appl. Sci.* **2022**, *12*, 3583. https://doi.org/10.3390/app12073583

Received: 28 March 2022
Accepted: 30 March 2022
Published: 1 April 2022

Publisher's Note: MDPI stays neutral with regard to jurisdictional claims in published maps and institutional affiliations.

1. Introduction

The significant advances in last decade in the research and technology of multi-rotor design, modeling and control, supported by the increasing variety of commercially available platforms, components and manufacturers, have facilitated a rise in the novel applications of aerial robots, capable of not only perceiving, but also interacting with the environment, allowing the realization of diverse operations and tasks in areas and workspaces that are difficult to access by human operators or ground vehicles. Although the use of drones in surveillance and monitoring, or in aerial filming (both professional and personal) is quite extended nowadays due to the affordable cost of commercial platforms, the use of aerial robots for the inspection and maintenance of infrastructures is currently the subject of research and development, particularly when involving physical interaction during flight. The integration of sensors and robotic manipulators in these platforms, capable of easily and quickly reaching high-altitude workspaces and maintaining their position while hovering, allows us to obtain measurements of interest from cracks in bridges or corrosion in viaducts, and conduct diverse operations such as the installation of bird flight diverters on power lines. Aerial robots can also be used for the maintenance of healthy environmental conditions in urban or rural areas.

Despite the significant maturity level reached with these platforms, new research and technological challenges arise from applications demanding the benefits of aerial robots, particularly in outdoor environments, where regulation aspects must be considered. The current paradigm is to develop methods and technologies driven by specific applications and operational conditions imposed by the sensors or devices involved in the inspection and maintenance tasks, also considering general goals such as reducing costs, improving performance time, safety, and precision, or reducing energy consumption. Therefore, considerable effort is still being devoted to the mechatronic development and integration of new robotic manipulators, mechanisms, and sensor devices required to accomplish the intended task in flight, extending the capabilities of conventional multi-rotors in terms of autonomy and force interaction.

This Special Issue presents several research works focused on the use of aerial robots to conduct inspection and maintenance operations on infrastructures such as power lines, bridges, viaducts, or walls involving physical interactions.

2. Contribution and Advances

This Special Issue collects eleven papers from different research groups from Spain, Croatia, Italy, Japan, the USA, the Netherlands and Denmark, focused on the design, development and experimental validation methods and technologies of aerial robotics for inspection and maintenance. The main contributions and innovation of these works are summarized below.

The lightweight and compliant dual-arm aerial manipulation robot presented in [1] employs one of the arms to estimate the position of the multirotor platform relative to the grabbing point, while the other arm is intended to conduct the manipulation operation in flight. This is motivated by the necessity to improve the positioning accuracy outdoors during the installation of devices on power lines, allowing the estimation and control of the interaction forces exerted by the arm, relying on the mechanical flexibility of the joints with deflection feedback. Reference [2] introduces the Cartesian aerial manipulator, a new morphology of aerial robot that exploits the benefits of a two degrees of freedom (DoF) Cartesian base (XY-axes) in terms of low weight/inertia and positioning accuracy, with a single DoF compliant joint, exploiting the deflection feedback from the flexible joint and a linear elastic link for contact force control and collision detection and reaction. The kinematic configuration of the manipulator is also applied in pick-and-store operations.

In order to increase the capability of conventional multi-rotors to generate forces along the horizontal plane, required in many contact-based inspection operations, the authors in [3] describe an add-on mechanism for multi-rotors, consisting of three ducted fans arranged in a Y-shaped structure that can be easily integrated in different types of multi-rotors, in such a way that the horizontal thrust allows the decoupling of the translational control from the attitude control. The developed prototype demonstrates how a relatively simple concept design can extend the functionalities and control capabilities of aerial robots.

The inspection and maintenance of power lines is the scope of the work in [4–6], in the context of the AERIAL-CORE H2020 project. The high altitude, high voltage, and difficult access of this essential infrastructure, comprising thousands of kilometers in any country, makes the realization of operations such as the installation of bird diverters (imposed by regulation to protect bird species from collision or electrocution) particularly risky for human operators, who have to climb the towers, using elevated lift platforms or even manned helicopters to reach the cables. This motivates the use of multi-rotor-based aerial robots and the development of new capabilities specifically for this application. In this context, reference [4] details the mathematical formulation for estimating the position and orientation of a multi-rotor platform from the magnetic field generated by the power line, measured by three magnetometers. Analytical expressions for the position of the aerial robot relative to the power line are derived, along with an exhaustive analysis of the different solutions and possible arrangements of the sensors.

Several designs and preliminary results in control, planning, and manipulation for the installation of devices on power lines, such as bird flight diverters and electrical spacers, are presented in [5], covering also the design of cognitive human–machine interfaces and the use of aerial manipulators for fast and safe tool delivery to human operators working on the power lines. The paper provides an overview of the different solutions explored as part of the AERIAL-CORE project in terms of safe local aerial manipulation. A particular technological solution for the installation of approved bird flight diverters is detailed in [6]. The main challenge here is to exert very high forces to install the device on the cable, using for this purpose a linear actuator with a customized clamp mechanism that holds the device and supports the reaction forces, so the aerial platform is isolated during the realization of the operation on flight.

The inspection of bridges [7], viaducts [8], and other civil infrastructures [9] using aerial robots requires the integration of specific sensors and devices, as well as the implementation of perception and navigation methods to conduct the operations in flight with a sufficient level of autonomy and positioning accuracy. The authors in [7] propose the deployment of a team of cooperative aerial robots to install inspection devices on bridges by spraying a resin onto the surface and applying a pushing force to keep the device attached. The paper presents the design of the control framework, covering the attitude/position and impedance controllers, the path planning, and the detection and estimation of the marked point. Two different aerial robotic solutions are presented in [8] for the visual and contact inspection of viaducts, where GNSS (Global Navigation Satellite System) positioning is not possible, requiring the integration of 3D LiDAR (Light Detection and Ranging) combined with robotic stations to generate accurate maps of the environment. The presented work illustrates the development and integration efforts to increase the TRL (Technology Readiness Level) of aerial robots applied in real inspection scenarios.

Specific methods based on vibration monitoring for the assessment of civil infrastructure using aerial robots equipped with deployable sensor units are described in [9]. Different sensor technologies for Structural Health Monitoring (SHM) are identified, presenting the design of a sensor unit based on an accelerometer with a docking mechanism that can be attached to metallic structures using the aerial platform. Damages on monitored structures are detected as changes in their Dynamic Signature Response (DSR), relying on the B-Spline Impulsive Response Function (BIRF) for representing the time-variable system dynamics. The paper illustrates, in laboratory conditions, the deployment of the inspection sensor in a metallic structure subject to vibrations using the aerial robot, including the data acquisition, analysis and interpretation of results.

Traditional contact-based inspection methods are carried out by human operators, typically consisting of placing the sensor device in the point of interest. In this sense, the operator is responsible for determining the desired inspection point according to the observations of the environment and specific knowledge of the assessment. Therefore, in some cases it is not convenient or feasible to implement a fully autonomous inspection operation with an aerial robotic system, but it is convenient to allow the human operator to interact with the environment through the aerial robot. This is the scope of the research work presented in [10], in which a multi-rotor equipped with a passive and compliant end effector is teleoperated to exert contact forces using a haptic device that provides the user with feedback to improve his/her situation awareness. The paper is focused on the design of the bilateral teleoperation scheme and stability analysis to maintain contact forces in flight.

Although the inspection and maintenance of industrial and civil infrastructures is one of the most immediate application areas of aerial robots, reference [11] extends the scope by proposing the application of drones for releasing sterilized insects, in order to prevent the increase in insect populations that could become vectors of disease transmission in urban and rural environments. Given the capabilities of Unmanned Aircraft System (UAVs), either fixed wing or rotary wing, to cover large areas following accurately desired paths in an autonomous way, the paper considers the current regulation challenges in Europe and risk assessment when flying over populated environments, particularly when the drone operates beyond the visual line of sight (BVLOS).

3. Conclusions

The execution of inspection and maintenance operations in illustrative scenarios such as power lines, bridges, viaducts and other civil infrastructures, consisting typically of the appropriate collocation of a sensor or device in the point of interest, presents considerable risk when conducted at high altitude by human operators, and becomes a technological challenge when it is intended to be performed by an aerial robot operating outdoors. Overcoming the gap between research in indoor laboratories and practical application in realistic outdoor conditions also introduces significant difficulties, particularly in terms of positioning accuracy and reliability, requiring the integration of additional systems that reduce the effective payload. New mechatronic designs, estimation and measurement methods, control frameworks and technological solutions emerge from demanding applications that have been shown increasing interest in terms of the application of aerial robots to reduce the time, cost and risk compared with traditional procedures.

Since most of the inspection and maintenance devices currently employed in the realization of these operations are intended to be used by human operators, aerial robotics researchers and engineers face the problem of integrating or adapting these devices to aerial platforms, where the limited payload and flight time capacity are still the main constraints. It is foreseeable, however, that manufacturers will develop new inspection solutions adapted for aerial robots given the benefits and potential uses evidenced by multi-rotors.

Author Contributions: Conceptualization, A.S., J.C. and M.O.; investigation, A.S., J.C. and M.O.; writing—original draft preparation, A.S.; writing—review and editing, A.S., J.C. and M.O.; visualization, J.C. and M.O.; supervision, J.C. and M.O.; All authors have read and agreed to the published version of the manuscript.

Funding: This research was partially funded by the European Commission grant number 871479 through the AERIAL-CORE H2020 project (AERIAL COgnitive integrated multi-task Robotic system with Extended operation range and safety).

Acknowledgments: We want to thank Anibal Ollero from the University of Seville, coordinator of the AERIAL-CORE project, for his contribution in the dissemination of this Special Issue.

Conflicts of Interest: The authors declare no conflict of interest.

References

1. Suarez, A.; Sanchez-Cuevas, P.J.; Heredia, G.; Ollero, A. Aerial Physical Interaction in Grabbing Conditions with Lightweight and Compliant Dual Arms. *Appl. Sci.* **2020**, *10*, 8927. [CrossRef]
2. Suarez, A.; Perez, M.; Heredia, G.; Ollero, A. Cartesian Aerial Manipulator with Compliant Arm. *Appl. Sci.* **2021**, *11*, 1001. [CrossRef]
3. Miyazaki, R.; Paul, H.; Shimonomura, K. Development of Add-On Planar Translational Driving System for Aerial Manipulation with Multirotor Platform. *Appl. Sci.* **2021**, *11*, 1462. [CrossRef]
4. Martinović, D.; Bogdan, S.; Kovačić, Z. Mathematical Considerations for Unmanned Aerial Vehicle Navigation in the Magnetic Field of Two Parallel Transmission Lines. *Appl. Sci.* **2021**, *11*, 3323. [CrossRef]
5. Cacace, J.; Orozco-Soto, S.M.; Suarez, A.; Caballero, A.; Orsag, M.; Bogdan, S.; Vasiljevic, G.; Ebeid, E.; Rodriguez, J.A.A.; Ollero, A. Safe Local Aerial Manipulation for the Installation of Devices on Power Lines: AERIAL-CORE First Year Results and Designs. *Appl. Sci.* **2021**, *11*, 6220. [CrossRef]
6. Rodriguez-Castaño, A.; Nekoo, S.R.; Romero, H.; Salmoral, R.; Acosta, J.Á.; Ollero, A. Installation of Clip-Type Bird Flight Diverters on High-Voltage Power Lines with Aerial Manipulation Robot: Prototype and Testbed Experimentation. *Appl. Sci.* **2021**, *11*, 7427. [CrossRef]
7. Ivanovic, A.; Markovic, L.; Car, M.; Duvnjak, I.; Orsag, M. Towards Autonomous Bridge Inspection: Sensor Mounting Using Aerial Manipulators. *Appl. Sci.* **2021**, *11*, 8279. [CrossRef]
8. Caballero, R.; Parra, J.; Trujillo, M.Á.; Pérez-Grau, F.J.; Viguria, A.; Ollero, A. Aerial Robotic Solution for Detailed Inspection of Viaducts. *Appl. Sci.* **2021**, *11*, 8404. [CrossRef]
9. Carroll, S.; Satme, J.; Alkharusi, S.; Vitzilaios, N.; Downey, A.; Rizos, D. Drone-Based Vibration Monitoring and Assessment of Structures. *Appl. Sci.* **2021**, *11*, 8560. [CrossRef]
10. Mohammadi, M.; Bicego, D.; Franchi, A.; Barcelli, D.; Prattichizzo, D. Aerial Tele-Manipulation with Passive Tool via Parallel Position/Force Control. *Appl. Sci.* **2021**, *11*, 8955. [CrossRef]
11. Garcia, M.; Maza, I.; Ollero, A.; Gutierrez, D.; Aguirre, I.; Viguria, A. Release of Sterile Mosquitoes with Drones in Urban and Rural Environments under the European Drone Regulation. *Appl. Sci.* **2022**, *12*, 1250. [CrossRef]

Article

Aerial Physical Interaction in Grabbing Conditions with Lightweight and Compliant Dual Arms

Alejandro Suarez *, Pedro J. Sanchez-Cuevas, Guillermo Heredia and Anibal Ollero

GRVC Robotics Labs, University of Seville, 41092 Seville, Spain; psanchez16@us.es (P.J.S.-C.); guiller@us.es (G.H.); aollero@us.es (A.O.)
* Correspondence: asuarezfm@us.es

Received: 4 November 2020; Accepted: 12 December 2020; Published: 14 December 2020

Abstract: This paper considers the problem of performing bimanual aerial manipulation tasks in grabbing conditions, with one of the arms grabbed to a fixed point (grabbing arm) while the other conducts the task (operation arm). The goal was to evaluate the positioning accuracy of the aerial platform and the end effector when the grabbing arm is used as position sensor, as well as to analyze the behavior of the robot during the aerial physical interaction on flight. The paper proposed a control scheme that exploits the information provided by the joint sensors of the grabbing arm for estimating the relative position of the aerial platform w.r.t. (with respect to) the grabbing point. A deflection-based Cartesian impedance control was designed for the compliant arm, allowing the generation of forces that help the aerial platform to maintain the reference position when it is disturbed due to external forces. The proposed methods were validated in an indoor testbed with a lightweight and compliant dual arm aerial manipulation robot.

Keywords: aerial manipulation; dual arm; compliance

---

1. Introduction

The reliability in the realization of an aerial manipulation task on flight strongly depends on the positioning accuracy of the aerial robot, which mainly depends on the accuracy of the position sensors, the performance of the multirotor controller, and the effect of endogenous/exogenous forces raised during the execution of the operation. On the one hand, it is desirable that the accuracy in the position estimation of the aerial vehicle is below the 10% of the reach of the manipulator [1], being capable of compensating the undesired deviations while the multirotor hovers within the workspace. Different positioning systems have been employed in the literature. Motion capture systems such as Vicon or Opti-Track have been extensively used in indoor testbeds [2–6] due to their high accuracy (<1 cm) and high update rates (100–200 Hz), as well as because no additional devices have to be integrated in the aerial platform, but only the passive markers. Similarly, the laser tracking systems used in [7] only require the addition of a reflective marker or a prism to the multirotor, although this solution imposes that the marker is not occluded by any obstacle in the line of the laser. Several on-board perception systems have been developed for multirotor platforms, including optical flow [8], stereo vision [9], live 3D dense reconstruction [10], or laser scanners [11]. However, these solutions reduce the payload capacity of the aerial platform, as additional devices such as cameras and on-board computers have to be added and complicate the system integration. Not only that, but each of these technologies presents certain limitations relative to the operation range, accuracy, reliability, or the update rate. The docking system described in [12] is an alternative solution that exploits the proximity of the aerial platform to the workspace during the manipulation phase, using an articulated link for obtaining the relative position from the encoders of the joints.

An aerial manipulation robot operating on flight will be affected by three types of perturbations: the reaction wrenches induced over the multirotor platform due to the motion of the arms [13,14], the contact forces associated to the physical interactions on flight [4,6,15–17], as well as the aerodynamic effects [7]. As consequence, the realization of certain manipulation tasks requiring the correct positioning of the end effector, such as object grasping [2,5], valve turning [3], or inspection by contact [7,15,18], may be compromised or become unfeasible. In order to overcome these problems, several methods and strategies have been proposed, such as the estimation and control of the external wrenches acting over the aerial platform [19,20], the development of multilayer control architectures [13], or the design of lightweight and compliant robotic arms [21,22]. In this sense, it is desirable to improve the accommodation of the aerial platform to the position deviations when it is operating in contact with the environment, exploiting for this purpose the mechanical compliance of the arms [22].

The main contribution of this paper is the design, modelling, and validation of a lightweight and compliant dual arm system that allows for the estimation and control of the position of an aerial robotic manipulator relative to a fixed grabbing point, using one of the arms for grabbing and as position sensor (grabbing arm), while the other is intended to conduct the operation while flying (operation arm). Figure 1 illustrates the application of the dual arm aerial manipulator for the installation of clip-type bird diverters on a power line [23]. Two methods are proposed and evaluated. Firstly, a zero-torque controller was implemented in the grabbing arm, and thus the reaction wrenches induced over the multirotor were relatively low, using the position estimation obtained from the joint servos to control the deviations in the position of the platform with respect to the reference pose. Secondly, a force controller based on Cartesian deflection was developed to achieve the desired impedance behavior during the aerial physical interaction, using the grabbing arm to exert a force that helps the multirotor controller to reach the reference pose relative to the grabbing point. Experimental results carried out in test-bench and indoor flight tests (Figure 1) demonstrated the performance of both approaches.

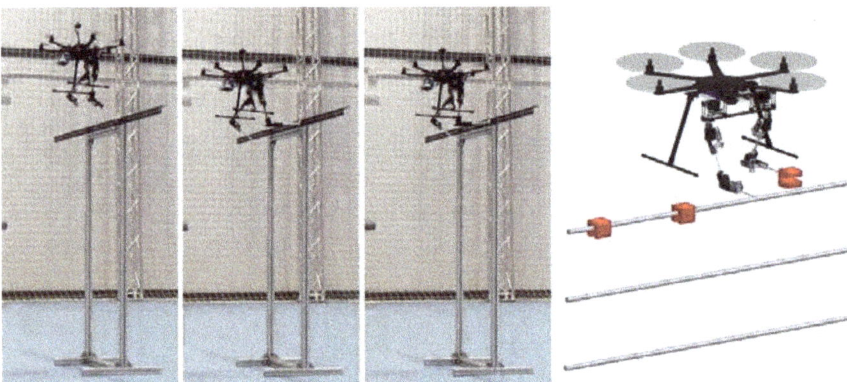

Figure 1. Dual arm aerial manipulation robot grabbed a linear structure with the right (grabbing) arm.

The innovative aspects of this paper with respect to our previous published works [17,21,22] can be summarized in the following points:

1. The development and testing of a new functionality for the dual arm aerial manipulator: estimating the position of the robot relative to the grabbing point with one of the arms while the other is intended to conduct the operation on flight.
2. The evaluation of the positioning accuracy in the estimation provided by the grabbing arm compared to the ground truth given by an Opti-Track system.
3. The combination of passive (mechanical) and active (control) compliance methods in the grabbing arm to facilitate the accommodation of the aerial robot to sustained grabbing forces.
4. The experimental evaluation and qualitative analysis of the effects of the grabbing arm and the injected disturbances over the stability of the multirotor controller.

The rest of the paper is organized as follows. The prototype employed in the experiments is firstly described in Section 2. Section 3 covers the kinematics and position estimation, with the mentioned control methods described in Section 4. The experimental results are presented in Section 5, and the conclusions are summarized in Section 6.

2. System Description

2.1. Compliant Dual Arm

The manipulator used for validating the methods described in Section 4 is a lightweight and compliant dual arm system developed at the GRVC Robotics Labs. A picture of the arms can be seen in Figure 2, indicating in Table 1 its main features. Each arm provides three degrees of freedom for end effector positioning in the following kinematic configuration [22]: shoulder yaw at the base (q_1), shoulder pitch (q_2), and elbow pitch (q_3). The arms are built with the Herkulex DRS-0201 smart servos, and a customized frame structure manufactured in carbon fiber and aluminum, providing full servo protection at the shoulder yaw joint with a pair of polymer bearings, and partial servo protection in the other two joints [21]. In order to estimate and control the torques and forces, the grabbing arm (right arm) integrates 14-bit resolution magnetic encoders that are interfaced through a STM32F303 microcontroller board, sending the deflection measurement to the main computer at 200 Hz with 1 ms latency. Each of the arms and the sensors are connected to the Raspberry Pi 3B+ through USB-to-USART interfaces, where control program of the arms is executed. The manipulator is fed with a 2S, 650 mAh LiPo battery, providing an operation time of around 20 min.

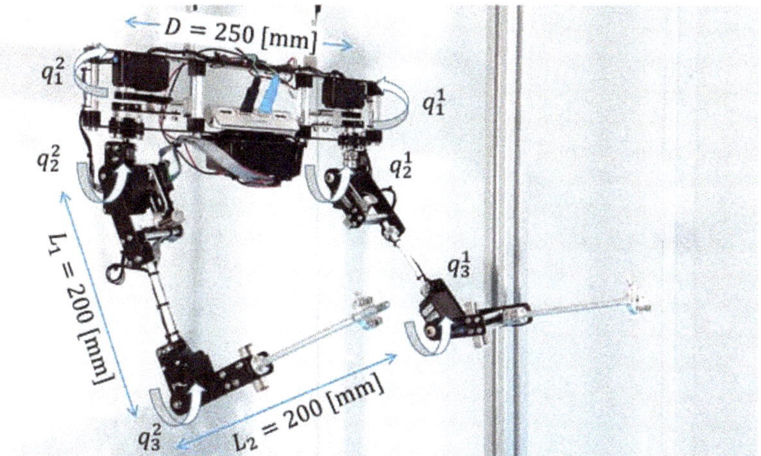

Figure 2. Compliant dual arm used in the experiments.

Table 1. Main features of the lightweight and compliant dual arm.

Total Weight	1.0 (kg)
Maximum lift load (1 s playtime)	At elbow: 0.3 (kg) At shoulder: 0.12 (kg)
Joint stiffness	5 (Nm/rad)—all joints
Maximum joint speed	300 (°/s)

The tip of the forearm link includes an aluminum flange to facilitate the integration of the end effector in the manipulator. For safety reasons, the grabbing operation is conducted with a magnetic gripper (around 5 N force), using the linear metallic structure shown in Figure 1 for grabbing.

2.2. Aerial Manipulation Robot

The dual arm system is integrated in an S550 platform, a hexarotor similar to a DJI F550. This is equipped with six DJI 2312E brushless motors with 9×4.5 inch propellers. Figure 3 shows a picture of the aerial robot, identifying its components, including the Pixhawk 2.1 autopilot, the Raspberry Pi 3B+ computer board, and the 4S 4400 mAh LiPo battery (0.5 kg) used as counterweight of the arms. The setup is similar to the one described in [14], constraining the motion of the arms to prevent the collision with the landing gear. The hardware and software architecture of the system is represented in Figure 4. The control program of the arms, developed in C/C++, as well as the different software modules that control the aerial platform, based on ROS and the UAV Abstraction Layer [24], are executed in the Raspberry Pi and interfaced by the Ground Control Station (GCS) through the wireless link. Four USB-to-USART devices are connected to the computer board: (1) the Pixhawk autopilot, (2) the left arm, (3) the right arm, and (4) the microcontroller board that reads the sensors. The position and orientation of the multirotor are obtained from an Opti Track system, used as ground truth.

Figure 3. Dual arm aerial manipulation robot.

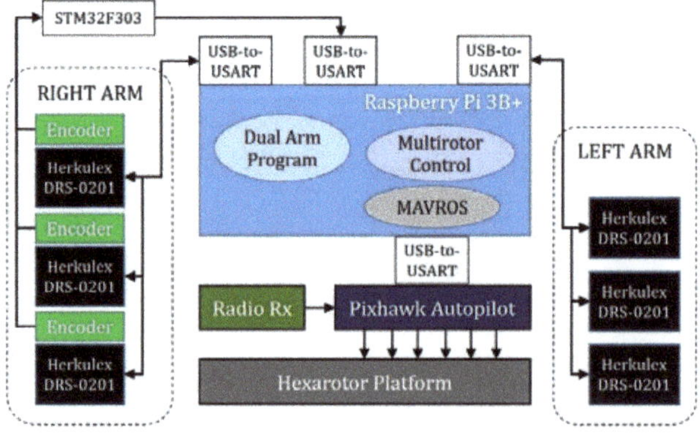

Figure 4. Hardware architecture of the aerial manipulation robot.

3. Modelling

3.1. Kinematics

As usual, three reference frames are defined for the aerial manipulation system, as illustrated in Figure 5: the Earth fixed frame {E} (inertial), the multirotor body frame {B}, and the manipulator frame {i}, with $i = \{1, 2\}$ for the left/right arms. In the following, $^A r_B$ denotes the position of a certain point B w.r.t. (with respect to) reference frame {A}. In this way, $^E r_B = [x,\ y,\ z]^T$ and $^E \eta_B = [\phi,\ \theta,\ \psi]^T$ represent the multirotor position and orientation relative to {E}, and $^i r_{TCP,i}$ is the position of the tool center point (TCP) of the i-th manipulator expressed in its own frame. The origin of {i} is located at the intersection of the shoulder joints, with the x-axis pointing forwards, the y-axis parallel to the baseline of the two arms, and the z-axis pointing upwards.

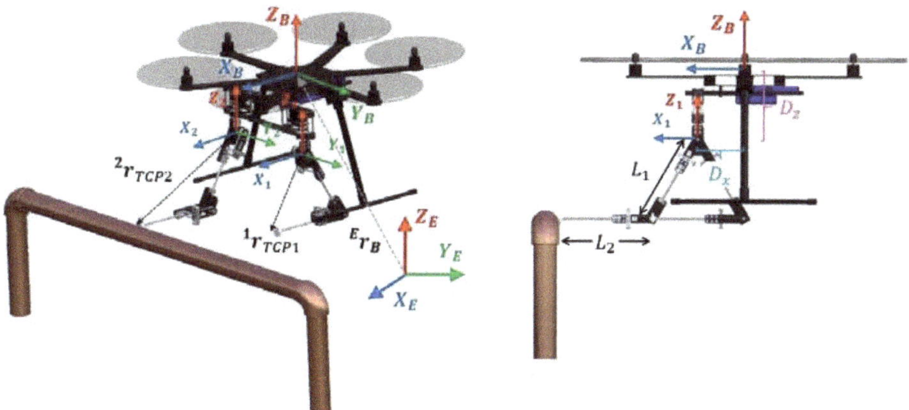

Figure 5. Kinematic model of the dual arm aerial manipulator.

The three reference frames are related through the corresponding transformation matrices:

$$^E T_B = \begin{bmatrix} ^E R_B(\phi,\ \theta,\ \psi) & ^E r_B \\ 0_{1\times3} & 1 \end{bmatrix} \quad ; \quad ^B T_i = \begin{bmatrix} I_{3\times3} & ^B r_i \\ 0_{1\times3} & 1 \end{bmatrix} \quad ; \quad ^E T_i = {}^E T_B \cdot {}^B T_i \tag{1}$$

where $^E R_B$ is the multirotor rotation matrix; $^B r_i = [D_x,\ \pm D/2,\ D_z]^T$ is the origin of {i} relative to {B}, with D_x and D_z being the displacement of the arms with respect to {B} in the x- and z-axes, respectively; and D is the separation distance between the arms in the y-axis.

The arms implement the 3-DOF (degrees of freedom) configuration considered in our previous work [14,22], with three joints for TCP positioning: shoulder yaw (base), shoulder pitch, and elbow pitch. The wrist joints are not considered due to the convenience to simplify the mechanical construction and reduce the weight of the arms. The rotation angle of the j-th joint of the i-th arm is denoted as q^i_j, whereas θ^i_j is the corresponding servo shaft position. The difference between these two variables is the deflection angle, $\Delta\theta^i_j = \theta^i_j - q^i_j$, measured by the encoders [17,21]. The forward and inverse kinematic models are computed as follows (superscript i is omitted for clarity reasons):

$$^i r_{TCP,i} = FK\left(q^i\right) = \begin{bmatrix} r(q_2, q_3) \cdot \cos(q_1) \\ r(q_2, q_3) \cdot \sin(q_1) \\ L_1 \cos(q_2) + L_2 \cos(q_2 + q_3) \end{bmatrix} \tag{2}$$

$$q^i = IK\left({}^i r_{TCP,i}\right) = \begin{bmatrix} atan2(y, x) \\ \cos^{-1}\left(\dfrac{x^2+y^2+z^2-L_1^2-L_2^2}{2L_1\sqrt{x^2+y^2}}\right) \\ \cos^{-1}\left(\dfrac{x^2+y^2+z^2-L_1^2-L_2^2}{2L_1 L_2}\right) \end{bmatrix} \tag{3}$$

where L_1 and L_2 are the upper arm/forearm link lengths, and $r(q_2, q_3)$ is given by

$$r(q_2, q_3) = L_1 \sin(q_2) + L_2 \sin(q_2 + q_3) \tag{4}$$

3.2. Relative Position Estimation

If the end effector of the grabbing arm is firmly attached to a fixed point, then it is possible to estimate the position of the multirotor relative to this grabbing point just applying the homogeneous transformation from {2} to {B}, taking into account that the position of the TCP referred to {2} is directly obtained from the forward kinematic model given by Equation (2). The accuracy in the position estimation can be obtained multiplying the joint position error (including the deflection error) by the Jacobian of the arm. The wrist joints are not essential for this purpose, as the multirotor orientation can be obtained from the inertial measurement unit (IMU) of the aerial platform. However, a certain level of accommodation is required at the wrist so the robotic arm can follow the position and orientation deviations of the aerial platform while it is grabbed.

Two similar mechanisms have been proposed in previous works for estimating the pose of an aerial manipulator relative to a contact point. Reference [12] presents a docking tool consisting of an articulated arm with passive joints that is deployed over a pipe with a stiff-joint dual arm, whereas reference [25] relies on a passive spherical wrist joint and an IMU integrated at the end effector of a 3-DOF arm. In this paper, we combine the passive/active compliance methods for estimating the multirotor position while controlling the interaction force.

3.3. Dynamics

The dynamic model of the compliant joint dual arm aerial manipulation robot is derived from the Lagrangian and the generalized equations of the forces and torques:

$$L = K - V \tag{5}$$

$$\frac{d}{dt}\left\{\frac{\partial L}{\partial \dot{\xi}}\right\} - \left\{\frac{\partial L}{\partial \xi}\right\} = \Gamma + \Gamma_{ext} \tag{6}$$

where L is the Lagrangian; K and V are the kinetic and potential energies, respectively; ξ is the vector of generalized coordinates; and Γ and Γ_{ext} respectively represent the generated and external wrenches acting on the aerial robot. The vector of generalized coordinates includes the multirotor position and orientation, as well as the servo shaft and output link angular position vectors, $\theta^i = \begin{bmatrix} \theta_1^i & \theta_2^i & \theta_3^i \end{bmatrix}$ and $q^i = \begin{bmatrix} q_1^i & q_2^i & q_3^i \end{bmatrix}$, respectively, and thus ξ is defined as follows.

$$\xi = \begin{bmatrix} {}^E r_B & {}^E \eta_B & \theta^1 & q^1 & \theta^2 & q^2 \end{bmatrix}^T \in \mathfrak{R}^{18} \tag{7}$$

Analogously, the vector of generalized forces comprises the forces and torques acting over the multirotor and the joints of the manipulator.

$$\Gamma = \begin{bmatrix} F_B & \tau_B & \tau_m^1 & \tau^1 & \tau_m^2 & \tau^2 \end{bmatrix}^T \in \mathfrak{R}^{18} \tag{8}$$

The vector of external forces Γ_{ext} models the disturbance wrenches exerted on the multirotor base [18–20], the contact forces at the end effector [4,6,17], as well as the aerodynamic forces raised when the thrust of the rotors is affected by close surfaces in the environment [7,26].

The kinetic energy of the aerial manipulator can be expressed as the sum of the kinetic energy of the aerial platform and the kinetic energy of the robotic arms:

$$K = K_{UAV} + K_{arms} \tag{9}$$

Each of these components comprises two terms corresponding to the translation and rotation of the masses with respect to the inertial frame $\{E\}$:

$$K_{UAV} = \frac{1}{2} m_{UAV} \|{}^E \dot{r}^B\|^2 + \frac{1}{2} {}^E \omega_B^T I_{UAV} {}^E \omega_B \tag{10}$$

$$K_{arms} = \sum_{i=1}^{2} \sum_{j=1}^{4} \left(\frac{1}{2} m_j^i \|{}^E \dot{r}_i^j\|^2 + \frac{1}{2} {}^E \omega_j^{i,T} I_j^i {}^E \omega_j^i \right) \tag{11}$$

where m_{UAV} and I_{UAV} are the mass and inertia tensor of the aerial platform, respectively, whereas m_j^i and I_j^i are mass and inertia of the j-th joint of the i-th arm, respectively. The potential energy of the aerial manipulator also includes two terms, the gravity and elastic potential of the compliant joints:

$$V = g \left(m_{UAV} z_{UAV} + \sum_{i=1}^{2} \sum_{j=1}^{4} m_j^i z_j^i \right) + \sum_{i=1}^{2} \sum_{j=1}^{4} k_j^i \left(\theta_j^i - q_j^i \right)^2 \tag{12}$$

Here, g is the gravity constant, and k_j^i is the corresponding joint stiffness. After some work, it is possible to express the dynamic model in the usual compact matrix form [14]:

$$M\ddot{\xi} + C(\xi, \dot{\xi}) + G(\xi) + K(\xi) + D(\dot{\xi}) = \Gamma + \Gamma_{ext} \tag{13}$$

where $M \in \mathfrak{R}^{18 \times 18}$ is the generalized inertia matrix; C and $G \in \mathfrak{R}^{18}$ represent the centrifugal, Coriolis, and gravity terms; and K and $D \in \mathfrak{R}^{18}$ correspond to the stiffness and damping terms of the compliant manipulator, respectively [17]. The dynamic coupling between the arms and the aerial platform is associated with the cross terms in the generalized inertia matrix, which can be decomposed in three groups of submatrices identified in [14]: multirotor translation (decoupled from rotation), multirotor rotation with coupling terms, and dual arm manipulator with coupling terms. Since the grabbing arm will be held to a fixed point, the corresponding inertia, Coriolis, and gravity terms will be negligible compared to the operation arm. Moreover, if the sensitivity of the zero-torque controller described in next section is good enough, then the magnitude of the wrenches associated to the stiffness and damping terms will be relatively low, and with it, the influence over the multirotor controller.

As stated in the introduction and illustrated in Figure 1, the operation arm is intended to perform the manipulation operation (the installation of a bird flight diverter) while the grabbing arm provides the position estimation. This can be assimilated to a close kinematic chain [27,28] with floating base [29], in which the pushing/pulling force exerted by the operation arm will cause a reaction torque on the aerial platform that should be cancelled by the multirotor controller with the help of the grabbing arm in order to prevent undesired position deviations. This motivates the implementation of an active impedance control scheme with the grabbing arm, as explained below.

4. Control

4.1. Definition of the Control Task

As stated in the introduction, a dual arm system allows the realization of aerial manipulation tasks in grabbing conditions, using one arm for grabbing and for estimating the position of the aerial platform relative to the grabbing point (Section 3.2), whereas the other takes care of conducting the task, for example, the installation of a sensor device [21]. Since the grabbing arm is actuated and mechanically compliant, it is possible to estimate and control the forces and torques acting over the manipulator from the deflection of the joints [17,21]. The idea is that the arm helps the multirotor to reach the desired position when it is disturbed by an external force, exerting a pushing/pulling force in the opposite direction of the position error, as Figure 6 illustrates.

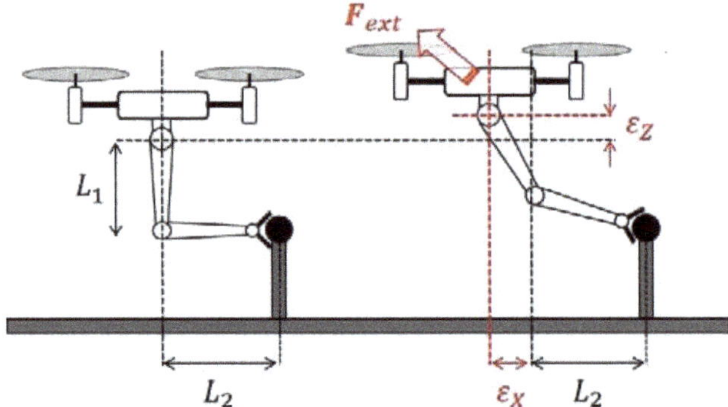

Figure 6. Model considered in grabbing conditions. Nominal operation pose (left), and displacement due to external force (right). The grabbing arm compensates the disturbance exerting a reaction force.

Therefore, the grabbing arm can be used in two ways:

- As relative position sensor, with zero torque control, so the reaction wrenches induced over the multirotor are relatively small.
- As an active impedance link, exerting a controlled force over the aerial platform to compensate external forces and guide the multirotor towards the reference position.

In the first case, the joints of the grabbing arm implement a PI (proportional-integral) controller to maintain a zero deflection (torque) reference, acting over the servo position as follows:

$$\theta^2_{j,ref} = \theta^2_j + \left(K_p\left(\theta^2_j - q^2_j\right) + K_i \int \left(\theta^2_j - q^2_j\right)dt \right) \tag{14}$$

Here, θ^2_j is the position of the j-th servo of the right arm, whereas the term on the right side is the incremental position correction. The proportional and integral gains, K_p and K_i, can be tuned experimentally, taking into account the nominal values of the deflection ($\approx$5 degrees). Now, we impose that the nominal operation position of the aerial robot relative to the grabbing point is the L-shaped configuration of the arm, since this is far enough from the joint limits and the kinematic singularities (although any other could be considered):

$$^2\mathbf{r}^{ref}_{TCP2} = \begin{bmatrix} L_2 \\ 0 \\ -L_1 \end{bmatrix} = FK\left(q^2_{ref}\right) = FK\left(\begin{bmatrix} 0 \\ 0 \\ -\pi/2 \end{bmatrix} \right) \tag{15}$$

The position deviation of the aerial platform is then defined as the displacement of the TCP of the grabbing arm w.r.t. the reference position. That is,

$$\varepsilon = \begin{bmatrix} \varepsilon_X & \varepsilon_Y & \varepsilon_Z \end{bmatrix}^T = FK\left(q^2_{ref}\right) - FK\left(q^2\right) \tag{16}$$

Note that the maximum deviation is limited by the reach of the arm, $\|\varepsilon\| < L_1 + L_2 - \sqrt{L_1^2 + L_2^2}$. If the influence of the grabbing arm over the attitude controller is relatively low due to the zero-torque controller, this measurement can then be taken as input by the position controller of the multirotor platform, as represented in upper part of Figure 7, replacing the GPS (Global Positioning System) used for navigating to achieve better accuracy during the manipulation phase.

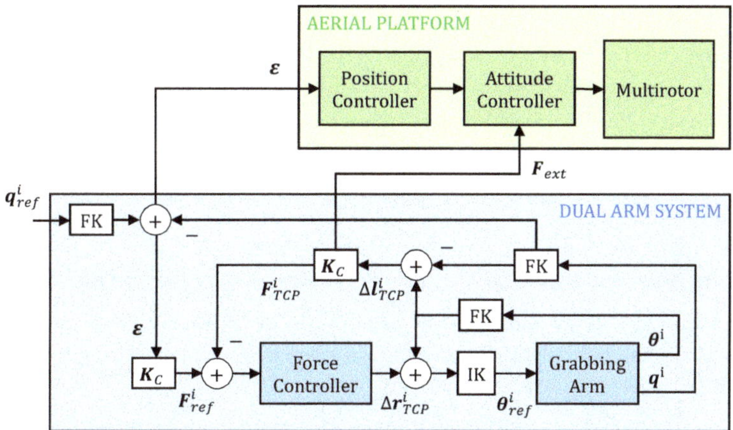

Figure 7. Control scheme of the aerial manipulator with grabbing arm ($i = 2$).

4.2. Impedance Control in Grabbing Conditions

Assuming that the aerial platform is close to the hover state during the grabbing maneuver ($\phi \cong \theta \cong 0$), it is desired that the position deviation of the aerial platform is assimilated to an impedance behavior characterized as follows:

$$M_d\ddot{\varepsilon} + D_d\dot{\varepsilon} + K_d\varepsilon = F_{ext} \tag{17}$$

where M_d, D_d, and K_d are the desired inertia, damping, and stiffness, respectively, and F_{ext} is the external force exerted over the multi-rotor. This force can be estimated from the Cartesian deflection of the grabbing arm, as it will be seen in next subsection, and taken as input by the attitude controller of the aerial platform so it can be partially compensated with the wrenches generated by the rotors.

The grabbing arm will react to the external force exerting a pushing/pulling force at the end effector in the direction of the position deviation, relying on the Cartesian force controller described in the next subsection. The impedance control is then achieved, generating a force reference that cancels the dynamic behavior described by Equation (17). Figure 7 represents the case of desired stiffness, $F^i_{e,TCP} = K_d\varepsilon$.

4.3. Force Control Based on Cartesian Deflection

The force control of the compliant arm is formulated in the Cartesian space and based on the Cartesian deflection, defined as the position deviation of the TCP of the compliant arm w.r.t. the same point in an equivalent stiff joint arm. That is,

$$\Delta l_{TCPi} = FK(\theta^i) - FK(q^i) \tag{18}$$

This definition is useful for expressing the force at the end effector directly in the task space:

$${}^iF_{TCPi} = K^i_C\Delta l_{TCPi} + D^i_C\dot{\Delta l}_{TCPi} \tag{19}$$

where K_C^i and D_C^i are the Cartesian stiffness and damping matrices, respectively, whose values can be obtained from the physical joint stiffness and damping through the Jacobian:

$$K_C^i = \left(J^T\right)^{-1} K_p^i \, J^{-1} \tag{20}$$

where $K_p^i = diag\left\{k_j^i\right\}$ is the physical joint stiffness matrix [17]. The Cartesian damping can be obtained analogously. The force controller, represented in Figure 7 relies on the inverse kinematics, giving as output an incremental position correction term for the TCP:

$$^i r_{TCPi}^{ref} = {}^i r_{TCPi} + \Delta r_{TCP}^i = FK\left(q^i\right) + \left(K_P^F F_{e,TCP}^i + K_I^F \int F_{e,TCP}^i dt\right) \tag{21}$$

Here, Δr_{TCP}^i is the position increment that should be applied in the grabbing arm to achieve the desired force reference; $F_{e,TCP}^i = F_{ref}^i - F_{TCP}^i$ is the force control error at the TCP; whereas $K_P^F = diag\left\{K_{Px}^F, K_{Py}^F, K_{Pz}^F\right\}$ and $K_I^F = diag\left\{K_{Ix}^F, K_{Iy}^F, K_{Iz}^F\right\}$ are the proportional and integral gain matrices, respectively, whose value is tuned experimentally knowing the nominal values of the forces (around 1 $[N]$) and the Cartesian deflection (around 0.02 $[m]$).

5. Experimental Results

5.1. Position Estimation with Zero Deflection/Torque Control

The goal of this experiment was to evaluate the accuracy of the position estimation provided by the grabbing arm when this was used as position sensor, enabling the zero-torque controller to reduce the reaction wrenches induced over the aerial platform on flight (see Section 4.1). The experiment, shown in the video provided as Supplementary Material, consisted of four phases:

(1)   The multirotor takes off.
(2)   The arms adopt the nominal operation pose given by Equation (15).
(3)   The aerial robot approaches the contact point using the Opti Track system for navigating.
(4)   Once the arm grabs the support structure, the zero-torque controller and the estimator are enabled while the multirotor is controlled in position with the Opti Track system.

Figures 8–10 represent the signals of interest during the grabbing maneuver, that is, the multirotor position (arm estimation and ground truth) relative to the nominal pose, the multirotor orientation, the estimation error, and the joint position and torque in the grabbing arm. In order to appreciate the ability of accommodation of the arm and the accuracy in the estimation, we intentionally deviated the aerial platform in the x- and z-axes, with small deviations in the y-axis. Note that in the experiments the end-effector is supposed to be firmly attached to the grabbing point in such a way that the displacement of the multirotor platform will force the displacement of the grabbing arm. However, the magnetic gripper may slip around the contact area, introducing errors in the position estimation. It was also assumed that the multirotor heading (yaw angle) was almost constant, with the Y_B-axis parallel to the linear structure. Otherwise, additional degrees of freedom should be integrated in the wrist joint to estimate the relative orientation, although this is out of the scope of this work. The influence of these two effects over the positioning accuracy can be observed in Figure 9, where the estimation error increased as the multirotor orientation in the yaw angle changed with respect to its initial value, causing mainly errors in the y-axis estimation.

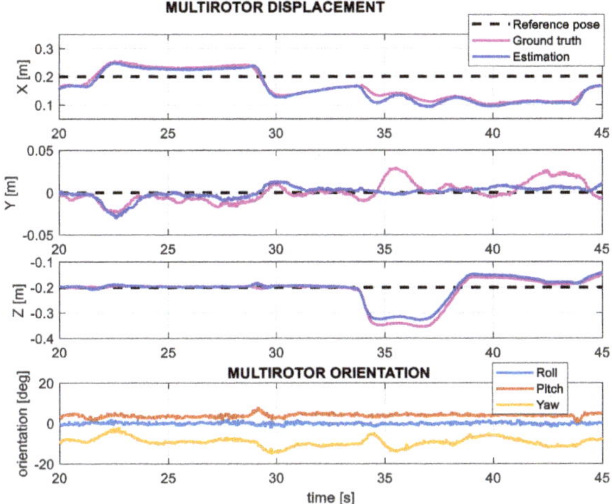

Figure 8. Multirotor displacement in the grabbing maneuver with zero torque control. Nominal pose (black), displacement measured by OptiTrack (magenta), and arm estimation (blue).

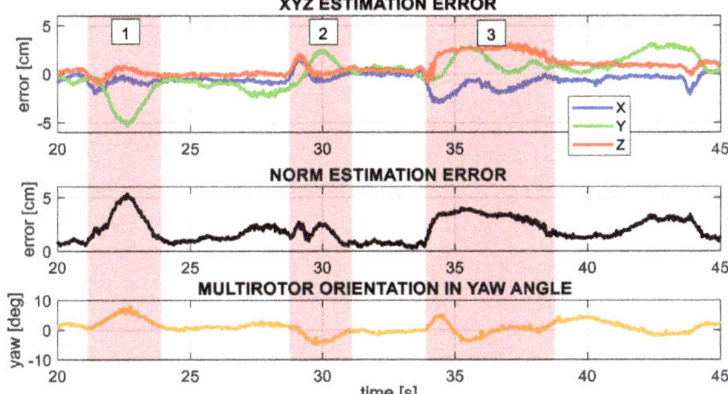

Figure 9. Position estimation error, using Opti-Track as ground truth. x-, y-, and z-axes error (up) and error norm (middle). The three shaded areas correspond to situations in which the platform is rotated in yaw and the error increased since the magnetic gripper was not firmly attached to the point.

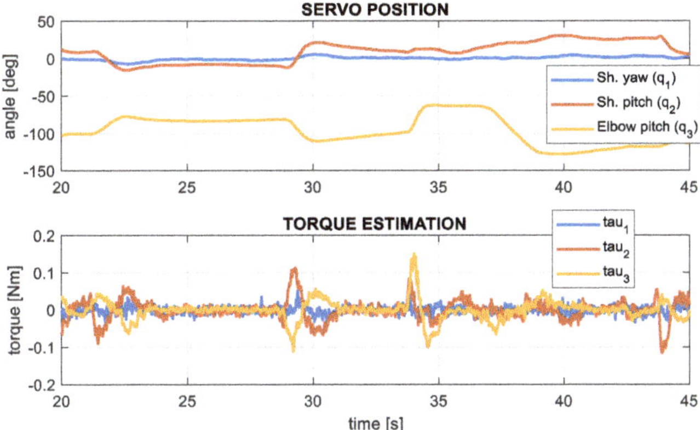

Figure 10. Evolution of the servo position and torque estimation during the grabbing maneuver.

5.2. Impedance Control in Test-Bench

Before its evaluation on flight, the impedance controller of the grabbing arm was firstly validated in test-bench in order to evidence more clearly the variation in the desired stiffness, as defined in Equation (17). During the experiment, the arm adopted the L-shaped configuration (Figure 6 and Equation (15)), manually exerting an external force at the end effector that was estimated and compensated from the Cartesian deflection, as expressed by Equations (19) and (21). The desired stiffness was set to $K_d = 50\ Nm$ for the experiment illustrated in Figure 11, and to $K_d = 100\ Nm$ for the experiment represented in Figure 12. As it can be seen, the position deviation of the end effector was lower in the second case, as expected, and the instantaneous stiffness ($K_x = F_x / \varepsilon_x$) was similar to the desired value. Note that the stiffness is affected by the singularity associated to the deflection when this tends to zero.

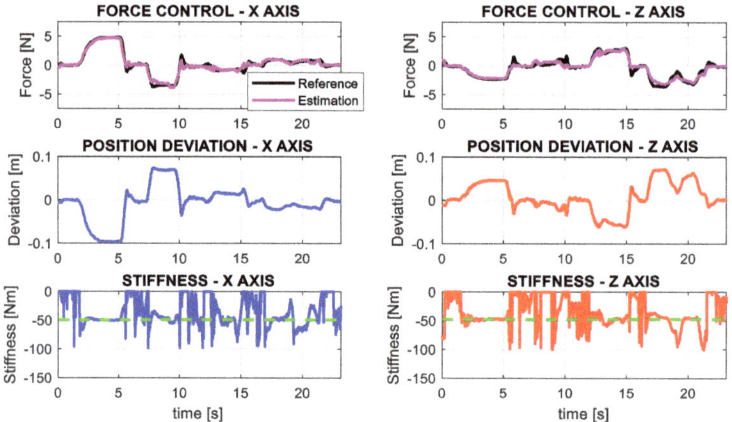

Figure 11. Force reference and estimation (up), position deviation (middle), and instantaneous stiffness (down) in the *x*- and *z*-axes for a desired stiffness $K_d = 50\ [Nm]$.

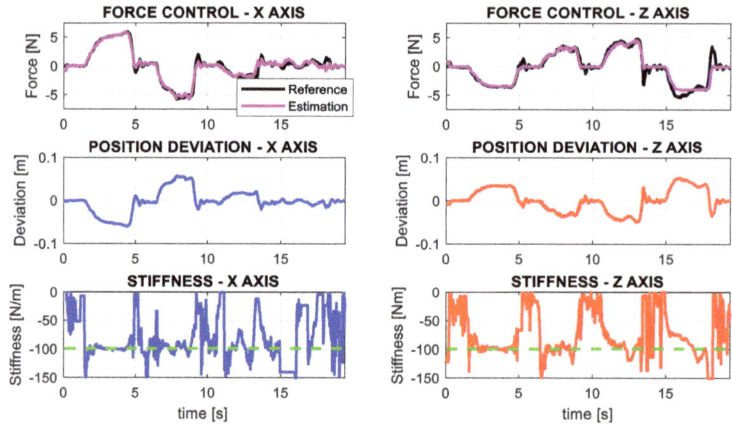

Figure 12. Force reference estimation (up), position deviation (middle), and instantaneous stiffness (down) in the x- and z-axes for a desired stiffness $K_d = 100\ [Nm]$.

5.3. Grabbing Interaction on Flight with Impedance Control

The goal of this experiment was to analyze the behavior of the aerial manipulation robot on flight when the grabbing arm was actively controlling the impedance in the x- and z-axes, considering a desired Cartesian stiffness $K_{d,X} = K_{d,Z} = 50 \ [N/m]$. The performance of the controller was evaluated by injecting a disturbance on the multirotor position that emulates the effect of an external force, such as a wind gust. The zero-torque controller of the first joint (shoulder yaw) was disabled, relying only on the passive compliance to support the lateral displacements (y-axis) and heading variations (yaw angle) of the multirotor. The flight test consists of eight phases:

1. The aerial manipulator takes off with the arms resting in landing configuration.
2. The grabbing arm adopts the nominal pose (L-shaped) while the operation arm is retracted.
3. The impedance controller is enabled, imposing a zero reference for the first joint ($\theta_1^2 = 0$).
4. The aerial manipulator approaches to the linear structure until the magnetic gripper grabs it.
5. The aerial manipulator hovers while the impedance controller of the grabbing arm is active.
6. A position disturbance is injected through the radio controller to observe the response of the aerial manipulator and the reaction of the grabbing arm.
7. The torque control of the servos is disabled to release the grabbing arm.
8. The arms adopt the landing pose, and the platform goes back and lands.

Figures 13 and 14 show the evolution of the system in the time intervals corresponding to phase 5 and phase 6, respectively. On the one hand, Figure 13 covers the time interval from t = 25 to t = 35 s, in which the multirotor was hovering at fixed position while the grabbing arm exerted a small force in the x-axis, around 0.5 N, due to the slight displacement of the platform with respect to the nominal pose. On the other hand, Figure 14 covers the interval from t = 48 to t = 70 s. At t = 49.6 [s], we intentionally applied a 10 cm displacement in the x-axis position of the multirotor, emulating a wind gust. The grabbing arm reacted, exerting a pushing force with a peak of 2.2 [N] that counteracted the disturbance, recovering the nominal operation pose in 2 s. The instantaneous stiffness varied w.r.t. the desired value (50 N/m) due to the singularity in the displacement ($K_{d,x} = F_x / \varepsilon_x$). Note also that, since the grabbing arm was not aligned with the $\mathbf{X_B Z_B}$ plane of the base but it was displaced a distance $D/2$ (see Section 3.1), then the force exerted during the interaction would cause a reaction torque in the yaw angle of the multirotor, and with it, a position deviation in the y-axis, as can be seen in Figure 14. The effect was accentuated as the pilot disturbed the multirotor controller between t = 52 and t = 60 s. Although the passive deflection of the shoulder yaw joint provided a certain level of accommodation, the motion constraint associated with the grabbing condition had a more significant effect on the y-axis and yaw angle, in accordance with the results shown in Figure 9.

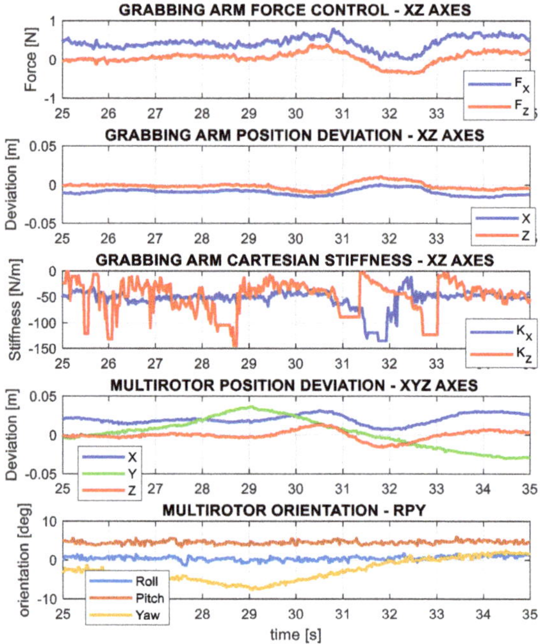

Figure 13. Evolution of the grabbing arm while the multirotor hovers with no disturbance during the grabbing maneuver (phase 5) with impedance control in the x- and z-axes ($K_x = K_z = 50\,N/m$).

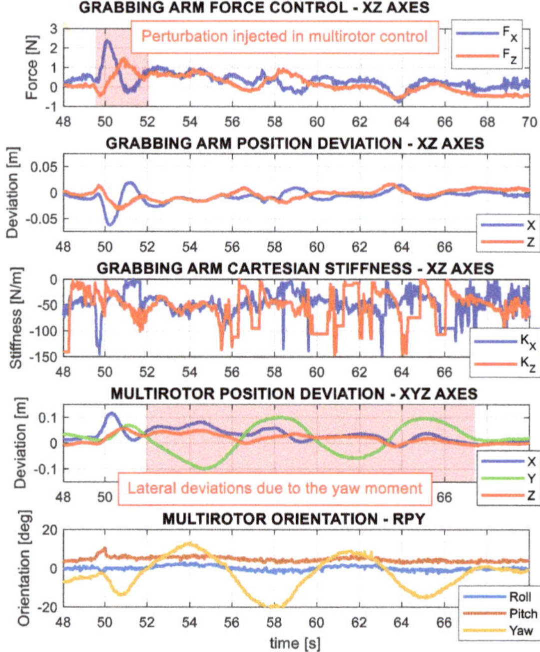

Figure 14. Evolution of the grabbing arm and the multirotor, disturbed at t = 49.6 seconds, during the grabbing maneuver (phase 6) with impedance control in the x- and z-axes ($K_x = K_z = 50\,N/m$).

6. Conclusions

The paper presented two approaches for the estimation and control of the physical interactions of a compliant dual arm aerial manipulation robot operating in grabbing conditions, in which one of the arms was used as position sensor relative to a grabbing point, implementing a zero-torque controller so the wrenches induced over the multirotor base were relatively low. The mechanical joint compliance of the arm was also exploited for the development of active compliance methods, achieving desired impedance behaviors that improved the response of the aerial manipulator when it was affected by external disturbances while interacting physically with the environment. The experimental results presented here validated the concepts, allowed us to evaluate the accuracy of the position estimation, and evidenced the convenience of combining the passive/active compliance on the three Cartesian axes.

Although the aerial manipulation robot will require a navigation system to reach the workspace, the positioning accuracy required to perform the manipulation task (which should be around 10% of the reach of the arm) cannot be achieved with typical sensors employed in outdoor environments, such as GPS, LIDAR (light detecting and ranging), or vision systems. In this sense, the estimation methods described in this work result from special interest to avoid the integration of additional positioning systems, taking into account the limited payload capacity of the aerial platform, being also an effective and reliable solution. In terms of future work, we propose the application of this system for the installation of bird flight diverters on power lines, or for inspection and maintenance of railways, chemical plants, and other linear infrastructure.

Supplementary Materials: The following are available online at http://www.mdpi.com/2076-3417/10/24/8927/s1. Video S1: Aerial_Physical_Interaction.mp4.

Author Contributions: Conceptualization, A.O. and A.S.; software, A.S. and P.J.S.-C.; validation, A.S. and P.J.S.-C.; data curation, A.S.; writing—original draft preparation, A.S.; writing—review and editing, G.H. and A.S.; supervision, G.H.; project administration, A.O.; funding acquisition, A.O. All authors have read and agreed to the published version of the manuscript.

Funding: This work was funded by the AERIAL-CORE project (H2020-2019-871479) funded by the European Commission; the European Research Council Advanced Grant GRIFFIN project (Action 788247); and the ARM-EXTEND (DPI2017-89790-R) and ARTIC (RTI2018-102224-B-I00) projects funded by the Spanish Ministerio de Economia, Industria, y Competitividad.

Conflicts of Interest: The authors declare no conflict of interest.

References

1. Suarez, A.; Vega, V.M.; Fernandez, M.; Heredia, G.; Ollero, A. Benchmarks for Aerial Manipulation. *IEEE Robot. Autom. Lett.* **2020**, *5*, 2650–2657. [CrossRef]
2. Kim, S.; Choi, S.; Kim, H.J. Aerial manipulation using a quadrotor with a two DOF robotic arm. In Proceedings of the IEEE/RSJ International Conference on Intelligent Robots and Systems, Tokyo, Japan, 3–7 November 2013; pp. 4990–4995.
3. Orsag, M.; Korpela, C.; Bogdan, S.; Oh, P. Dexterous aerial robots—Mobile manipulation using unmanned aerial systems. *IEEE Trans. Robot.* **2017**, *33*, 1453–1466. [CrossRef]
4. Cataldi, E.; Muscio, G.; Trujillo, M.A.; Rodriguez, Y.; Pierri, F.; Antonelli, G.; Caccavale, F.; Viguria, A.; Chiaverini, S.; Ollero, A. Impedance control of an aerial-manipulator: Preliminary results. In Proceedings of the 2016 IEEE/RSJ International Conference on Intelligent Robots and Systems (IROS), Daejeon, Korea, 9–14 October 2016; pp. 3848–3853.
5. Santamaria-Navarro, A.; Grosch, P.; Lippiello, V.; Solà, J.; Andrade-Cetto, J. Uncalibrated visual servo for unmanned aerial manipulation. *Trans. Mechatron.* **2017**, *22*, 1610–1621. [CrossRef]
6. Tognon, M.; Chavez, H.A.T.; Gasparin, E.; Sable, Q.; Bicego, D.; Mallet, A.; Lany, M.; Santi, G.; Revaz, B.; Cortess, J.; et al. A truly-redundant aerial manipulator system with application to push-and-slide inspection in industrial plants. *IEEE Robot. Autom. Lett.* **2019**, *4*, 1846–1851. [CrossRef]
7. Jimenez-Cano, A.E.; Sanchez-Cuevas, P.J.; Grau, P.; Ollero, A.; Heredia, G. Contact-based bridge inspection multirotors: Design, modeling, and control considering the ceiling effect. *IEEE Robot. Autom. Lett.* **2019**, *4*, 3561–3568. [CrossRef]

8.	Rhudy, M.B.; Chao, H.; Gu, Y. Wide-field optical flow aided inertial navigation for unmanned aerial vehicles. In Proceedings of the IEEE/RSJ Int. Conference on Intelligent Robots and Systems, Chicago, IL, USA, 14–18 September 2014; pp. 674–679.

9.	Schmid, K.; Tomic, T.; Ruess, F.; Hirschmüller, H.; Suppa, M. Stereo vision based indoor/outdoor navigation for flying robots. In Proceedings of the 2013 IEEE/RSJ International Conference on Intelligent Robots and Systems, Tokyo, Japan, 3–7 November 2013; pp. 3955–3962.

10.	Faessler, M.; Fontana, F.; Forster, C.; Mueggler, E.; Pizzoli, M.; Scaramuzza, D. Autonomous, vision-based flight and live dense 3D mapping with a quadrotor micro aerial vehicle. *J. Field Robot.* **2016**, *33*, 431–450. [CrossRef]

11.	Huh, S.; Shim, D.H.; Kim, J. Integrated navigation system using camera and gimbaled laser scanner for indoor and outdoor autonomous flight of UAVs. In Proceedings of the IEEE/ International Conference on Intelligent Robots and Systems, Tokyo, Japan, 3–7 November 2013; pp. 3158–3163.

12.	Ramon Soria, P.; Arrue, B.C.; Ollero, A. A 3D-printable docking system for aerial robots: Controlling aerial robotic manipulators in outdoor industrial applications. *IEEE Robot. Autom. Mag.* **2019**, *26*, 44–53. [CrossRef]

13.	Ruggiero, F.; Trujillo, M.A.; Cano, R.; Ascorbe, H.; Viguria, A.; Perez, C.; Lippiello, V.; Ollero, A.; Siciliano, B. A multilayer control for multirotor UAVs equipped with a servo robot arm. In Proceedings of the IEEE Int. Conference on Robotics and Automation (ICRA), Seattle, WA, USA, 26–30 May 2015; pp. 4014–4020.

14.	Suarez, A.; Jimenez-Cano, A.E.; Vega, V.M.; Heredia, G.; Rodriguez-Castaño, A.; Ollero, A. Design of a lightweight dual arm system for aerial manipulation. *Mechatronics* **2018**, *50*, 30–44. [CrossRef]

15.	Fumagalli, M.; Naldi, R.; Macchelli, A.; Forte, F.; Keemink, A.Q.L.; Stramigioli, S.; Carloni, R.; Marconi, L. Developing an aerial manipulator prototype: Physical interaction with the environment. *IEEE Robot. Autom. Mag.* **2014**, *21*, 41–50. [CrossRef]

16.	Hamaza, S.; Georgilas, I.; Fernandez, M.; Sanchez, P.; Richardson, T.; Heredia, G.; Ollero, A. Sensor installation and retrieval operations using an unmanned aerial manipulator. *IEEE Robot. Autom. Lett.* **2019**, *4*, 2793–2800. [CrossRef]

17.	Suarez, A.; Heredia, G.; Ollero, A. Physical-virtual impedance control in ultralightweight and compliant dual-arm aerial manipulators. *IEEE Robot. Autom. Lett.* **2018**, *3*, 2553–2560. [CrossRef]

18.	Bodie, K.; Brunner, M.; Pantic, M.; Walser, S.; Pfändler, P.; Angst, U.; Siegwart, R.; Nieto, J. An omnidirectional aerial manipulation platform for contact-based inspection. In Proceedings of Robotics: Science and Systems. *arXiv* **2019**, arXiv:1905.03502.

19.	Ruggiero, F.; Cacace, J.; Sadeghian, H.; Lippiello, V. Impedance control of VToL UAVs with a momentum-based external generalized forces estimator. In Proceedings of the IEEE International Conference on Robotics and Automation (ICRA), Hong Kong, China, 31 May–7 June 2014; pp. 2093–2099.

20.	Tomić, T.; Haddadin, S. A unified framework for external wrench estimation, interaction control and collision reflexes for flying robots. In Proceedings of the IEEE/RSJ International Conference on Intelligent Robots and Systems, Chicago, IL, USA, 14–18 September 2014; pp. 4197–4204.

21.	Suarez, A.; Real, F.; Vega, V.M.; Heredia, G.; Rodriguez-Castaño, A.; Ollero, A. Compliant bimanual aerial manipulation: Standard and long reach configurations. *IEEE Access* **2020**, *8*, 88844–88865. [CrossRef]

22.	Suarez, A.; Heredia, G.; Ollero, A. Lightweight compliant arm with compliant finger for aerial manipulation and inspection. In Proceedings of the IEEE/RSJ International Conference on Intelligent Robots and Systems (IROS), Daejeon, Korea, 9–14 October 2016; pp. 4449–4454.

23.	AERIAL-CORE Project Home Page. Available online: https://aerial-core.eu/ (accessed on 12 October 2020).

24.	Real, F.; Torres-González, A.; Ramon-Soria, P.; Capitan, J.; Ollero, A. UAL: An abstraction layer for unmanned aerial vehicles. In Proceedings of the 2nd International Symposium on Aerial Robotics, Philadelphia, PA, USA, 11–12 June 2018.

25.	Perez-Jimenez, M.; Montes-Grova, M.A.; Ramon-Soria, P.; Arrue, B.C.; Ollero, A. POSITRON: Lightweight active positioning compliant joints robotic arm in power lines inspection. In Proceedings of the 2020 International Conference on Unmanned Aircraft Systems (ICUAS), Athens, Greece, 1–4 September 2020; pp. 729–736.

26.	Sanchez-Cuevas, P.J.; Heredia, G.; Ollero, A. Characterization of the aerodynamic ground effect and its influence in multirotor control. *Int. J. Aerosp. Eng.* **2017**, *2017*, 1823056. [CrossRef]

27.	Liu, T.; Lei, Y.; Han, L.; Xu, W.; Zou, H. Coordinated resolved motion control of dual-arm manipulators with closed chain. *Int. J. Adv. Robot. Syst.* **2016**, *13*, 80. [CrossRef]

Appl. Sci. **2020**, *10*, 8927

28. Lee, J.; Chang, P.H.; Jamisola, R.S. Relative impedance control of dual-arm robots performing asymmetric bimanual tasks. *Trans. Ind. Electron.* **2014**, *61*, 3786–3796. [CrossRef]

29. Stolfi, A.; Gasbarri, P.; Sabatini, M. A combined impedance-PD approach for controlling a dual-arm space manipulator in the capture of a non-cooperative target. *Acta Astronaut.* **2017**, *139*, 243–253. [CrossRef]

Publisher's Note: MDPI stays neutral with regard to jurisdictional claims in published maps and institutional affiliations.

Article

Cartesian Aerial Manipulator with Compliant Arm

Alejandro Suarez *, Manuel Perez, Guillermo Heredia and Anibal Ollero

GRVC Robotics Labs, University of Seville, 41092 Sevilla, Spain; manup1192@gmail.com (M.P.);
guiller@us.es (G.H.); aollero@us.es (A.O.)
* Correspondence: asuarezfm@us.es

Abstract: This paper presents an aerial manipulation robot consisting of a hexa-rotor equipped with a 2-DOF (degree of freedom) Cartesian base (XY–axes) that supports a 1-DOF compliant joint arm that integrates a gripper and an elastic linear force sensor. The proposed kinematic configuration improves the positioning accuracy of the end effector with respect to robotic arms with revolute joints, where each coordinate of the Cartesian position depends on all the joint angles. The Cartesian base reduces the inertia of the manipulator and the energy consumption since it does not need to lift its own weight. Consequently, the required torque is lower and, thus, the weight of the actuators. The linear and angular deflection sensors of the arm allow the estimation, monitoring and control of the interaction wrenches exerted in two axes (XZ) at the end effector. The kinematic and dynamic models are derived and compared with respect to a revolute-joint arm, proposing a force-position control scheme for the aerial robot. A battery counterweight mechanism is also incorporated in the X–axis linear guide to partially compensate for the motion of the manipulator. Experimental results indoors and outdoors show the performance of the robot, including object grasping and retrieval, contact force control, and force monitoring in grabbing situations.

Keywords: aerial manipulation; Cartesian manipulator; compliance; hexa-rotor

Citation: Suarez, A.; Perez, M.; Heredia, G.; Ollero, A. Cartesian Aerial Manipulator with Compliant Arm. *Appl. Sci.* **2021**, *11*, 1001. https://doi.org/10.3390/app11031001

Received: 28 December 2020
Accepted: 18 January 2021
Published: 22 January 2021

Publisher's Note: MDPI stays neutral with regard to jurisdictional claims in published maps and institutional affiliations.

1. Introduction

1.1. Aerial Manipulation

Aerial manipulation robots (drones equipped with robotic arms) are extending the potential applications of multirotor platforms, allowing the realization of tasks involving the manipulation of objects and tools in flight, as well as interaction with the environment with a certain level of dexterity and control capability. The development of this technology is motivated by interest in reducing the time, cost and risk for human workers associated with the realization of certain tasks in high altitude or difficult access workspaces such as power lines [1,2], chemical plants [3], oil and gas refineries [4], and other infrastructures [5,6]. Recent works in this field have demonstrated the possibility to conduct operations such as object grasping [7–9], valve turning [10], sensor installation and retrieval [2,11], contact-based inspection [3,12,13], insulation of cracks and leaks [14], or the realization of other tasks with grippers and other tools [15,16]. Several prototypes and morphologies of manipulators have been specifically developed for their integration in multi-rotors, including multi-joint arms [17,18], dual arm systems [7,10], linear actuators [11], delta manipulators [14], compliant joint arms [2,19,20], long reach aerial manipulators [2,21], or three-arm manipulators used for object grasping and as reconfigurable landing gear [22].

Compliance, either mechanical [7,19–23] or at control level [11,24–26], is a highly desirable feature for an aerial manipulation robot operating in contact with the environment since the stability of the aerial platform may be compromised due to the interaction wrenches exerted on flight [24,25,27]. The design and development of compliant manipulators aim to increase safety by exploiting the energy storage capacity and passivity properties of springs [7,19,26] and elastomers [23] while protecting the aerial robot against impacts and overloads [23]. Unlike most industrial manipulators that use expensive force-torque sensors in the joints or in the end effector, the servo actuators (Herkulex, Dynamixel) typically employed for building lightweight robotic arms for aerial manipulation do not provide a direct measurement of the torque; or this is based on the current, so its accuracy is low due to the friction of the gearbox. This has motivated the development of methods for estimating and controlling the forces and torques from the measurement of the deflection in the elastic element, either at joint level [7,19] or in the Cartesian space [26]. Despite the evident benefits of mechanical compliance during the physical interactions of the aerial robot while flying, the main drawback of these mechanisms is the reduction in position accuracy due to the deflection of the joints [19] or links [2], a consequence of the second order dynamics associated with the mass-spring-damper system [26]. As a result, the realization of tasks requiring accurate positioning tends to become more difficult.

1.2. Positioning Accuracy

Positioning accuracy is one of the most relevant and challenging requirements for an aerial manipulator operating on flight since it determines the reliability and success rate in the realization of tasks like grasping [7] or the installation of devices on power lines [1,26], especially outdoors where Vicon or OptiTrack systems are not available, whereas sensors like 3D laser scanners [28] and vision-based systems [29] are more suited. This feature is directly related to the performance of the multirotor position controller [30,31], affected by the endogenous and exogenous wrenches exerted through the manipulator [25,27]. On the one hand, tilted-rotor hexa-rotors, also known as fully actuated aerial platforms [32,33], allow the translation of the platform as well as the application of wrenches [34,35] without changing the orientation, which contributes to increasing the accuracy in the realization of manipulation tasks compared to coplanar multi-rotors. On the other hand, the error propagation at the end effector is closely related with the kinematic configuration and the physical construction of the manipulator. This is, in most cases, a multi-link arm [7,8,17,18,24] with revolute joints. However, this configuration present two main drawbacks: (1) the joint limits (especially in the elbow joint) constraint or reduce the effective reach and workspace of the arm, (2) a significant amount of torque and energy is needed to lift the weight of the manipulator itself (forearm-upper arm configuration), requiring high torque actuators that tend to increase the weight. Alternatively, other works propose the use of simple manipulators [3,5,11,14,15] in order to reduce the weight, at expenses of reducing the reach and dexterity.

1.3. Contribution of This Work

The main contribution of this paper is the design, development and experimental validation of a new morphology of aerial manipulation robot consisting of a hexa-rotor platform integrating a 2-DOF (degree of freedom) Cartesian base (XY–axes) that carries a 1-DOF compliant joint arm equipped with a gripper and a deflection-based force-torque sensor. Figure 1 shows a picture of the prototype. The proposed morphology is intended to improve the positioning accuracy of the end effector while reducing the inertia, torque and weight of the actuators, since the Cartesian base supports most of the weight of the robotic manipulator (0.85 kg weight, 1 m reach, and 10 cm/s speed). The elastic deflection sensors integrated in the compliant joint and in the fore-arm link are used to estimate and monitor the force (XZ–axes) and torque (pitch) exerted by the manipulator over the aerial platform. This allows the realization of a number of tasks such as sensor installation with contact force control, force monitoring in grabbing situations, or object grasping and retrieval. The kinematic and dynamic models of the aerial manipulator are derived, while analyzing in more detail the error propagation at the end effector from the Jacobian, comparing the proposed morphology with respect to that of a revolute-joint arm in terms of positioning accuracy, reach, and workspace. This paper also proposes a force-position control scheme for the aerial robot that takes into account the wrench estimation provided by the arm.

Experimental results conducted in test-bench as well as indoor/outdoor flight tests validate the functionalities and capabilities of the developed Cartesian aerial robot.

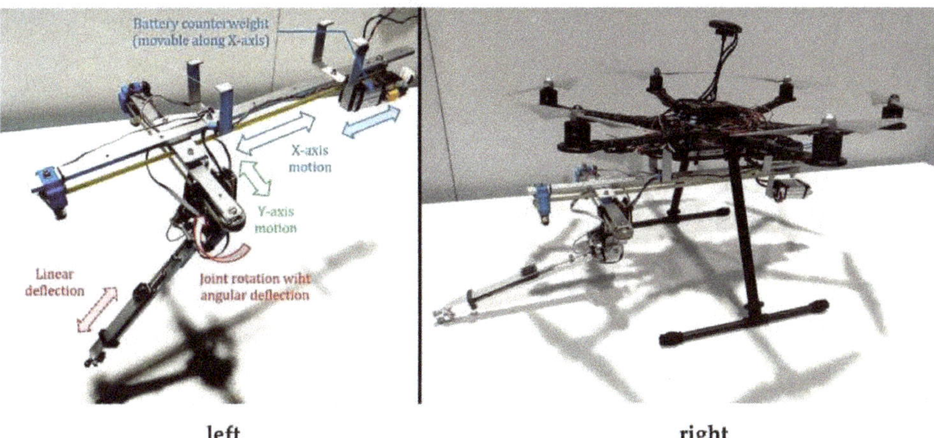

Figure 1. Cartesian manipulator (**left**) integrated in S550 hexarotor platform (**right**).

The rest of the paper is organized as follows. Section 2 presents the motivation for and describes the prototype of a Cartesian aerial manipulator with compliant joint arm. Section 3 presents the kinematic and dynamic models of the aerial robot, analyzing and comparing the features of the Cartesian manipulator with respect to a revolute-joint arm. Section 4 covers the control of the aerial robot, the experimental results are shown in Section 5, and the conclusions in Section 6.

2. System Description

2.1. Motivation

The Cartesian manipulator aims to avoid three inconveniences of the human-like robotic arms developed in our previous work [7,19]. First of all, several manipulation tasks like grasping or peg-in-hole are defined in the Cartesian space rather than in the joint space, so it is more convenient that the mapping between both spaces is straightforward in order to improve the positioning accuracy. According to the forward kinematic model, the XYZ position of the end effector depends simultaneously on the rotation angle of the three [19] or four [7] joints of the manipulator (shoulder and elbow). Therefore, the joint position errors, as well as errors associated with the manufacturing and assembly of the arm (clearance, misalignment), will be propagated to the three Cartesian coordinates of the end effector. Additionally, the mechanical deflection of the joints in a compliant arm [7] may cause deviations of several centimeters in the position of the end effector with respect to an equivalent stiff-joint arm. In a Cartesian manipulator, instead, the position of the end effector on each axis is controlled with its own actuator independently from the other joints, so the position errors are decoupled and the trajectory control in the task space is simplified.

Secondly, the effective reach of a human-like manipulator (~50 cm reach) is in practice relatively small, around 30 cm, taking into account that the nominal operation position of the arm is typically an L-shaped configuration (forearm lifted 90°) and due to the limit in the rotation angle of the elbow joint. Note that humans extend the reach of the hands with the torso and legs, whereas in an aerial manipulator the motion of the arms is usually constrained by the landing gear and the perimeter of the propellers. In this sense, the Cartesian base avoids dead-zones in its workspace due to joint limits, allowing access to points close to the multirotor base, which can be useful, for example, in object retrieval and storage. Finally, a two link manipulator in the upper arm-forearm morphology typically requires a significant amount of torque and energy to lift its own weight, whereas in the developed prototype the Cartesian manipulator only has to compensate for the inertia and friction of the carriage, reducing in this way the required torque of the actuators, and thus, their weight. Not only that, but the linear guide also isolates the actuators from impacts and loads, avoiding the design of bearing-based protection mechanisms, as in [7].

The Cartesian manipulator shown in Figure 1 implements a Prismatic-Prismatic-Revolute (PPR) configuration, introducing a revolute joint attached at the Cartesian base, since the landing gear and central body of the hexa-rotor platform do not allow the placement of a linear actuator in the vertical axis, and because this configuration extends the reach of the end effector in the X–axis (forward direction). No additional degrees of freedom are considered for end effector positioning due to the convenience of reducing the weight, complexity and size of the manipulator, and because the capabilities of the robot can be extended with the additional degrees of freedom provided by the aerial platform [13,32].

2.2. Mechanical Design

The Cartesian aerial manipulator shown in Figure 1 consists of two systems. The aerial platform is built from a Banggood S550 hexa-rotor (similar to the DJI F550) that integrates the DJI 2312E brushless motors with 9 × 4.5 inch propellers, and DJI 430 LITE electronic speed controllers (ESC). The onboard controller comprises a Pixhawk autopilot and a Raspberry Pi model 3B+ that implements the software architecture described in the next subsection. The Cartesian manipulator is divided into the five parts identified in Figure 2: X–axis linear guide (base), Y–axis linear guide, servo actuator, compliant arm and battery counterweight mechanism. The mechanical specifications of the manipulator are indicated in Table 1, representing in Table 2 the mass and inertia parameters obtained from the 3D model.

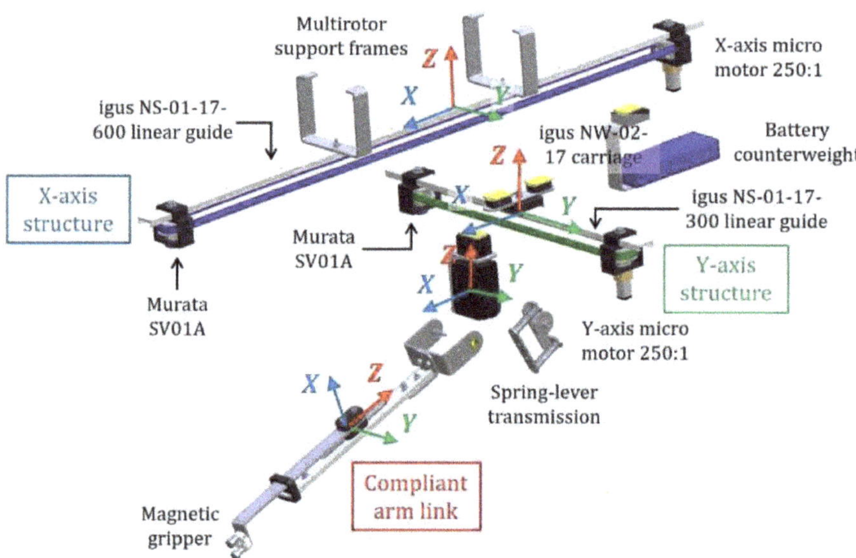

Figure 2. Components of the Cartesian manipulator, indicating the center of mass of each structure (XYZ axes).

Table 1. Specifications of the Cartesian manipulator and the compliant joint arm.

Cartesian Manipulator		Compliant Joint Arm	
Actuator features	Stall torque: 0.3 Nm Mass: 10 g, 120 RPM	Actuator features	Stall torque: 1.17 Nm Stiffness: 1.4 Nm/rad
Linear range	X–axis: 500 mm	Rotation range	$\pm120°$
	Y–axis: 250 mm	Joint deflection	$30°$
Max. speed	100 mm/s	Link length	250 mm
Total mass	0.85 kg (including battery–2S LiPo, 1500 mAh)		

Table 2. Mass and inertia parameters of the hexa-rotor platform and the different structures of the Cartesian manipulator relative to the center of mass. The cross inertial terms (I_{xy}, I_{xz}, I_{yz}) are omitted since their value is negligible.

Structure	Mass [kg]	I_{xx} [kg·m^2]	I_{yy} [kg·m^2]	I_{zz} [kg·m^2]
Hexa-rotor	1.27	1.9×10^{-2}	1.8×10^{-2}	3.0×10^{-2}
X–axis	0.2	7.9×10^{-5}	6.8×10^{-3}	6.8×10^{-3}
Y–axis	0.14	1.0×10^{-3}	2.1×10^{-5}	1.0×10^{-3}
Servo	0.08	3.4×10^{-5}	3.0×10^{-5}	1.0×10^{-5}
Link	0.09	6.0×10^{-4}	6.0×10^{-4}	1.1×10^{-5}

The 2-DOF Cartesian base is built from the igus® NS-01-17-600 (X–axis) and NS-01-17-300 (Y–axis) linear guides, using three NW-02-17 carriages to support the mobile parts. Two 22 mm Ø timing pulleys are placed at the tip of both linear guides to transmit the motion of the Pololu micro metal gear motors (10 g weight, 250:1 reduction ratio, 2.7 kg·cm stall torque) to the timing belts, measuring the rotation angle with a Murata SV01A potentiometer and with the magnetic encoder of the micro-motors. The linear displacement of the Cartesian base relative to an initial position is obtained from the rotation angle and the number of turns given by the encoder (6 pulses per revolution at the micro motor shaft, 1500 pulses per turn at the output shaft of the gearbox). Figure 3 illustrates the timing pulley-belt actuator in the X and Y linear guides. The battery counterweight is supported by a C-shaped aluminum frame that can slide along the X–axis linear guide and move in synchronization with the Y–axis structure, but in the opposite direction when it is attached to the X–axis timing belt (see Section 5.2).

The integration of the Cartesian manipulator in the hexa-rotor platform and the battery counter-weight mechanism are shown in Figure 4. The X–axis linear guide is supported by the carbon fiber tubes attached to the multirotor base through the two U-shaped aluminum frames identified in Figure 2. As can be seen in the front view of Figure 4, the legs of the landing gear should be separated to avoid collision of the Y–axis linear guide. However, this additional separation is not needed if the battery counter-weight mechanism is enabled. As depicted in Figure 4, this mechanism consists of a C–shaped aluminum frame that supports the 2S LiPo battery and is actuated by the X–axis timing pulley, but in the opposite direction to the Y–axis linear guide. Since both structures share the same linear guide, the effective range is reduced by half. The counterweight mechanism can be set to a fixed position by simply removing the zip tie that connects the timing belt with the linear carriage.

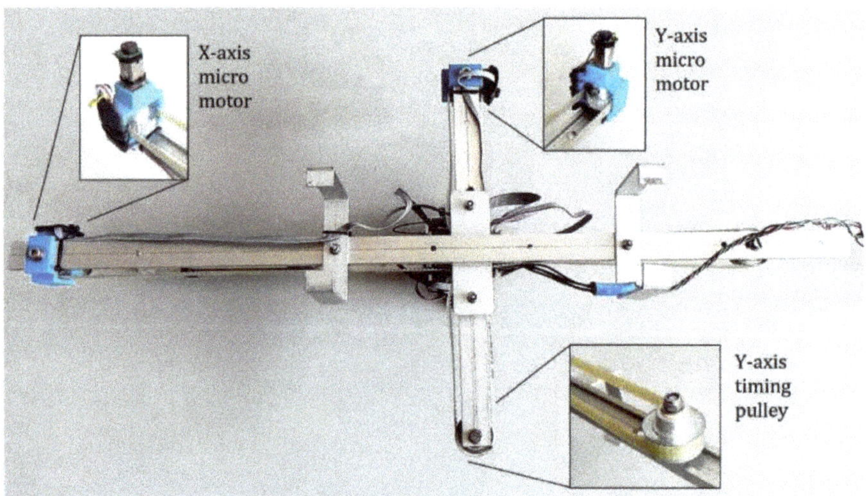

Figure 3. Pololu micro metal gear motors employed in the XY axes' timing pulley-belt transmission.

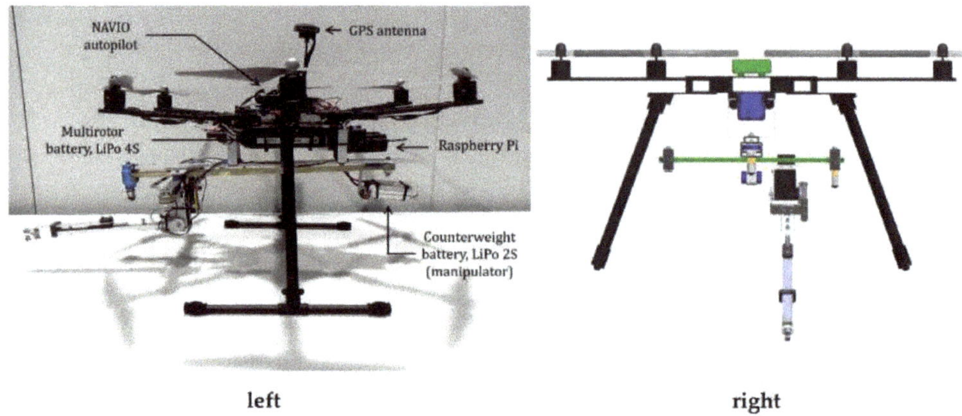

Figure 4. Side (**left**) and front (**right**) views of the Cartesian aerial manipulator, indicating the components.

The compliant arm depicted in Figure 5 employs a Herkulex DRS-0101 servo and a spring-lever transmission mechanism (1.4 Nm/rad stiffness) that allows the estimation and control of the torque from the deflection measurement [19], integrating an AS5048 magnetic encoder to obtain the deflection angle at 500 Hz with 0.2° accuracy. The estimator described in Section 4.2 employs this signal for collision detection and controls the torque/force/impedance [26] exerted at the end effector. The link of the arm consists of a NS–01–17–200 linear guide and a NW–02–17 carriage that supports the displacement of the end effector ($\pm$30 mm range), using a pair of extension springs (3.2 N/mm stiffness) and a linear potentiometer to estimate the pushing/pulling forces in the XZ–axes in grabbing conditions, as will be explained in Section 4.2.

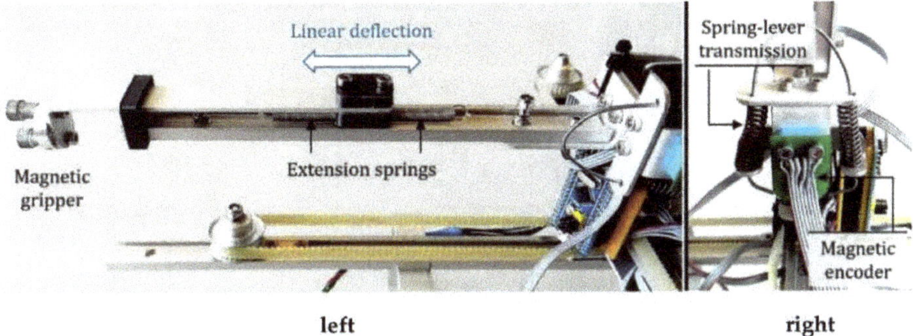

Figure 5. Magnetic gripper supported by the compliant linear joint (**left**). Compliant joint actuator (**right**).

The end effector of the arm can be reconfigured according to the intended task and type of object to be manipulated. This can be a simple hook to retrieve a device with handles, a magnetic gripper for catching ferromagnetic objects as hand tools, or a servo-gripper for grabbing objects of different shapes.

As indicated in the previous subsection, one of the main benefits of the Cartesian morphology with respect to the human-like dual arm systems developed in our previous work [7,26] is that the manipulator wastes lesser energy in lifting its own weight or reaching any point in its workspace, as the displacement of the end effector along the XY–axes is supported by the linear guide system. Note that in a forearm-upper arm configuration, the shoulder actuators must lift a 50 cm length frame structure, including the servo actuators placed in the links. Thus, the proposed solution is more energy efficient in this sense.

Finally, it is necessary to remark that the purpose of the Cartesian manipulator and the compliant arm is to provide a certain level of dexterity (3–DOFs) and sense capability (F_x, F_z, τ_θ) to the aerial platform, sufficient to implement a number of functionalities and tasks without unnecessarily increasing the weight with additional sensors or actuators.

2.3. Hardware Architecture

The components and architecture of the Cartesian aerial manipulator are represented in Figure 6. The aerial platform consists of the S550 hexa-rotor integrating the NAVIO autopilot and a Raspberry Pi 3B+ board running the PX4 firmware [36] as well as the control program of the manipulator, developed in C/C++. This program is built around the Task Manager class [2], which implements the different functionalities used in the experiments: go to rest position, go to operation position, teleoperation, force control and monitoring, and Cartesian base calibration. The manipulator is interfaced through a USB–to–USART (Universal Synchronous-Asynchronous Receiver Transmitter) device connecting the Raspberry Pi board with the STM32F303K8 Nucleo microcontroller that takes care of the low level control of the XY–axes' micro motors and reads the measurements from the three sensors:

- The two potentiometers used to estimate the displacement of the linear base.
- The magnetic encoder attached to the servo that measures the joint deflection.
- The linear potentiometer in the link used to estimate the pushing/pulling force.

The microcontroller generates as output the PWM (Pulse Width Modulation) signals taken as input by the DRV 8833 dual H-bridge that controls the micro motors (Section 4.1), providing another PWM signal to control a gripper servo. The Herkulex servo of the compliant joint is controlled in position through a dedicated USART interface [7,26]. The control references generated by the Task Manager and the feedback measurements collected by the microcontroller are sent at 100 Hz through the USART as data packets codified as C-type data structures.

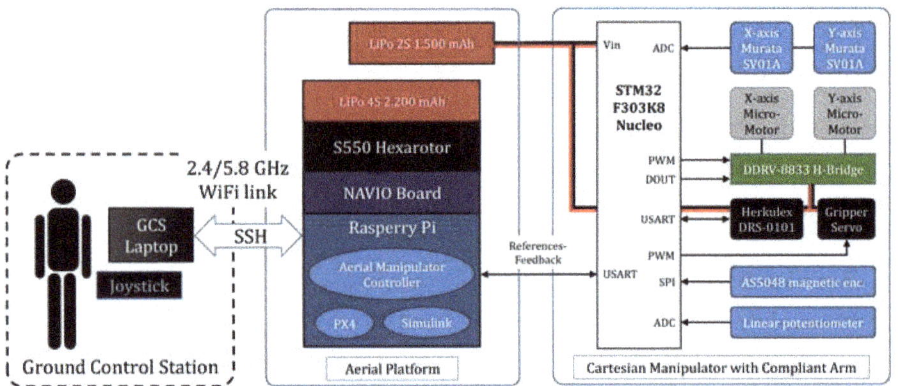

Figure 6. Components and architecture of the Cartesian aerial manipulator with ground control station (GCS).

The Cartesian aerial manipulator is operated from the Ground Control Station (GCS) through a wireless link, typically a Wi–Fi network, that supports the SSH (Secure Shell) session used to launch and manage remotely the aerial manipulator control programs. The commands introduced by the operator to select the particular task to be executed or to tele-operate the manipulator using a joystick are sent as C–type data packets through UDP (User Datagram Protocol) sockets.

3. Modelling

3.1. Kinematic Model

In the definition of an aerial manipulation task, three reference frames can be identified: the Earth fixed frame $\{E\}$ (inertial), the multirotor base frame $\{B\}$, and the manipulator base frame $\{0\}$. The position and orientation of the multirotor, ${}^{E}r_B = [x,y,z]^T$ and ${}^{E}\eta_B = [\phi,\theta,\psi]^T$, are referred to the Earth fixed frame, whereas the position of the tool center point (TCP) of the manipulator, ${}^{0}r_{TCP} = [x_{TCP}, y_{TCP}, z_{TCP}]^T$, is expressed in $\{0\}$ for convenience, considering that the onboard perception system usually gives the reference positions in the local axes. Figure 7 depicts the reference frames, position vectors and the joint variables in the Cartesian aerial manipulator. Here d_1 and d_2 denote the displacement of the Cartesian base along the $X_0 Y_0$ axes, respectively, q_3 is the rotation angle of the joint in the pitch angle, whereas D_1, D_2 and L are the corresponding link lengths.

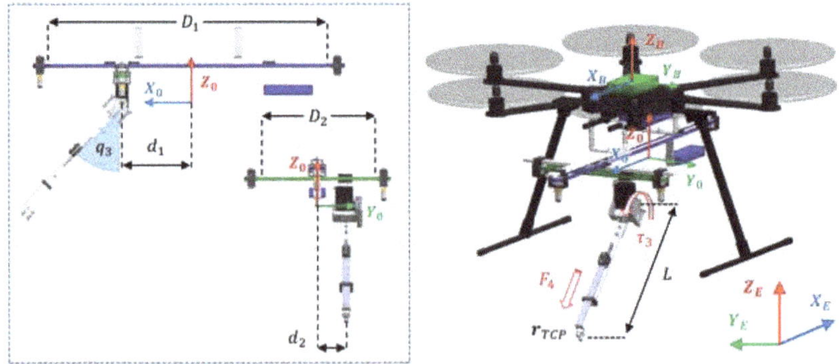

Figure 7. Kinematic model of the Cartesian aerial manipulator.

The position of the tool center point referred to $\{0\}$ can be computed easily from the joint variables applying the forward kinematic model $FK : \Re^3 \rightarrow \Re^3$:

$$
{}^0r_{TCP} = \begin{bmatrix} x_{TCP} \\ y_{TCP} \\ z_{TCP} \end{bmatrix} = FK\left(\begin{bmatrix} d_1 \\ d_2 \\ q_3 \end{bmatrix}\right) = \begin{bmatrix} d_1 + L\cdot\sin(q_3) \\ d_2 \\ -L\cdot\cos(q_3) \end{bmatrix} \tag{1}
$$

Note that x_{TCP} depends on two joint variables (d_1 and q_3), although q_3 is determined by z_{TCP}. The inverse kinematic model, $IK : \Re^3 \rightarrow \Re^3$, provides two analytical solutions for q_3 if $|x_{TCP}| < D_1/2$, and only one solution if $|x_{TCP}| \geq D_1/2$. The double solution is determined by the sign of q_3, and indicates the pointing direction of the link, either forwards or backwards:

$$
\begin{bmatrix} d_1 \\ d_2 \\ q_3 \end{bmatrix} = IK\left(\begin{bmatrix} x_{TCP} \\ y_{TCP} \\ z_{TCP} \end{bmatrix}\right) = \begin{bmatrix} x_{TCP} \mp L\cdot\sin(\alpha) \\ y_{TCP} \\ \pm\alpha \end{bmatrix} \tag{2}
$$

where angle α is defined in the following way:

$$
\alpha = \left| \cos^{-1}\left(\frac{z_{TCP}}{L}\right) \right| \tag{3}
$$

3.2. Positioning Accuracy and Workspace

It is interesting to compare the workspace, joint limits and position errors of the proposed Cartesian manipulator with compliant joint arm with respect to other kinematic configurations, like the 3–DOF robotic arm described in [19], where the TCP position is given by:

$$
{}^0r_{TCP}^* = \begin{bmatrix} r(q_2, q_3)\cdot\cos(q_1) \\ r(q_2, q_3)\cdot\sin(q_1) \\ L_1\cos(q_2) + L_2\cos(q_2 + q_3) \end{bmatrix} \tag{4}
$$

Here L_1 and L_2 are the upper arm and forearm link lengths, respectively, and:

$$
r(q_2, q_3) = L_1\sin(q_2) + L_2\sin(q_2 + q_3) \tag{5}
$$

As stated in the introduction and in Section 2.1, the Cartesian base reduces the propagation of positioning errors in the TCP due to angular deviations (joint deflection) or manufacturing errors (misalignments, clearance). This can be quantitatively evaluated through the Jacobian matrix:

$$
\varepsilon_{TCP} = J(q)\cdot\varepsilon_q \tag{6}
$$

where ε_{TCP} is the position deviation of the TCP in the XYZ axes due to a variation in the vector of joint variables ε_q, which depends on the Jacobian of the arm particularized to the nominal pose, $J(q)$. Although it is omitted for space reasons, the Jacobian of the manipulator described by Equation (4) introduces coupled terms in the XYZ axes, i.e., each of the Cartesian coordinates depends on the three joint variables simultaneously, whereas in the kinematic configuration considered in this paper, the Jacobian matrix is significantly simplified, resulting in the following expression.

As can be seen, the positioning accuracy in the X–axis depends on the accuracy of the X–axis actuator and on the compliant joint, whereas the accuracy on the YZ axes dep

$$
\varepsilon_{TCP} = \begin{bmatrix} \varepsilon_Y \\ \varepsilon_Y \\ \varepsilon_Z \end{bmatrix} = \begin{bmatrix} 1 & 0 & L\cdot\cos(q_3) \\ 0 & 1 & 0 \\ 0 & 0 & L\cdot\sin(q_3) \end{bmatrix} \cdot \begin{bmatrix} \varepsilon_{d_1} \\ \varepsilon_{d_2} \\ \varepsilon_{q_3} \end{bmatrix} = \begin{bmatrix} \varepsilon_{d_1} + L\cdot\cos(q_3)\cdot\varepsilon_{q_3} \\ \varepsilon_{d_2} \\ L\cdot\sin(q_3)\cdot\varepsilon_{q_3} \end{bmatrix} \tag{7}
$$

ends on the accuracy of the d_2 and q_3 coordinates.

The volume of operation of the Cartesian manipulator is a parallelepiped of length $D_1 = 0.5$ [m], depth $D_2 = 0.2$ [m], and height $L = 0.2$ [m], with two additional circular sections generated by the revolution of the compliant joint with maximum rotation. q_3^{max} is around 120°:

$$V = D_1 D_2 L + D_2 L^2 q_3^{max} = 0.037 \left[\mathrm{m}^3 \right] \tag{8}$$

whereas the volume of operation of a 3–DOF robotic arm corresponding to a hollow semi-sphere with upper arm link length $L_1 = 0.2$ [m] and forearm link length $L_2 = 0.2$ [m] is given by:

$$V^* = \frac{4}{3} \pi \left[(L_1 + L_2)^3 - R_{min}^3 \right] = 0.234 \left[\mathrm{m}^3 \right] \tag{9}$$

This is represented graphically in Figure 8, and compared with respect to the workspace of the 3–DOF manipulator described in [19]. The hollow area is determined by the maximum rotation angle of the elbow joint, q_3^{max}, defining a minimum radius R_{min} from the origin of the manipulator frame to the end effector. Although the difference is significant, the dexterity and performance of configurations with revolute joints are typically reduced when integrated in multirotor platforms due to the motion constraints associated with the landing gear and the effect of the kinematic singularities when the arm tends to be stretched. The effective workspace of the Cartesian prototype described here can be extended by exploiting the rotation of the multirotor in the yaw angle, equal to $V = 0.13$ $\left[\mathrm{m}^3 \right]$ for a full rotation. The combination of this morphology with tilted or fully actuated platforms like [32–34] is especially convenient in applications related to object retrieval and storage in order to extend the workspace of the aerial robot while providing a high positioning accuracy.

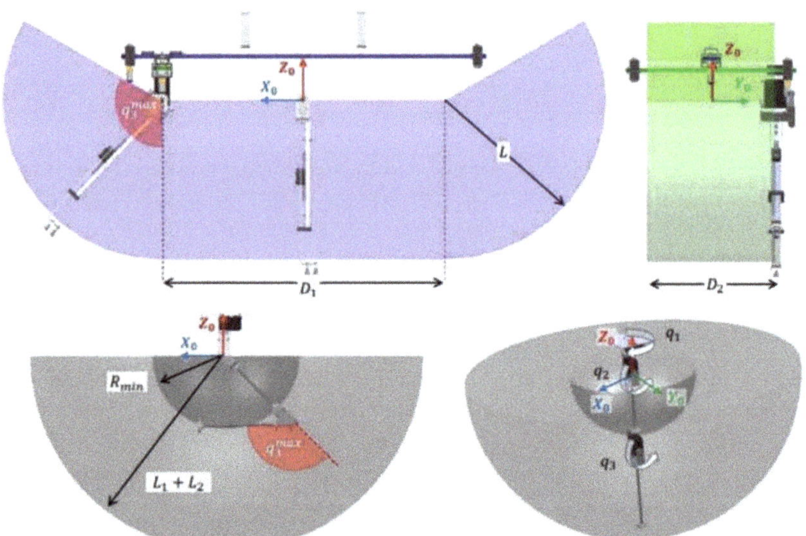

Figure 8. Volume of operation in the Cartesian manipulator (**up**) and in a 3-degrees of freedom (3-DOF) revolute joint arm (**down**).

3.3. Dynamic Model

The equations of the dynamic model of the Cartesian aerial manipulator with compliant arm can be derived following the Euler-Lagrange formulation based on the Lagrangian and the generalized equation of the forces and torques:

$$\frac{d}{dt} \left\{ \frac{\partial L}{\partial \dot{\xi}} \right\} - \frac{\partial L}{\partial \xi} = \boldsymbol{\Gamma} + \boldsymbol{\Gamma}_{ext} \; ; \; L = K - V \tag{10}$$

Here L is the Lagrangian, which is the difference between the kinetic and potential energies, K and V, respectively, ξ is the vector of generalized coordinates, whereas Γ and Γ_{ext} are the vectors of endogenous and exogenous (external) generalized forces, defined as follows.

$$\xi = \begin{bmatrix} {}^E r_B^T & {}^E \eta_B^T & d_1 & d_2 & q_3 & \theta_3 & \Delta l_4 \end{bmatrix}^T \in \Re^{11} \tag{11}$$

$$\Gamma = \begin{bmatrix} F_B^T & \tau_B^T & F_1 & F_2 & \tau_3 & \tau_{m,3} & F_4 \end{bmatrix}^T \in \Re^{11} \tag{12}$$

where ${}^E r_B = [x, y, z]^T$ and ${}^E \eta_B = [\phi, \theta, \psi]^T$ represent the position and orientation of the aerial platform in the Earth fixed frame, d_1, d_2, and q_3 are the joint variables defined earlier, and Δl_4 is the linear elongation of the end effector. The forces and torques acting over the hexa-rotor are denoted by F_B and τ_B, whereas F_1 and F_2 are the force exerted by the Cartesian manipulator in the X and Y axes, respectively, τ_3 is the torque applied over the compliant joint, and F_4 is the force exerted or supported in the direction of the arm link.

The kinetic energy is the sum of two terms corresponding to the aerial platform, considered as a rigid solid, and the masses of the manipulator:

$$K = \frac{1}{2} m_{AP} \parallel {}^E \dot{r}_B \parallel^2 + \frac{1}{2} {}^E \omega_B^T I_{AP} {}^E \omega_B + \frac{1}{2} \sum_{i=1}^{4} \left(m_i \parallel {}^E \dot{r}_i \parallel^2 + {}^E \omega_i^T I_i {}^E \omega_i \right) \tag{13}$$

where m_{AP} and I_{AP} are the mass and inertia tensor of the aerial platform (including the X–axis support structure), m_i and I_i are the mass and inertia of the i-th link of the manipulator (Y–axis linear structure, servo structure, arm link and forearm, as indicated in Figure 2 and Table 2), ${}^E \dot{r}_B$ and ${}^E \omega_B$ are the translational and rotational speed of the aerial platform, whereas ${}^E \dot{r}_i$ and ${}^E \omega_i$ are the speed of the i-th link of the manipulator.

The potential energy includes the gravity potential and elastic potential of the compliant joints:

$$V = g e_Z^T \left\{ m_{AP} {}^E r_B + \sum_{i=1}^{4} m_i {}^E r_i \right\} + \frac{1}{2} k_3 (\theta_3 - q_3)^2 + \frac{1}{2} K_S \Delta l^2 \tag{14}$$

Here g is the gravity constant, $e_Z^T = [0,\ 0,\ 1]$ is the Z–axis unit vector, k_3 is the torsional stiffness of the compliant joint, and K_S is the stiffness of the extension springs that support the end effector, as shown in Figure 5. Applying the Lagrangian in Equation (10) over the kinetic and potential energies given by Equations (13) and (14), the usual compact matrix form is derived, given by:

$$M(\xi)\ddot{\xi} + C\left(\xi, \dot{\xi}\right) + G(\xi) + K(\xi) = \Gamma + \Gamma_{ext} \tag{15}$$

where $M \in \Re^{11 \times 11}$ is the generalized inertia matrix, $C \in \Re^{11}$ represents the centrifugal and Coriolis terms, $G \in \Re^{11}$ corresponds to the gravity wrenches, and $K \in \Re^{11}$ are the elastic potential terms. It is worth noting that, since the analytical expression of the kinematics of the Cartesian manipulator is simpler compared to a revolute joint arm due to the reduction in sine/cosine terms and the decoupling of the Cartesian coordinates, the complexity of the dynamic model is reduced, and consequently so is the computational time, facilitating real-time implementation.

4. Estimation and Control

4.1. Cartesian Base Control

The Cartesian manipulator described in Section 2.2 is built with DC micro-motors controlled at low level through a PWM signal (1 kHz frequency) generated by the microcontroller board. The motion of the linear guide system depicted in Figure 3 can be modelled as follows:

$$m_i \ddot{d}_i + f_i \dot{d}_i = r_i \cdot pwm_i; \ i = \{1,\ 2\} \tag{16}$$

where m_i is the displaced mass, f_i is the friction coefficient of the linear carriage, $pwm_i \in [-1, 1]$ represents the mean current injected to the motor relative to the stall current, whereas r_i is a constant that depends on the motor current-torque constant and the radius of the timing pulley.

Two control modes are implemented on the microcontroller. On the one hand, the open-loop velocity control consists of directly applying the PWM reference provided by the higher level controller. According to Equation (16), if the acceleration term is neglected, then the linear velocity can be regulated directly through the PWM signal. On the other hand, a PI position controller executed at 100 Hz is used for moving the Cartesian base at desired absolute positions:

$$pwm_i^{ref} = \left\{ \begin{array}{ll} sign(e_{d_i}) \cdot PWM_i^{max} & if\ |e_{d_i}| \geq e_{d_i}^{th} \\ k_{P_i} \cdot e_{d_i} + k_{I_i} \cdot \int e_{d_i} dt & if\ |e_{d_i}| < e_{d_i}^{th} \end{array} \right\} \tag{17}$$

Here $e_{d_i} = d_{i,ref} - d_i$ is the position error of the i-th Cartesian coordinate, $PWM_i^{max} \in [0,1]$ is the maximum PWM of the micro motor ($PWM_i^{max} < 0.8$ to prevent damaging the motor winding when it is stalled), whereas k_{P_i} and k_{I_i} are the proportional and integral gains. To avoid the discontinuity of the control signal, it is imposed that $k_{P_i} \cdot e_{d_i}^{th} = PWM_i^{max}$. The integral term is reset when $|e_{d_i}| \geq e_{d_i}^{th}$ to prevent the saturation of the controller due to the wind-up effect.

4.2. Force-Torque Estimation with Compliant Arm

The linear and angular deflection sensors integrated in the compliant arm allow the estimation of the force and torque exerted on the manipulator in the corresponding directions, as represented in Figure 7. On the one hand, the spring-lever transmission mechanism introduced in the servo joint allows the estimation and control of the torque from the deflection measurement [7,19]. If k_3 and d_3 denote the joint stiffness and damping, the torque delivered by the motor can be estimated as follows:

$$\tau_{m,3} = k_3 \Delta\theta_3 + d_3 \dot{\Delta\theta_3} \tag{18}$$

This torque is spent to compensate three terms: gravity, output link inertia, and the torque due to external forces:

$$\tau_{m,3} = J_3 \ddot{q}_3 + m_3 g l_3 \sin(q_3) + F_{ext} \sin(\beta) L \tag{19}$$

Here m_3 and J_3 are the mass and inertia of the output link, l_3 is the distance from the rotation axis to the center of mass, and F_{ext} is the external force acting over the tool center point at distance L with an angle β. The transmitted torque can be controlled through the servo position reference [19]:

$$\theta_3^{ref} = \theta_3 + \left(k_{P_3} \tau_{e,3} + k_{I_3} \int \tau_{e,3} dt \right) \tag{20}$$

where θ_3 is the current servo position, $\tau_{e,3} = \tau_3^{ref} - \tau_{m,3}$ is the torque error, and k_{P_3} and k_{I_3} are the gains in PI control. This controller will be applied in the experiment shown in Section 5.4. On the other hand, the force acting along the direction of the arm link is estimated from the elongation of the extension springs (see Figure 5), measured with a linear potentiometer:

$$F_4 = K_4 \Delta l_4 \tag{21}$$

where K_4 is the spring stiffness and Δl_4 is the elongation. The experiment described in Section 5.3 (force monitoring in grabbing situations) exploits this estimation to detect over-loads and release the gripper in order to prevent crashes. The torque exerted by the compliant joint and the force along the link are supported by the thrust generated by the multirotor. These wrenches are depicted in Figure 7. However, note that the arm does not provide an estimation of the Y–axis wrenches.

4.3. Cartesian Aerial Manipulator Control Scheme

The control scheme of the Cartesian aerial manipulator is shown in Figure 9. The Task Manager [2,7] is the high-level functional block that implements the functionalities of the robot, generating the appropriate control references for the manipulator and the aerial platform, relying on the low level controllers of the Cartesian manipulator (Sections 4.1 and 4.2) and the autopilot. Some of the tasks are:

- For the multirotor: take-off, land, follow trajectory, go to workspace, approach reference position, interaction force control [25,27,33–35,37].
- For the manipulator: move manipulator to rest/operation position, object grasping [7,8] and store, sensor installation/retrieval [2,11], teleoperation.

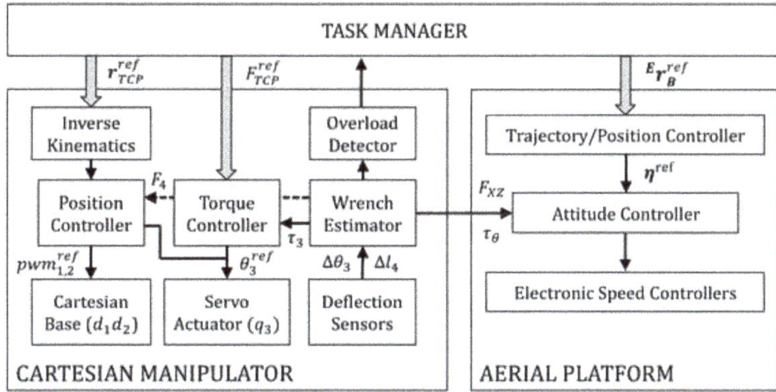

Figure 9. Functional blocks of the Cartesian aerial manipulator controller.

Since the aerial robot is intended to operate in contact with objects and the environment, it is necessary to estimate and monitor the interaction wrenches exerted by the manipulator to ensure the stability of the aerial platform, exploiting for this purpose the linear and angular deflection sensors of the compliant arm (see Figure 5). The Overload Detector, evaluated in Section 5.3, will report the Task Manager in case the arm is not able to retrieve a grabbed object, whereas the Wrench Estimator [25,27] can be used by the Attitude Controller to compensate for the external forces (see Section 5.4). In this way the attitude control is partially coupled with the manipulator control. The stability of the interaction control in the Cartesian configuration has been analyzed in terms of Lyapunov in [25,37].

5. Experimental Results

5.1. Cartesian Base Control

As stated in the introduction, the Cartesian base that supports the compliant arm enhances the positioning accuracy of the end effector compared to a revolute joint manipulator, in such a way that the mapping defined by the forward kinematics given by Equation (1) is almost straightforward. In order to evaluate this feature and the low level position control of the Cartesian base described in Section 4.1, a calibration test is conducted, executing a stair sequence from $d_1 = -100$ to $d_1 = 100$ (mm)and from $d_2 = -50$ to $d_2 = 50$ (mm), with increments of $\Delta d_1 = 25$ and $\Delta d_2 = 10$ (mm), respectively. Figure 10 represents the evolution of the two variables along with the position control error. The Murata SV01 potentiometers used to measure the rotation angle and the number of turns of the micro motors provide an effective rotational angle of $333°$. Note that a pair of potentiometers is required to overcome the dead zone. The 12-bit resolution ADC limits the accuracy to $0.08°$, which corresponds to a linear displacement of 0.014 mm for a 21 mm Ø timing pulley.

However, in practice, the accuracy is limited by the backlash of the gearbox of the micro motors (around 2°) and the clearance in the timing belt.

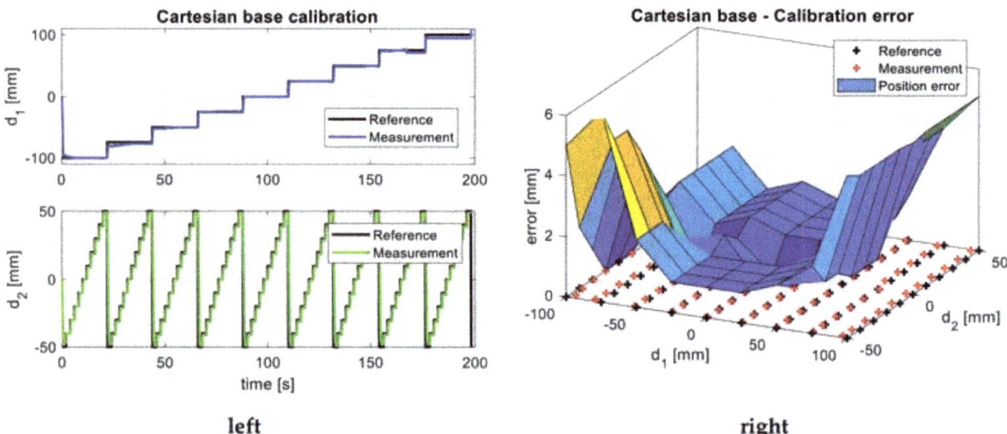

Figure 10. Cartesian base calibration: reference and measured displacement (**left**), and control error (**right**) in $X_0 Y_0$.

5.2. Compensation of Mass Displacement

The positioning accuracy of the end effector is also affected by the dynamic coupling between the aerial platform and the manipulator, as the movement of the Cartesian base causes a variation of the center of mass and a reaction torque over the multirotor. The X–axis linear guide in Figure 4 incorporates a counterweight mechanism that moves the manipulator battery synchronized with the Y–axis guide, so the mass distribution is more symmetric, but at expenses of reducing the range of d_1 by half. The goal of this experiment is to evaluate the effectiveness of this mechanism in a flight test, showing the evolution of the position and velocity of the multirotor when the Y–axis structure and the compliant arm move in the forward direction while the multirotor is controlled in stabilize mode. The position controller is disabled in order to observe more clearly the desired effect, using a laser tracker to measure the multirotor position. Figure 11 ilustrates the execution of the experiment, representing the signals of interest in Figure 12.

The manipulator is initially at rest position, $d_1 = 0$, $d_2 = 0$, $q_3 = 0$. Between $t = 0$ and $t = 1$ [s], the Cartesian base is displaced to $d_1 = 95$ [mm], and between $t = 4$ and $t = 5$ [s] q_3 is rotated from 0 to 90°. Note that the rotation of the joint cannot be compensated with the counter-weight mass. As can be seen on the lower part of Figure 12, the perturbation in the position and velocity of the multirotor along the X–axis (blue line) is more evident when the counterweight mechanism is disabled.

5.3. Monitoring Linear Deflection in Grabbing Conditions

The link of the compliant arm is equipped with an elastic force sensor consisting of a pair of extension springs and a linear potentiometer that measures the deflection along the link (Figure 5). This mechanism is used in this experiment to detect and react against overloads in grabbing situations, for example when the arm tries to retrieve an object that is too heavy to be grasped or it is anchored to a fixed point. This situation is depicted in Figure 13. In order to avoid the destabilization of the attitude controller due to the interaction forces, the linear deflection is monitored in such a way that the gripper is released if the deflection exceeds a threshold of 10 (mm), corresponding to a force of around 3 (N). The compliant joint is controlled imposing a zero torque reference, according to the controller given by Equation (20), so the reaction torque in the pitch angle of the multirotor is relatively low. Figure 13 shows the monitoring-release capability on flight, representing the position of the Cartesian base, along with the linear deflection, the joint torque and the orientation of the multirotor. Once the support structure is grabbed at $t = 3$ [s], the X–axis guide moves backwards trying to retrieve the object. At $t = 3.6$ [s], the linear deflection threshold is exceeded, so the Cartesian base reacts by moving the gripper forwards while the compliant joint is rotated to release the bar. The process is repeated at $t = 10$ [s]. The hexa-rotor was controlled in stabilize mode.

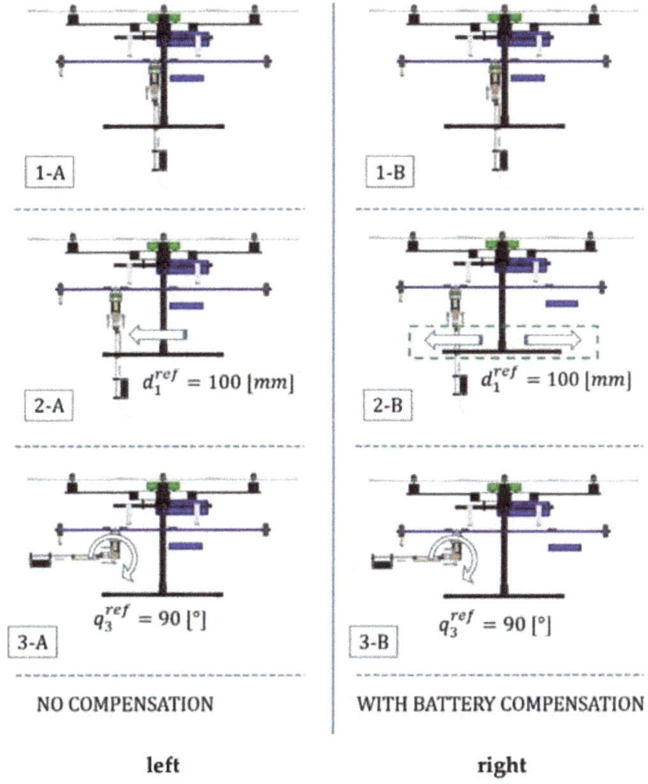

Figure 11. Phases involved in the battery counterweight experiment: no battery (**left**), battery counterweight (**right**).

5.4. Contact Force Control

This experiment is intended to evaluate the performance of the contact force controller described in Section 4.2 and the effect over the attitude controller of the hexa-rotor platform. A sequence of images taken from the video is depicted in Figure 14. The Cartesian base and the compliant joint arm are initially at rest position ($d_1 = 0$, $d_2 = 0$, $q_3 = 0$). Once the hexa-rotor is close to the contact point, the Cartesian base moves the output link in the forward direction until the collision with the obstacle is detected at $t = 3.36$ [s]. The deflection detection threshold is set to $\Delta\theta_3^{th} = 2°$. Then, the PI force controller is activated, generating the force stair reference represented in Figure 15. As it can be seen, the pushing force is compensated by the attitude controller increasing the reference in the pitch angle, so the thrust of the multirotor is projected in the X–axis. The experiment was repeated again in order to validate the results, represented in Figure 16. The ability to accommodate the compliant joint arm is more evident here, considering the deviation in the position of the multirotor. The contact force control is useful in tasks like sensor installation and retrieval, inspection by contact, or for moving objects in the environment.

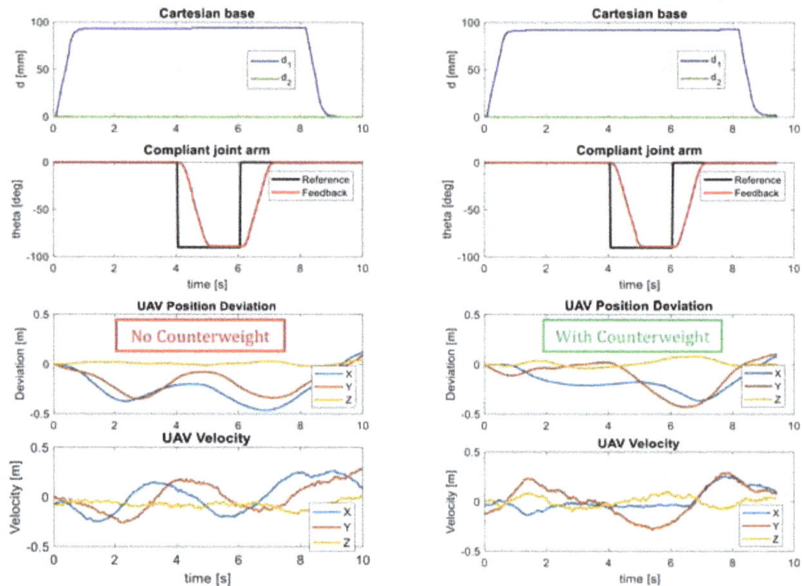

Figure 12. Multirotor position deviation when the manipulator moves in the forward direction (d_1, q_3).

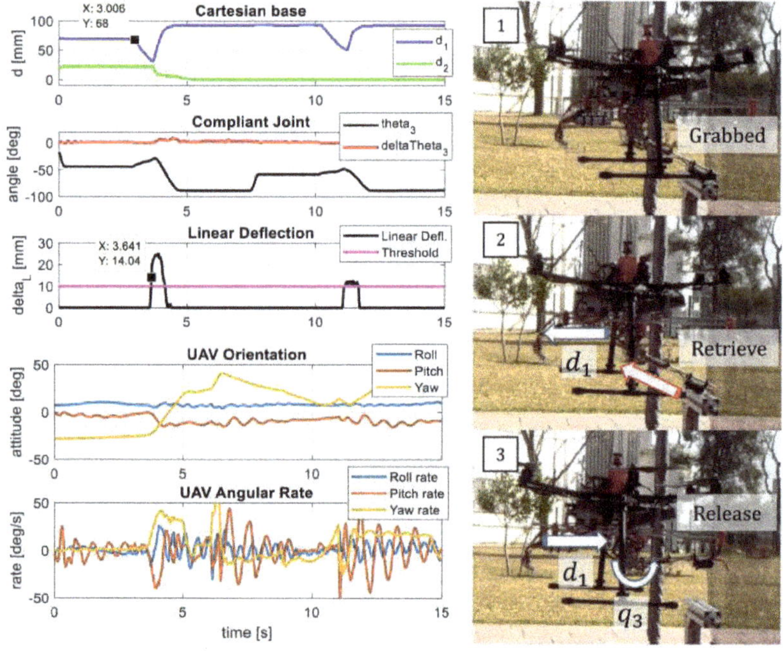

Figure 13. Variables of interest (**left**) and sequence of images (**right**) in the force monitoring experiment.

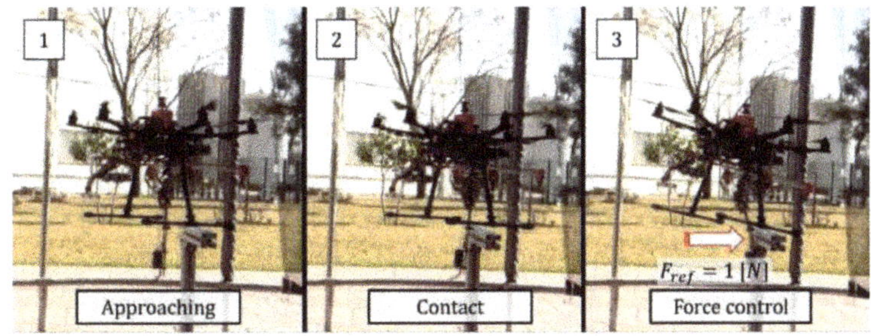

Figure 14. Sequence of images from the contact force control experiment.

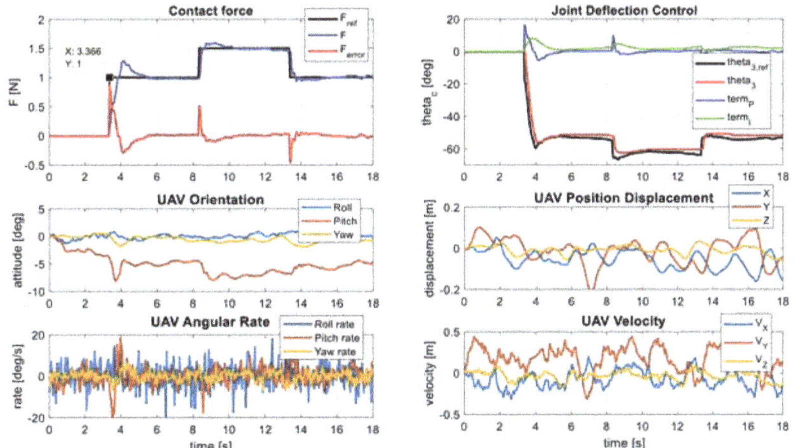

Figure 15. Contact force control experiment. Force reference and estimation (**up, left**), servo position and PI correction terms (**up, right**), multirotor position, velocity, orientation and angular rate (**down**).

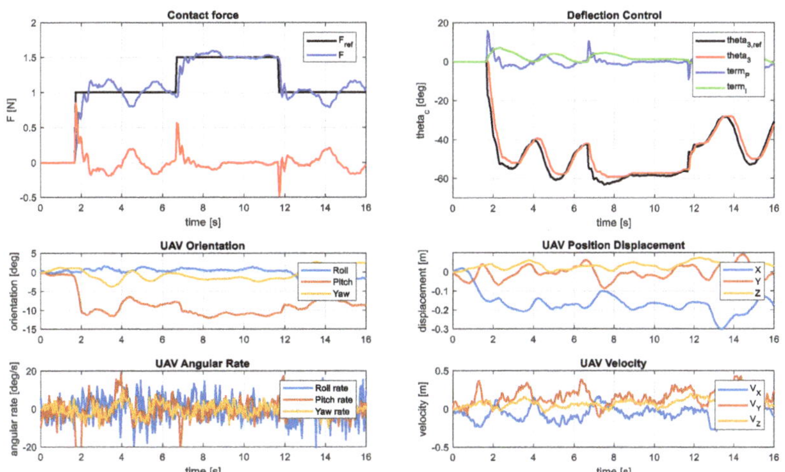

Figure 16. Contact force control experiment. Second test.

6. Conclusions

This paper described an aerial manipulator consisting of a multirotor equipped with a 2–DOF Cartesian base (XY–axes) and a compliant arm with angular and linear deflection sensors that allow the estimation, monitoring and control of the interaction wrenches exerted on flight, improving safety in grabbing conditions. This new morphology presents several benefits with respect to revolute joint manipulators in terms of positioning accuracy, energy consumption, and effective workspace, thanks to mechanical construction with two transversal linear guide systems that support the translation of the compliant arm. The forward kinematic model shows that the map between the joint space and the Cartesian space, in which the intended task is usually defined, is almost straightforward, so the Cartesian coordinates (XYZ) can be controlled with the corresponding actuators, reducing in this way the kinematic coupling between the joints and, with it, the propagation of errors at the end effector. The lack of dead-zones around the base of the manipulator extends the effective workspace, particularly useful for object retrieval and store operations. In this sense, the experiments presented in the paper showed how the linear deflection sensor can be used to detect overloads when the arm tries to grab an object too heavy to be lifted, moving the Cartesian base to release it and preventing the stability of the aerial platform from being compromised. The experiments also show the possibility of exerting sustained contact forces with the compliant joint arm, and the effect of the battery counterweight over positioning accuracy (at the expense of reducing the motion range in the X–axis by half). Although the mechanism is able to compensate partially for the reaction wrenches caused by the displacement of the Y–axis structure in the pitch angle, its effectiveness is reduced when the compliant arm is rotated, so this may not be practical in some cases taking into account the limitation it imposes on the range of motion of the X–axis structure.

Author Contributions: Conceptualization, A.S.; validation, A.S and M.P.; writing—original draft preparation, A.S.; writing—review and editing, A.S. and G.H.; supervision, G.H. and A.O.; project administration, A.O.; funding acquisition, A.O. All authors have read and agreed to the published version of the manuscript.

Funding: This work was funded by the AERIAL-CORE (H2020-2019-871479) and HYFLIERS (H2020-ICT-25-2017-779411) projects funded by the European Commission; and the ARTIC (RTI2018-102224-B-I00) projects funded by the Spanish Ministerio de Economia, Industria, y Competitividad.

Informed Consent Statement: Not applicable.

Acknowledgments: Authors want to thank Antonio Enrique Jimenez-Cano for his support in the realization of the experiments.

Conflicts of Interest: The authors declare no conflict of interest.

References

1. The AERIAL-CORE Project. Available online: https://aerial-core.eu/ (accessed on 28 December 2020).
2. Suarez, A.; Real, F.; Vega, V.M.; Heredia, G.; Rodriguez-Castano, A.; Ollero, A. Compliant Bimanual Aerial Manipulation: Standard and Long Reach Configurations. *IEEE Access* **2020**, *8*, 88844–88865. [CrossRef]
3. Trujillo, M.A.; Dios, J.R.M.-D.; Martín, C.; Viguria, A.; Ollero, A. Novel Aerial Manipulator for Accurate and Robust Industrial NDT Contact Inspection: A New Tool for the Oil and Gas Inspection Industry. *Sensors* **2019**, *19*, 1305. [CrossRef] [PubMed]
4. Ollero, A.; Cortés, J.; Santamaria-Navarro, Àngel; Trujillo, M.A.; Balachandran, R.; Andrade-Cetto, J.; Rodriguez, A.; Heredia, G.; Franchi, A.; Antonelli, G.; et al. The AEROARMS Project: Aerial Robots with Advanced Manipulation Capabilities for Inspection and Maintenance. *IEEE Robot. Autom. Mag.* **2018**, *25*, 12–23. [CrossRef]
5. Ikeda, T.; Yasui, S.; Fujihara, M.; Ohara, K.; Ashizawa, S.; Ichikawa, A.; Okino, A.; Oomichi, T.; Fukuda, T. Wall contact by octo-rotor UAV with one DoF manipulator for bridge inspection. In Proceedings of the 2017 IEEE/RSJ International Conference on Intelligent Robots and Systems (IROS), Vancouver, BC, Canada, 24–28 September 2017; pp. 5122–5127.
6. Fumagalli, M.; Naldi, R.; Macchelli, A.; Forte, F.; Keemink, A.Q.L.; Stramigioli, S.; Carloni, R.; Marconi, L. Developing an Aerial Manipulator Prototype: Physical Interaction with the Environment. *IEEE Robot. Autom. Mag.* **2014**, *21*, 41–50. [CrossRef]
7. Suarez, A.; Heredia, G.; Ollero, A. Design of an Anthropomorphic, Compliant, and Lightweight Dual Arm for Aerial Manipulation. *IEEE Access* **2018**, *6*, 29173–29189. [CrossRef]

8. Seo, H.; Kim, S.; Kim, H.J. Aerial grasping of cylindrical object using visual servoing based on stochastic model predictive control. In Proceedings of the 2017 IEEE International Conference on Robotics and Automation (ICRA), Singapore, 29 May–3 June 2017; pp. 6362–6368.

9. Chen, H.; Quan, F.; Fang, L.; Zhang, S. Aerial Grasping with a Lightweight Manipulator Based on Multi-Objective Optimization and Visual Compensation. *Sensors* **2019**, *19*, 4253. [CrossRef] [PubMed]

10. Orsag, M.; Korpela, C.; Bogdan, S.; Oh, P. Valve turning using a dual-arm aerial manipulator. In Proceedings of the 2014 International Conference on Un-manned Aircraft Systems (ICUAS), Orlando, FL, USA, 27–30 May 2014; pp. 836–841.

11. Hamaza, S.; Georgilas, I.; Fernandez, M.; Sanchez-Cuevas, P.J.; Richardson, T.; Heredia, G.; Ollero, A. Sensor Installation and Retrieval Operations Using an Unmanned Aerial Manipulator. *IEEE Robot. Autom. Lett.* **2019**, *4*, 2793–2800. [CrossRef]

12. Wopereis, H.W.; Hoekstra, J.J.; Post, T.H.; Folkertsma, G.A.; Stramigioli, S.; Fumagalli, M. Application of substantial and sustained force to vertical surfaces using a quadrotor. In Proceedings of the 2017 IEEE International Conference on Robotics and Automation (ICRA), Singapore, 29 May–3 June 2017; pp. 2704–2709.

13. Tognon, M.; Chavez, H.A.T.; Gasparin, E.; Sable, Q.; Bicego, D.; Mallet, A.; Lany, M.; Santi, G.; Revaz, B.; Cortés, J.; et al. A Truly-Redundant Aerial Manipulator System with Application to Push-and-Slide Inspection in Industrial Plants. *IEEE Robot. Autom. Lett.* **2019**, *4*, 1846–1851. [CrossRef]

14. Chermprayong, P.; Zhang, K.; Xiao, F.; Kovac, M. An Integrated Delta Manipulator for Aerial Repair: A New Aerial Robotic System. *IEEE Robot. Autom. Mag.* **2019**, *26*, 54–66. [CrossRef]

15. Shimahara, S.; Leewiwatwong, S.; Ladig, R.; Shimonomura, K. Aerial torsional manipulation employing multi-rotor flying robot. In Proceedings of the 2016 IEEE/RSJ International Conference on Intelligent Robots and Systems (IROS), Daejeon, Korea, 9–14 October 2016; pp. 1595–1600.

16. Nguyen, H.N.; Park, S.; Lee, D. Aerial tool operation system using quadrotors as rotating thrust generators. In Proceedings of the IEEE International Conference on Intelligent Robots and Systems, Hamburg, Germany, 28 September–2 October 2015.

17. Bellicoso, C.D.; Buonocore, L.R.; Lippiello, V.; Siciliano, B. Design, modeling and control of a 5-DoF light-weight robot arm for aerial manipulation. In Proceedings of the 2015 23rd Mediterranean Conference on Control and Automation (MED), Torremolinos, Spain, 16–19 June 2015; pp. 853–858.

18. Cano, R.; Pérez, C.; Pruano, F.; Ollero, A.; Heredia, G. Mechanical design of a 6-DOF aerial manipulator for assembling structures using UAVs. In Proceedings of the 2nd RED-UAS 2013 Workshop on Research, Education and Development of Unmanned Aerial Systems, Compiegne, France, 20–22 November 2013.

19. Suarez, A.; Heredia, G.; Ollero, A. Lightweight compliant arm with compliant finger for aerial manipulation and inspection. In Proceedings of the 2016 IEEE/RSJ International Conference on Intelligent Robots and Systems (IROS), Daejeon, Korea, 9–14 October 2016; pp. 4449–4454.

20. Yuksel, B.; Mahboubi, S.; Secchi, C.; Bulthoff, H.H.; Franchi, A. Design, identification and experimental testing of a light-weight flexible-joint arm for aerial physical interaction. In Proceedings of the 2015 IEEE International Conference on Robotics and Automation (ICRA), Seattle, WA, USA, 26–30 May 2015; pp. 870–876.

21. Miyazaki, R.; Paul, H.; Kominami, T.; Shimonomura, K. Wire-Suspended Device Control Based on Wireless Communication with Multirotor for Long Reach-Aerial Manipulation. *IEEE Access* **2020**, *8*, 172096–172104. [CrossRef]

22. Paul, H.; Miyazaki, R.; Ladig, R.; Shimonomura, K. Landing of a Multirotor Aerial Vehicle on an Uneven Surface Using Multiple On-board Manipulators. In Proceedings of the 2019 IEEE/RSJ International Conference on Intelligent Robots and Systems (IROS), Venetian Macao, Macau, China, 4–8 November 2019; pp. 1926–1933.

23. Bartelds, T.; Capra, A.; Hamaza, S.; Stramigioli, S.; Fumagalli, M. Compliant Aerial Manipulators: Toward a New Generation of Aerial Robotic Workers. *IEEE Robot. Autom. Lett.* **2016**, *1*, 477–483. [CrossRef]

24. Cataldi, E.; Muscio, G.; Trujillo, M.A.; Rodriguez, Y.; Pierri, F.; Antonelli, G.; Caccavale, F.; Viguria, A.; Chiaverini, S.; Ollero, A. Impedance Control of an aerial-manipulator: Preliminary results. In Proceedings of the 2016 IEEE/RSJ International Conference on Intelligent Robots and Systems (IROS), Daejeon, Korea, 9–14 October 2016; pp. 3848–3853.

25. Ruggiero, F.; Cacace, J.; Sadeghian, H.; Lippiello, V. Impedance control of VToL UAVs with a momentum-based external generalized forces estimator. In Proceedings of the 2014 IEEE International Conference on Robotics and Automation (ICRA), Hong Kong, China, 31 May–5 June 2014; pp. 2093–2099.

26. Suarez, A.; Sanchez-Cuevas, P.J.; Heredia, G.; Ollero, A. Aerial Physical Interaction in Grabbing Conditions with Lightweight and Compliant Dual Arms. *Appl. Sci.* **2020**, *10*, 8927. [CrossRef]

27. Tomic, T.; Haddadin, S. A unified framework for external wrench estimation, interaction control and collision reflexes for flying robots. In Proceedings of the 2014 IEEE/RSJ International Conference on Intelligent Robots and Systems, Chicago, IL, USA, 14–18 September 2014; pp. 4197–4204.

28. Garcia-Cruz, X.; Sergiyenko, O.; Tyrsa, V.; López, M.R.; Hernández-Balbuena, D.; Rodriguez-Quiñonez, J.; Basaca-Preciado, L.; Mercorelli, P.; Quiñonez, J.C.R. Optimization of 3D laser scanning speed by use of combined variable step. *Opt. Lasers Eng.* **2014**, *54*, 141–151. [CrossRef]

29. Lindner, L.; Sergiyenko, O.; Rivas-Lopez, M.; Ivanov, M.; Quiñonez, J.C.R.; Hernandez-Balbuena, D.; Fuentes, W.F.; Tyrsa, V.; Muerrieta-Rico, F.N.; Mercorelli, P. Machine vision system errors for unmanned aerial vehicle navigation. In Proceedings of the 2017 IEEE 26th International Symposium on Industrial Electronics (ISIE), Edinburgh, Scotland, 19–21 June 2017; pp. 1615–1620.

30. Ruggiero, F.; Trujillo, M.A.; Cano, R.; Ascorbe, H.; Viguria, A.; Perez, C.; Lippiello, V.; Ollero, A.; Siciliano, B. A multilayer control for multirotor UAVs equipped with a servo robot arm. In Proceedings of the 2015 IEEE International Conference on Robotics and Automation (ICRA), Seattle, WA, USA, 26–30 May 2015; pp. 4014–4020.

31. Lippiello, V.; Cacace, J.; Santamaria-Navarro, A.; Andrade-Cetto, J.; Trujillo, M.A.; Esteves, Y.R.; Viguria, A. Hybrid Visual Servoing With Hierarchical Task Composition for Aerial Manipulation. *IEEE Robot. Autom. Lett.* **2016**, *1*, 259–266. [CrossRef]

32. Rajappa, S.; Ryll, M.; Bulthoff, H.H.; Franchi, A. Modeling, control and design optimization for a fully-actuated hexa-rotor aerial vehicle with tilted propellers. In Proceedings of the 2015 IEEE International Conference on Robotics and Automation (ICRA), Seattle, WA, USA, 26–30 May 2015; pp. 4006–4013.

33. Ryll, M.; Bicego, D.; Franchi, A. Modeling and control of FAST-Hex: A fully-actuated by synchronized-tilting hexa-rotor. In Proceedings of the 2016 IEEE/RSJ International Conference on Intelligent Robots and Systems (IROS), Daejeon, Korea, 9–14 October 2016; pp. 1689–1694.

34. Ryll, M.; Muscio, G.; Pierri, F.; Cataldi, E.; Antonelli, G.; Caccavale, F.; Bicego, D.; Franchi, A. 6D interaction control with aerial robots: The flying end-effector paradigm. *Int. J. Robot. Res.* **2019**, *38*, 1045–1062. [CrossRef]

35. Bodie, K.; Brunner, M.; Pantic, M.; Walser, S.; Pfandler, P.; Angst, U.; Siegwart, R.; Nieto, J. An omni-directional aerial manipulation platform for contact-based inspection. In Proceedings of the Robotics: Science and Systems, Breisgau, Germany, 22–26 June 2019.

36. Meier, L.; Honegger, D.; Pollefeys, M. PX4: A node-based multithreaded open source robotics framework for deeply embedded platforms. In Proceedings of the 2015 IEEE International Conference on Robotics and Automation (ICRA), Seattle, WA, USA, 26–30 May 2015; pp. 6235–6240.

37. Lippiello, V.; Ruggiero, F. Cartesian Impedance Control of a UAV with a Robotic Arm. In Proceedings of the IFAC Proceedings Volumes, Dubrovnik, Croatia, 5–7 September 2012; Volume 45, pp. 704–709.

 *applied sciences*

Article

Development of Add-On Planar Translational Driving System for Aerial Manipulation with Multirotor Platform

Ryo Miyazaki *, Hannibal Paul and Kazuhiro Shimonomura *

Department of Robotics, Ritsumeikan University, Kusatsu, Shiga 5258577, Japan; gr0340vs@ed.ritsumei.ac.jp
* Correspondence: rr0045vp@ed.ritsumei.ac.jp (R.M.); skazu@fc.ritsumei.ac.jp (K.S.)

Abstract: We propose an add-on planar translational driving system (ATD) which can be equipped on a multirotor platform for aerial manipulation. The device is lightweight and consists of three ducted fans controlled via an on-board CPU. It uses a simple control method and enables a multirotor to perform positioning and generate force in two dimensions while keeping the airframe horizontal. By translating the multirotor without changing attitude, it can more smoothly and easily perform many types of aerial manipulation tasks with higher positioning accuracy. In this paper, we mainly show the design, modeling, and control of the ATD. Several preliminary experiments were performed to verify the positioning accuracy and effectiveness of the system. In addition, we successfully performed the push and pull task using a rigid arm.

Keywords: aerial manipulation; multirotor UAV; translational driving system

1. Introduction

The multirotor type unmanned aerial vehicle (UAV), which is simply called a drone, has increased in popularity in the consumer industry and industrial applications in the last 10 years. Because of their flight stability, easier operation, and autonomy, they are not only useful for hobbyist applications, but also for industrial applications such as civil engineering surveys, security, agriculture, infrastructure inspections, logistics, remote sensing [1], and many other fields. In addition, in high-altitude tasks which can pose danger to human workers, such as maintenance of bridges, tunnel roofs, windmill blades, dam walls, and high-voltage electric lines, assistance by multirotor UAVs can be a promising application. The technique of physical interaction with the environment by using a manipulator attached to the aerial robot is called "aerial manipulation", and many studies are focused on it [2–4]. In previously reported studies, a robotic gripper or arm is equipped on the bottom of the multirotor to perform the manipulation task downward from the body frame [5], on its side for contact inspection [6], or on its top for perching on a high altitude place [7,8], torsional task [9], and contact inspection of the roof [10]. Although several aerial manipulation tasks are successfully performed in these studies, it is still challenging to realize both the positioning and manipulation task with high accuracy. Traditional multirotors are able to take-off and land vertically on the ground and can move vertically while keeping the attitude of the body horizontal. However, it requires tilting the airframe for translating and generating contact force in horizontal direction. This motion affects the multirotor to perform the task not so smoothly. In addition, to stop the multirotor in horizontal direction after it moves, it requires break-like motion by quickly tilting the airframe in opposite direction. These movements make it difficult for the multirotor to move finely to accurately position itself and often prevents performing smooth manipulation tasks. Therefore, to realize aerial manipulation task in higher accuracy, it is desirable that the translation and attitude control of the multirotor be separated.

To perform the position and attitude control of multirotor independently, which is impossible with a typical multirotor, several studies on novel platform development are reported in [11–17]. An omnidirectional aerial vehicle with unidirectional thrusters fixed

Citation: Miyazaki, R.; Paul, H.; Shimonomura, K. Development of Add-On Planar Translational Driving System for Aerial Manipulation with Multirotor Platform. *Appl. Sci.* **2021**, *11*, 1462. http://doi.org/10.3390/app11041462

Academic Editor: Alejandro Suarez
Received: 29 December 2020
Accepted: 2 February 2021
Published: 5 February 2021

Publisher's Note: MDPI stays neutral with regard to jurisdictional claims in published maps and institutional affiliations.

to its airframe is proposed in [11,12]. The control algorithm was designed by [11] and the aerial robot was developed to perform the translation and attitude control on SE(3) separately [12]. Another approach for realizing positing and attitude independently is by tilting the rotors of a multirotor [13,14]. In [13], a quad tilt rotor UAV is developed and successfully showed hovering of the quad rotor with a 90 deg pitch angle. In [14], the design and optimal control of a novel omnidirectional vehicle that can exert a wrench in any orientation while maintaining efficient flight configurations is presented. In [15,17], a novel multi-linked type of multirotor is developed and performed a manipulation task [16]. In all of these studies, new aerial robot hardware structures and control techniques are needed to put them to practical use.

Figure 1. Add-on planar translational driving system (ATD) with the multirotor: (**a**) the overview of ATD with multirotor; and (**b–d**) the case of assuming some aerial manipulation task in narrow region, roof, and wall surfaces.

On the other hand, many flight controllers that can fly typical multirotors with high stability are already in the market and are widely being used. The availability of these platforms, which are used in today's major applications such as aerial photography and transportation, is convenient and leads to greater reach. Therefore, we focus on utilizing a typical type of multirotor. To realize position and attitude control independently with typical type of multirotor, we propose an add-on planar translational driving system (hereafter, ATD), which consists of three ducted fans arranged to generate thrust in the horizontal direction, as shown in Figure 1a. Three ducted fans generate thrust in any direction on the horizontal plane, allowing the multi-rotor aircraft to move while keeping its attitude horizontal. As a result, fine position control becomes easy, and the positioning ability is improved.

In this study, we mainly designed ATD and its control laws. Then, several preliminary experiments were performed to verify the accuracy of positioning and effectiveness of the system in aerial manipulation. The concept description of the ATD is provided in Section 2. Section 3 presents the kinematics, control laws, and system of ATD. Section 4 presents the force measurement of ATD to verify how much force currently works on the multirotor. Section 5 discusses comparison between the ATD control with the multirotor control. Section 6 evaluates the experiments of position control and translational movement in different directions. Moreover, we show push and pull operations performed using a rigid arm for verifying the possibility of aerial manipulations. Section 7 concludes this paper and discusses about future work.

2. Concept of ATD

In order for a normal multirotor UAV to move horizontally, it is necessary to appropriately change the balance of the propellers rotation speed and tilt the aircraft. As a result, a propulsive force in the horizontal direction can be obtained. To move in the horizontal direction while keeping the attitude of the aircraft horizontal, it is sufficient to add a driving force that generates thrust in the horizontal direction. Furthermore, to move in any direction of the horizontal plane, it is necessary to be able to generate a thrust vector in any direction. For this, at least three drive sources are required. The main features of the proposed ATD are based on the concept as follows.

(1) The ATD allows the multi-rotor aircraft to move horizontally while keeping its attitude horizontal. This leads to easier and fine position control, and the positioning accuracy can be improved.
(2) When applying a force to an object sideways in aerial work, the attitude of the aircraft is horizontal, so the force can be applied continuously and stably. Moreover, the addition of ATD increases the force that can be generated compared to the case of a normal multirotor aircraft alone.
(3) The vertical thrust is generated by the multirotor aircraft, and the horizontal thrust is generated by ATD. Since these can be controlled independently, the structure makes it easy to generate horizontal force at any given height of the multi-rotor aircraft.
(4) While moving horizontally using ATD, the driving force of the multi-rotor aircraft can be devoted to keeping the altitude and attitude of the aircraft constant. Therefore, it may become robust due to disturbances such as wind.
(5) The proposed ATD is add-on type and can be retrofitted to a multirotor, which means it can work with any of typical multirotors, from hobby to industrial ones.

The system can be useful for various applications; for example, in some contact or non-contact inspection tasks, by keeping attitude of multirotor horizontal, the task can be done more easily and smoothly while staying nearer the target surfaces in a narrow space (Figure 1b), roof (Figure 1c), or walls (Figure 1d). Moreover, it also can be useful for increasing accuracy of some tasks which requires multirotor generating force in horizontal direction without changing its attitude, such as high pressure cleaning or painting tasks.

To construct ATD, to be equipped on a typical multirotor UAV, we use ducted fan as the main actuator unit, which is lightweight, compact, and able to generate relatively high thrust. The generated thrust from ATD acts on the multirotor to allow it to translate and generate the force in horizontal direction. At least three forces applied in a plane are required for positioning a body in that plane. Three or more ducted fans are required to generate thrust in any direction on the horizontal plane. In this research, we designed and manufactured ATD using three ducted fans, which is the minimum configuration.

3. Design and Implementation of the ATD

3.1. Mechanical Structure

To construct ATD, to be equipped on a typical multirotor UAV, we use ducted fan as the main actuator unit, which is lightweight, compact, and able to generate relatively high thrust. The generated thrust from ATD acts on the multirotor to allow it to translate and generate the force in horizontal direction. At least three forces applied in a plane are required for positioning a body in that plane. The structure of ATD is shown in Figure 2a. Three ducted fans are placed every 120 degrees and set as Y-configurations, so that a well-balanced force can be generated by the combination of any of two thrusts of ducted fans. To avoid exhaust from the ducted fan from being affected by the downwash of the multirotor's propellers, the ducted fans are mounted outside the propeller radius of the multirotor. ATD comprises three ducted fans, power module, PCA9685-I2C to PWM interface, ESC, and a CPU board (LattePanda Alpha 864). The ducted fan is 50 mm in diameter and can generate 0.95 kg of maximum thrust. The CPU board is used for processing feedback control. The distance L from ducted fan can be set depending on the size of multirotor platform. The specifications of the device are summarized in Table 1.

To control ATD, for any feasible control force F, the output thrust of three ducted fans should be calculated. According to the arrangement of ducted fans, the model of ATD is shown in Figure 2b. The coordination is defined as the same as the multirotor body frame. F_1, F_2, and F_3 show the output thrusts of the ducted fans. The blocked region in Figure 2b is the feasible control force volume which is mixed by thrust of three ducted fans. The region can be separated into S_1, S_2, and S_3 and given as follows.

$$S_1 := \{F \in \Re^2 \mid -kF_x \le F_y, \; kF_x - F_{max} \ge F_y, \; -kF_x - F_{max} \ge F_y, \; kF_x \le F_y\} \tag{1}$$

$$S_2 := \{F \in \Re^2 \mid kF_x \ge F_y, \; -\frac{k}{2}F_{max} \le F_x \le 0, \; kF_x + F_{max} \le F_y\} \tag{2}$$

$$S_3 := \{F \in \Re^2 \mid -kF_x \ge F_y, \; 0 \le F_x \le \frac{k}{2}F_{max}, \; -kF_x + F_{max} \le F_y\} \tag{3}$$

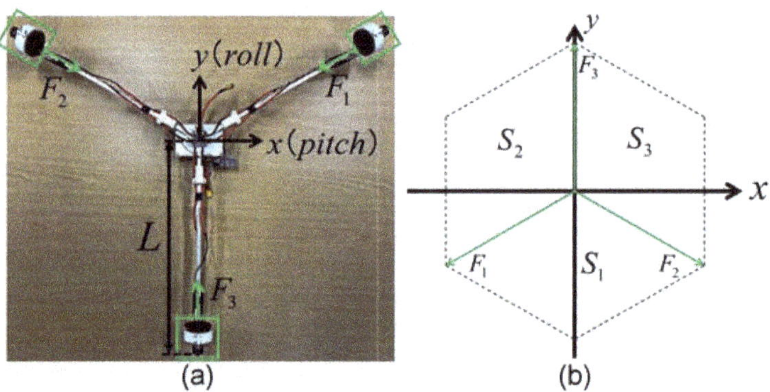

Figure 2. (**a**) Overview of ATD and its defined parameters; and (**b**) the model of ATD. The blocked region shows feasible control force volume and parameters corresponding to (**a**). The coordination is defined as the same as the body frame of multirotor.

Table 1. Specifications of ATD.

Gross weight	0.69 kg without battery
Size	0.7 m (W) × 0.6 m (D) × 0.12 m (H)
Number of ducted fan	3
Ducted fan size/motor	ϕ50 mm/4300 KV
Max thrust	950 g for each ducted fan
L	0.4 m
	L is adjustable depending on multirotor size.

In Equations (1)–(3), $\pm k$ represents the slope of F_1 and F_2 based on the coordination, and the value is $\pm\frac{\sqrt{3}}{3}$. F_{max} shows the maximum output thrust of ducted fan. Thus, for any feasible force F, output thrust of ducted fans $[F_1 \ F_2 \ F_3]^T$ can be given by as follows.

$$A = \begin{cases} \begin{bmatrix} k & 1 \\ k & -1 \\ 0 & 0 \end{bmatrix} & (F \in S_1) \\[2em] \begin{bmatrix} 2k & 0 \\ 0 & 0 \\ -k & 1 \end{bmatrix} & (F \in S_1) \ , \\[2em] \begin{bmatrix} 0 & 0 \\ 2k & 0 \\ k & 1 \end{bmatrix} & (F \in S_3) \end{cases} \qquad \begin{bmatrix} F_1 \\ F_2 \\ F_3 \end{bmatrix} = A \begin{bmatrix} F_x \\ F_y \end{bmatrix} \tag{4}$$

The input thrust of ducted fan u_i with its output thrust F_i is $u_i \propto F_i$, and we define the equation as follows.

$$F_i = \zeta u_i + C \qquad (0 \le u_i \le 100, \ i = 1, 2, 3) \tag{5}$$

In Equation (5), ζ and C are constants, and these parameters are estimated in Section 3.

3.2. Components of the System

The proposed system (shown in Figure 1a) consists of a multirotor UAV with ATD. The multirotor platform is constructed using DJI F550 frame (the diameters of the body and propellers are 550 and 238 mm, respectively) and DJI N3 flight controller. High level control for the multirotor navigation is achieved by the communication link between the flight controller and the on-board CPU. The pictures of ATD with multirotor system is shown Figure 3. The coordination is defined as the same as in Figure 2. RealSense T265 is used for estimating velocity of the UAV.

Through the communication between flight controller of the multirotor and the on-board CPU, we constructed the system, as shown in Figure 4, based on ROS(Robot Operating System). In the figure, the processes inside the CPU board block shows ROS nodes what we mainly developed and data flow shows the topics being published and subscribed in each nodes. Constructing the system using ROS improves the reusability of the system, and additional devices can be easily integrated into the system. To verify positioning accuracy of ATD with multirotor, a motion capture system is used for estimating the position. In the system, the state estimator summarizes converted sensor feedback and transmits necessary values to UAV and ATD controller. The I2C interface receives control input of three ducted fans and converts it to desired PWM signal using Python library of PCA9685. The UAV controller includes roll and pitch angle, yaw angle rate, throttle (vertical velocity/thrust), and enable/disable brake-like motion as control inputs. The brake-like motion of the multirotor will effect the flight, if it is enabled with the ATD control. Thus, we designed three different control modes, namely position control mode, manual ATD control mode, and manual UAV control mode, by making a switch based on CH5 position on the remote controller (RC) (shown in Figures 3 (side view) and 4). These control modes are described as follows.

- Position control mode allows the multirotor to translate automatically by keeping body frame horizontal. In the UAV controller, control inputs of roll and pitch angle are set to 0, yaw angle rate is controlled automatically, throttle is controlled manually, and break-like motion is disabled. In the ATD controller, the ducted fans are controlled automatically.

- Manual ATD control mode allows the operator to control the multirotor manually and translate it with fixed attitude of the airframe. In the UAV controller, the control inputs are set the same as in position control mode. The ATD controller is designed based on CH1~2 input values on the RC for operating the multirotor translating in horizontal, and a brake-like motion is designed for stopping the multirotor immediately after the operation.
- Manual UAV control mode allows the operator to control the multirotor manually which is same as typical operation. The UAV's roll and pitch angle, yaw rate, and throttle are set based on RC input of CH1~4 and brake-like motion is enabled. The ducted fans in the ATD controller are disabled.

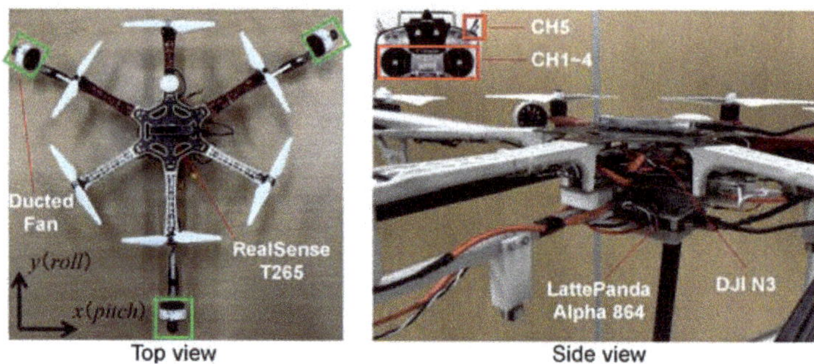

Figure 3. Pictures of ATD with multirotor system.

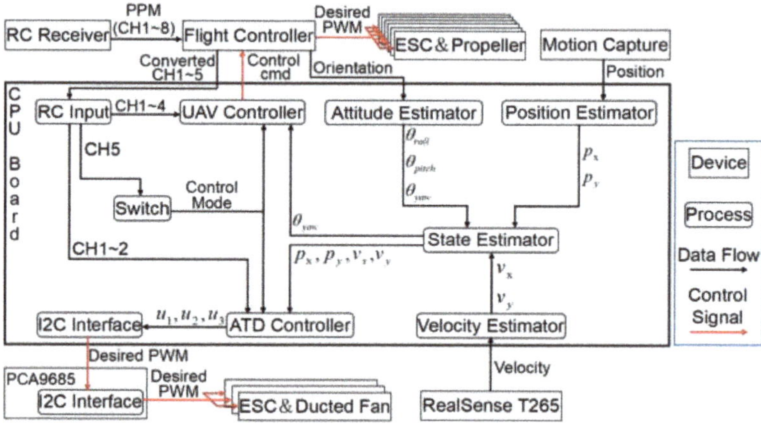

Figure 4. Block diagram of the ATD with multirotor system.

Through our designed system with three different control modes, the system allows the operator to easily switch desired control mode depending on the task. In the position control mode, the horizontal position and heading direction of multirotor are controlled automatically, and the operator only needs to adjust its height. It makes some inspection tasks much easier than when controlling the multirotor normally, such as the cases shown in Figure 1d. It can be realized by adding sensors such as LIDAR or RGB-D camera to our system for localization. On the other hand, it can be useful for performing some aerial manipulation tasks by switching to manual ATD control mode from UAV control mode. It allows an aerial manipulator to reach smoothly to the manipulated target.

3.3. Control

To design the controller of the multirotor and ATD corresponding to the three control modes described above, we defined control inputs of the multirotor as u_{roll}/u_{pitch} (roll and pitch angle), u_{yaw} (yaw angle rate), $u_{throttle}$ (vertical velocity), and u_{brake} (1 and 0 for enable/disable brake-like motion) and for ATD as u_x/u_y (thrust inputs in $x(roll)/y(pitch)$ axis), and u_1–u_3 (thrust inputs of three ducted fans). The RC input of CH1~CH5 values are set to $u_{ch1} \sim u_{ch5}$ ($u_{chi} \in \{1, -1\}$, $i = 1, \ldots, 5$).

In the case of position control mode, the control inputs of multirotor and ATD can be given by

$$u_{roll} = 0, \ u_{pitch} = 0, \ u_{yaw} = K_{pyaw}e_{yaw} + K_{dyaw}\dot{e}_{yaw}, \ u_{throttle} = K_{th}u_{ch4}, \ u_{brake} = 0$$
$$u = [u_x; u_y], \ u = K_p e_{pos} + K_d \dot{e}_{pos}, \quad (u_{ch5} = 1) \qquad (6)$$

In Equation (6), for the UAV controller, K_{pyaw} and K_{dyaw} are control gain of PD controller in yaw axis and e_{yaw} is the yaw angle error between target direction and its heading direction. K_{th} is the parameter for adjusting vertical velocity of the multirotor. In the ATD controller, K_p, K_p and K_d are the control gain of PD controller for position control by ATD, and e_{pos} is the position error in x and y axis.

In the case of manual ATD control mode, the control inputs can be given as follows:

$$u_{roll} = 0, \ u_{pitch} = 0, \ u_{yaw} = K_{pyaw}e_{yaw} + K_{dyaw}\dot{e}_{yaw}, \ u_{throttle} = K_{th}u_{ch4}, \ u_{brake} = 0$$
$$e_{yaw} = \theta_{taryaw} - \theta_{yaw}, \ \theta_{taryaw} = \theta_{yaw}(0) + K_{vyaw}\int_0^t u_{ch3}dt$$
$$u_{ch1,2} = [u_{ch1}; u_{ch2}], \ v = [v_x; v_y], \ u = \begin{cases} K_{fmax}u_{ch1,2}(u_{ch1}, u_{ch2} \neq 0) \\ -K_v v(u_{ch1}, u_{ch2} = 0) \end{cases} \quad (u_{ch5} = 0) \qquad (7)$$

In Equation (7), in the UAV controller, $\theta taryaw$, θyaw, and $\theta yaw(0)$ are target yaw angle, current yaw angle, and initial yaw angle of the airframe. K_{vyaw} is the parameter for adjusting yaw angle rate. Through calculation, yaw angle of multirotor can be controlled manually. In the ATD controller, v is the estimated velocity of multirotor. K_{fmax} is the control gain for the RC input values. K_v is the control gain for the damping control that allows multirotor to stop immediately after operating the ATD.

In the case of manual UAV control mode, the control inputs are given by

$$u_{roll} = K_{\theta max}u_{ch1}, \ u_{pitch} = K_{\theta max}u_{ch2}, \ u_{yaw} = K_{pyaw}e_{yaw} + K_{dyaw}\dot{e}_{yaw},$$
$$u_{throttle} = K_{th}u_{ch4}, \ u_{brake} = 1 \ ; \ u = [0; 0] \quad (u_{ch5} = -1) \qquad (8)$$

In Equation (8), in the UAV controller, $K_{\theta max}$ is the control gain for the RC input values, and it corresponds to the maximum feasible control input of roll and pitch angles. e_{yaw} is calculated the same as in Equation (7). In the ATD controller, the control input in x and y axis are given as 0 to disable the ATD control.

Through Equations (6)–(8), calculated control inputs for the multirotor are converted to desired PWM signal in the flight controller and sent to ESCs to control the propellers. In the ATD controller, control inputs of $u = [u_x; u_y]$ are converted to thrust inputs of ducted fans $[u_1 \ u_2 \ u_3]^T$ according to Equations (4) and (5), converted to the desired PWM signal, and sent to ESCs to control the ducted fans. To generate the PWM signal, an I2C to PWM interface (PCA9685 board) board is used with its Python library in the CPU. A range of input values were chosen for control by experimentally checking the change in the fan's speed from minimum to maximum for the given inputs. According to that, the minimum and maximum values were set as 150 and 535, respectively (e.g., the values 1000–2000 in Python control program are mapped to generate pulses in the range 1–2 ms at the output of PCA9685) at the frequency of 70 Hz. The process is shown in Figure 4.

4. Force Measurement of ATD

To verify the force generated by ATD on multirotor while it is flying, and to obtain the relation between thrust input with output force of ducted fans, we measured the force of ATD. The experimental setup is shown in Figure 5. To avoid the thrust of propellers from affecting the measurement results, we put the ATD with multirotor on the ground and fixed it to a force gauge using a wire. To reduce the friction between multirotor and ground, we fixed three ball-bearings (in this case, coefficient of friction is 0.05) under ATD and adjusted the thrust of the multirotor to slightly less than its total weight (3 kg including battery). Therefore, the friction between multirotor and ground can be ignored. In the experiment, we set the heading direction of the multirotor as the same as the wire by controlling yaw angle of the multirotor. The thrust input of ducted fan was set from 90% and reduced by 5% until 10%. The output force data were recorded for a duration of 5 s for each input.

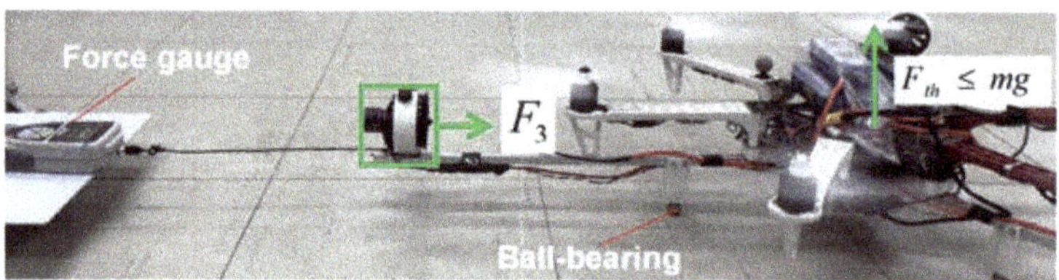

Figure 5. Experimental set up for force measurement.

The experimental result of the measurement is shown in Figure 6. The graph shows average value of the measured force for the thrust inputs and the dotted line shows its linear approximation, which is given by $y = 0.0826x + 0.1274$. In the graph, the small graph shows the recorded data of force during a 5-s interval, when the thrust input was at 60%. Accordingly, we verified if the output thrust is stable for the input thrust. The parameters ζ and C in Equation (5) can be decided based on this experiment.

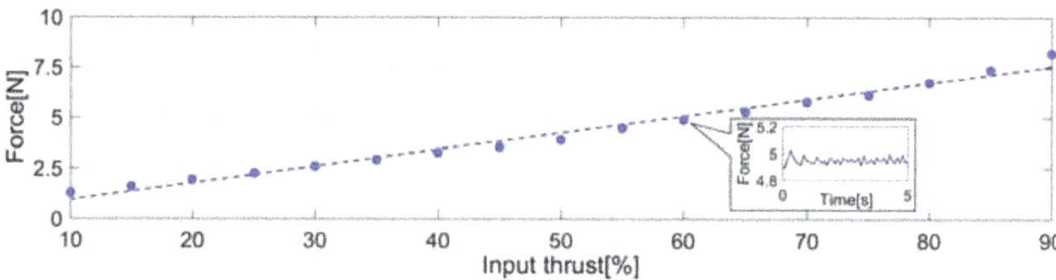

Figure 6. Experimental result on output force of the ATD for different input thrust. The linear approximation is $y = 0.0826x + 0.1274$.

5. Comparison of the Performance between ATD with Multirotor Control

To compare the performance of the ATD control and multirotor control, comparative experiments were performed. Both experiments were done with manual control. In the case of ATD control, if the input thrust in y axis (see Figure 3) is 50%, the output force can be estimated as 4.9 N (according to Figure 6). To generate the same force in y axis in the case of multirotor control, the pitch angle should be controlled to about 10 degrees. Thus, the experiment was done by translating the multirotor at the same force with a different control mode (ATD control mode and UAV control mode). To fix the control input to desired values, we reset the end points of the sticks in the RC transmitter. By limiting the maximum input value from the transmitter, the value can be fixed since we are fully bringing down the stick. The multirotor can perform brake-like motion after releasing the stick of RC transmitter in both control modes.

The experimental results are shown in Figure 7. The figure shows velocity in y axis and pitch angle change of the multirotor in different control modes. In the case of UAV control, the velocity was increased more quickly than in the case of ATD control, but, since we released the stick, the brake-like motion was performed autonomously by controlling its pitch angle in the opposite direction, and it was observed to be much larger than when translating the multirotor. Continuous images of the motion are shown in Figure 8a. In addition, it took 1.2 s to completely stop the body frame in horizontal direction. In the case of the ATD control, although velocity was increased more slowly than in the case of UAV control, it was increased more smoothly. It took 0.5 s to completely stop the body frame in horizontal direction from when the multirotor started to perform brake-like motion. Continuous images of the motion are shown in Figure 8b. During the flight, the attitude of multirotor remained stable. Through the experiments, we verified that the ATD control was successfully able to translate the multirotor by keeping the attitude horizontal including brake-like motion, and it can be helpful for the multirotor to move finely to accurate position and perform smooth aerial manipulation tasks.

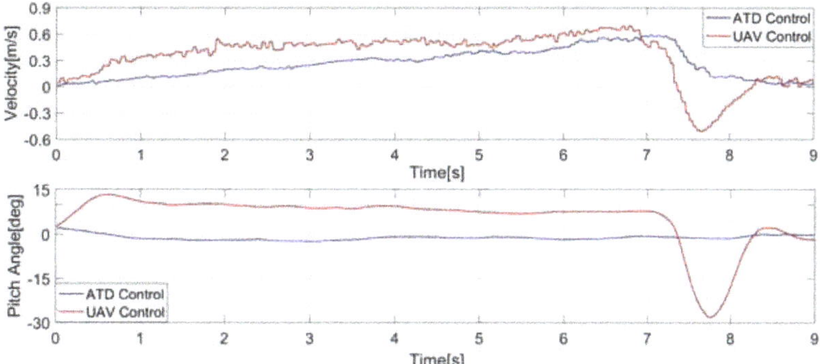

Figure 7. Experimental results of the ATD control and multirotor control. The result was recorded from the start of multirotor translating until it stopped: (top) the velocity change of the multirotor in y axis; and (bottom) the pitch angle change.

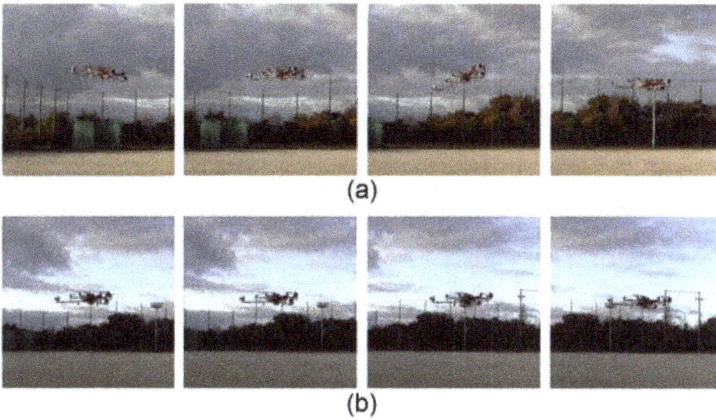

Figure 8. Continuous images of the brake-like motion in different control mode: (**a**) the multirotor control mode; and (**b**) ATD control mode. Both (**a**,**b**) follow from left to right, and they correspond to Seconds 7–9 in Figure 7. The video of the experiment is available from the supplementary materials.

6. Experiments

To assess efficiency of the ATD with multirotor system, several preliminary experiments were performed with motion capture system, and the experiment of push and pull operation was performed outdoors.

6.1. Position Control

The target position and initial position of multirotor was set to $[0,0]$ and $[-1,-1]$ on x and y axes. In the experiment, we switched to position control mode after the multirotor took off from the ground. The experimental result is shown in Figure 9. In the figure, the top graph shows position of the multirotor in x and y axis and the bottom graph shows roll and pitch angle of multirotor during the flight. The average position error was 0.0358 m on x-axis and 0.0188 m on y-axis, and it was in the region ±0.05 m. During the flight, the attitude of multirotor (Figure 9, bottom) was in the region $\pm$ 2 degree, therefore the body frame can be considered to be kept horizontal.

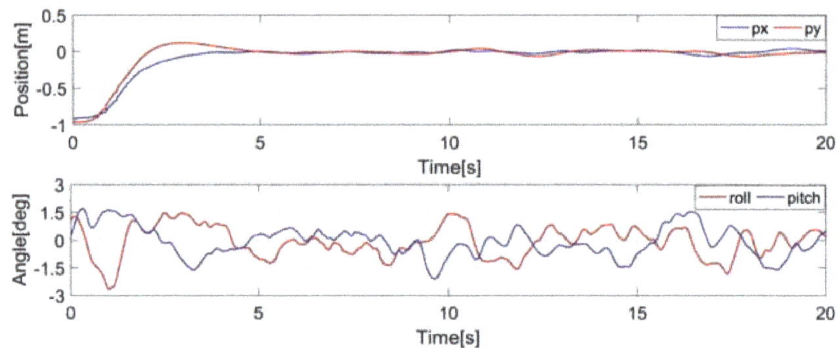

Figure 9. Experimental result of position control.

Moreover, to verify how fine of position control of the ATD with multirotor system is possible, another experiment was performed. According to experiment result of position control (Figure 9), it is possible to perform the positioning every 0.2 m. Thus, we set the initial position of the multirotor to $[0,0]$ and increased target position by 0.2 m in every 5-s interval. The experimental result is shown in Figure 10. In the figure, the top graph shows position of the multirotor in x and y axes and the bottom graph shows roll and pitch angle of multirotor during the flight. The target position was increased from the time Second 5 in the figure. The average position error was 0.037 m on x-axis and 0.0194 m on y-axis, and it was within the region of ±0.05 m for each target position. During the flight, attitude of the multirotor was also kept stable which was in the region ±2 degree. Through the experiment, we verified that it is possible to perform positioning at 0.2 m, and this is the minimum possible distance.

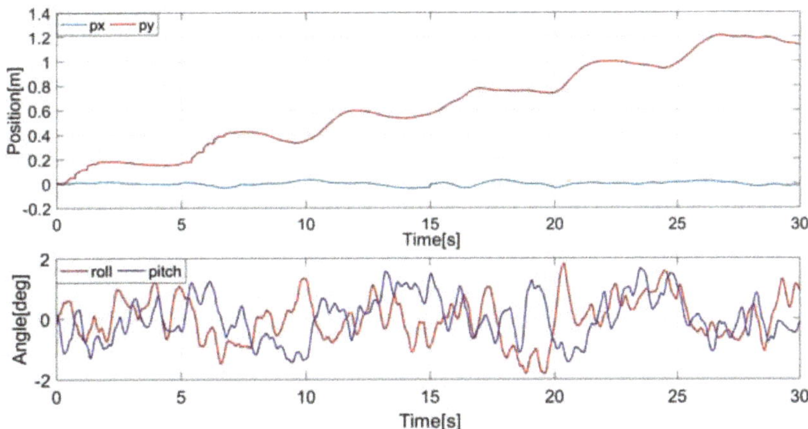

Figure 10. Experimental result of positioning by 0.2 m.

6.2. Experiment of Translational Movement in Different Directions

The experiment was performed for verifying how much the direction shifts while the ATD is translating the multirotor. The experimental setup is shown in Figure 11 (left). The target position was set 0.5 m away from the initial position at 30-degree intervals. The experimental result is shown in Figure 11 (right). The multirotor was observed to be translated almost straightly when the target position was on x and y axes. For the other target positions, it shifted about five degrees from the desired direction. It is caused by design error of ATD and control error from ESC of the ducted fans, but the direction error was absorbed since it was able to reach target position.

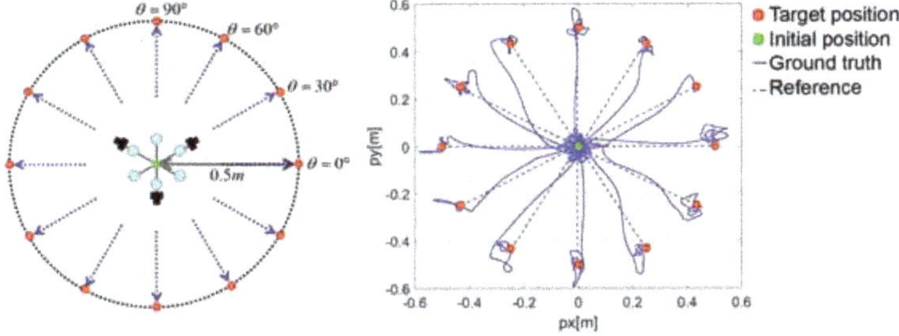

Figure 11. Experimental set up (**left**); and result (**right**) of translational movement in any direction.

6.3. Experiment of Push and Pull Operation

To verify if it possible to perform the aerial manipulation tasks with our system, we demonstrated a push and pull task. To perform the task, we equipped a rigid arm (the weight and length were 0.1 kg and 0.4 m, respectively) to the multirotor airframe. and a 3D printed disk-like object was fixed to the tip of the arm. The manipulated target was a nail-like object, and it was inserted to the vertical plate tightly. In the experiment, we used the manual ATD control mode to operate the multirotor. The external wind velocity during the experiment was 2–3 m/s. Continuous images of push and pull task are shown in Figure 12. Although wind disturbed the flight sometimes, both push and pull were successfully performed smoothly.

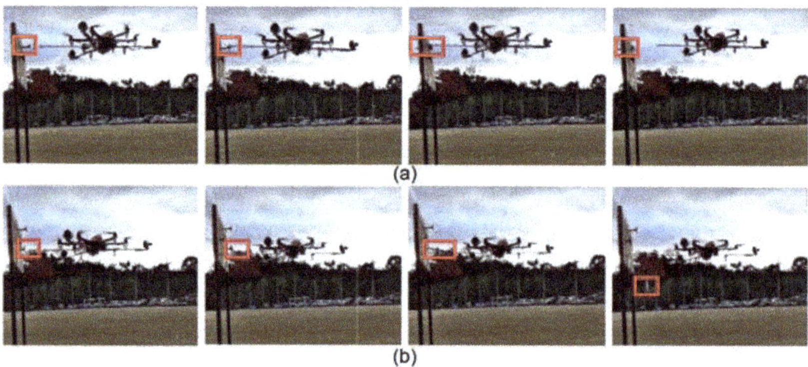

(a)

(b)

Figure 12. Continuous images of push (**a**) and pull (**b**) tasks. The images follow from left to right and the red region shows the target object. The video of the experiment is available from the supplementary materials.

During the flight, attitude of multirotor was kept horizontal (roll and pitch angles were in the region of ± 2 degree), as shown in Figure 13. In addition, we also tried the task by the UAV control mode, but it was challenging to operate the multirotor finely and impossible to push/pull the target object with such a rigid arm. Through the experiment, we verified that it was possible to perform the manipulation tasks, and it was much easier to perform the task than by controlling multirotor normally.

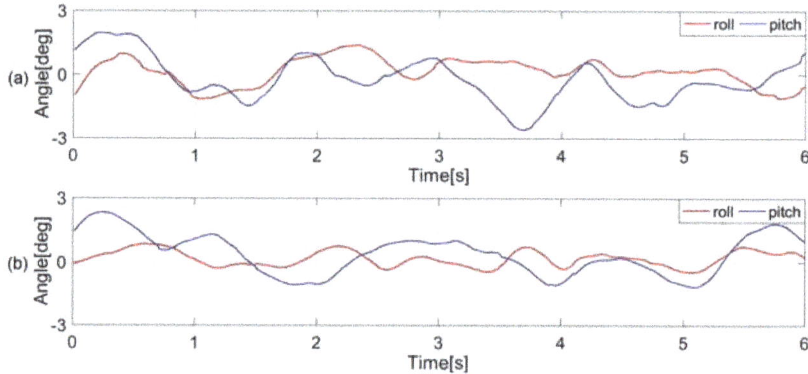

Figure 13. Experimental result of push (**a**) and pull (**b**) tasks. The graph shows roll and pitch angles of multirotor during the operation, and it corresponds to Figure 12.

7. Conclusions

In this study, we proposed and developed an add-on planar translational driving system (ATD) which is equipable to a typical type of multirotor for aerial manipulation. The device allows multirotor translating motion by keeping the attitude horizontal. This motion allows a multirotor to perform manipulation tasks easily, such as contact or non-contact inspections. To allow the operator to easily control the ATD with multirotor, three different control modes were designed, and the operator can switch among the control modes from RC transmitter. To verify the generated force from the ATD actually works on multirotor, we measured its force and verified relations between input thrust with output force. According to the measurement result, comparison of the performance between the ATD with multirotor control was performed, especially to observe the difference of brake-like motions. To verify positioning accuracy, i.e., to verify if it actually keeps the attitude of the multirotor stable while translating, how finely the position control can be performed, and if the translational movement can be performed in any direction, several preliminary experiments were performed. Finally, to verify the possibility of aerial manipulation by employing our system, a push and pull task was performed outdoors in an environment with wind velocity of 2–3 m/s. During the experiments, the ATD was observed to work as expected.

In future work, we are going to improve the design and control law of our system to allow for higher accuracy positioning and performing the task with much robustness. We also plan to develop several applications such as inspection of high-rise buildings and high pressure cleaning tasks by employing our system. Furthermore, implementing autonomous localization for our system will allow the multirotor to perform aerial manipulation at a higher precision.

Supplementary Materials: The following are available online at https://www.mdpi.com/2076-3417/11/4/1462/s1, Video S1: Supporting Video.mp4.

Author Contributions: R.M. developed the proposed device, controllers and system; designed and performed experiments; analyzed the data, and wrote the manuscript. H.P. advised on the design concepts, designed and performed experiments, and wrote the manuscript. K.S. provided general ideas about the work, advised on the design of experiments, and wrote the manuscript. All authors have read and agreed to the published version of the manuscript.

Funding: This research received no external funding.

Institutional Review Board Statement: Not applicable.

Informed Consent Statement: Not applicable.

Conflicts of Interest: The authors declare no conflict of interest.

References

1. Lippitt, C.D.; Zhang, S. The impact of small unmanned airborne platforms on passive optical remote sensing: A conceptual perspective. *Int. J. Remote Sens.* **2018**, *39*, 4852–4868. [CrossRef]
2. Ruggiero, F.; Lippiello, V.; Ollero, A. Aerial Manipulation: A Literature Review. *IEEE Robot. Autom. Lett.* **2018**, *3*, 1957–1964. [CrossRef]
3. Ding, X.; Guo, P.; Xu, K.; Yu, Y. A review of aerial manipulation of small-scale rotorcraft unmanned robotic systems. *Chin. J. Aeronaut.* **2019**, *32*, 200–214. [CrossRef]
4. Mendoza-Mendoza, J.; Gonzalez-Villela, V.J.; Aguilar-Ibanez, C.; Suarez-Castanon, M.; Fonseca-Ruiz, L. Snake Aerial Manipulators: A Review. *IEEE Access* **2020**, *8*, 28222–28241. [CrossRef]
5. Suarez, A.; Real, F.; Vega, V.M.; Heredia, G.; Rodriguez-Castaño, A.; Ollero, A. Compliant Bimanual Aerial Manipulation: Standard and Long Reach Configurations. *IEEE Access* **2020**, *8*, 88844–88865. [CrossRef]
6. Fumagalli, M.; Naldi, R.; Macchelli, A.; Carloni, R.; Stramigioli, S.; Marconi, L. Modeling and Control of a Flying Robot for Contact Inspection. In Proceedings of the IEEE/RSJ International Conference on Intelligent Robots and Systems (IROS), Vilamoura, Portugal, 7–12 October 2012; pp. 3532–3537.
7. Thomas, J.; Loianno, G.; Daniilidis, K.; Kumar, V. Visual Servoing of Quadrotors for Perching by Hanging From Cylindrical Objects. *IEEE Robot. Autom. Lett.* **2016**, *1*, 57–64. [CrossRef]
8. Paul, H.; Ono, K.; Ladig, R.; Shimonomura, K. A Multirotor Platform Employing a Three-Axis Vertical Articulated Robotic Arm for Aerial Manipulation Tasks. In Proceedings of the 2018 IEEE/ASME International Conference on Advanced Intelligent Mechatronics (AIM 2018), Auckland, New Zealand, 9–12 July 2018; pp. 478–485.
9. Shimahara, S.; Suphachart, L.; Ladig, R.; Shimonomura, K. Aerial torsional manipulation employing multirotor flying robot. In Proceedings of the IEEE/RSJ International Conference on Intelligent Robots and Systems (IROS), Daejeon, Korea, 9–14 October 2016; pp. 1596–1600.
10. Ikeda, T.; Yasui, S.; Minamiyama, S.; Ohara, K.; Ashizawa, S.; Ichikawa, A.; Okino, A.; Oomichi, T.; Fukuda, T. Stable impact and contact force control by UAV for inspection of floor slab of bridge. *Adv. Robot.* **2018**, *32*, 1061–1076. [CrossRef]
11. Tognon, M.; Franchi, A. Omnidirectional Aerial Vehicles With Unidirectional Thrusters: Theory, Optimal Design, and Control. *IEEE Robot. Autom. Lett.* **2018**, *3*, 2277–2282. [CrossRef]
12. Park, S.; Lee, J.; Ahn, J.; Kim, M.; Her, J.; Yang, G.; Lee, D. ODAR: Aerial Manipulation Platform Enabling Omnidirectional Wrench Generation. *IEEE/ASME Trans. Mechatron.* **2018**, *23*, 1907–1918. [CrossRef]

13. Oosedo, A.; Abiko, S.; Narasaki, S.; Kuno, A.; Konno, A.; Uchiyama, M. Large attitude change flight of a quad tilt rotor unmanned aerial vehicle. *Adv. Robot.* **2016**, *30*, 326–337. [CrossRef]
14. Allenspach, M.; Bodie, K.; Brunner, M.; Rinsoz, L.; Taylor, Z.; Kamel, M.; Siegwart, R.; Nieto, J. Design and optimal control of a tiltrotor micro-aerial vehicle for efficient omnidirectional flight. *Int. J. Robot. Res. (IJRR)* **2020**, *39*, 1305–1325. [CrossRef]
15. Zhao, M.; Anzai, T.; Shi, F.; Chen, X.; Okada, K.; Inaba, M. Design, Modeling, and Control of an Aerial Robot DRAGON: A Dual-Rotor-Embedded Multilink Robot With the Ability of Multi-Degree-of-Freedom Aerial Transformation. *IEEE Robot. Autom. Lett.* **2018**, *3*, 1176–1183. [CrossRef]
16. Zhao, M.; Shi, F.; Anzai, T.; Okada, K.; Inaba, M. Online Motion Planning for Deforming Maneuvering and Manipulation by Multilinked Aerial Robot Based on Differential Kinematics. *IEEE Robot. Autom. Lett.* **2020**, *5*, 1602–1609. [CrossRef]
17. Park, S.; Lee, Y.; Heo, J.; Lee, D. Pose and Posture Estimation of Aerial Skeleton Systems for Outdoor Flying. In Proceedings of the International Conference on Robotics and Automation (ICRA), Montreal, QC, Canada, 20–24 May 2019; pp. 704–710.

*applied
sciences*

MDPI

Article

Mathematical Considerations for Unmanned Aerial Vehicle Navigation in the Magnetic Field of Two Parallel Transmission Lines

Dean Martinović, Stjepan Bogdan * and Zdenko Kovačić

Faculty of Electrical Engineering and Computing, University of Zagreb, 10000 Zagreb, Croatia;
dean.martinovic@fer.hr (D.M.); zdenko.kovacic@fer.hr (Z.K.)
* Correspondence: stjepan.bogdan@fer.hr

Abstract: This publication deals with the navigation of unmanned aerial vehicles (UAVs) moving in the magnetic field of two long, straight, parallel conductors, which is of high interest for several new technical applications. How the position and orientation of the UAV can be calculated using a minimal number of only three three-axis magnetometers are discussed. It is shown that the angles can be determined without the knowledge of the conductor currents and the magnetic field equations, but only by combining the sensor measurements with the rotation matrix and exploiting a characteristic property of the magnetic field. Furthermore, different strategies were investigated to determine the respective sensor positions. An analytical solution was derived from the nonlinear magnetic field equations, which promises a low computational time. It is shown that for a given sensor, several solutions exist, from which the correct one has to be selected. Therefore, a specific detection method is introduced. Once the solution is known, the UAV location can be determined. Finally, the overall algorithm was tested by simulations far from and near the conductors with superimposed typical magnetic noise.

Keywords: magnetic field navigation; parallel conductors; transmission lines; unmanned aerial vehicles; aerial manipulation

Citation: Martinović, D.; Bogdan, S.; Kovačić, Z. Mathematical Considerations for Unmanned Aerial Vehicle Navigation in the Magnetic Field of Two Parallel Transmission Lines. *Appl. Sci.* **2021**, *11*, 3323. https://doi.org/10.3390/app11083323

Academic Editor: Alejandro Suarez

Received: 9 March 2021
Accepted: 2 April 2021
Published: 7 April 2021

Publisher's Note: MDPI stays neutral with regard to jurisdictional claims in published maps and institutional affiliations.

1. Introduction

Magnetic field-based navigation is an old discipline that had its beginnings in the 20th Century [1]. Magnetic fields have several advantages over other physical quantities. For example, unlike electromagnetic waves, they are not subject to multipath effects or fading because they do not reflect off surfaces. This allows for much higher accuracy. In addition, they are not attenuated when passing through magnetically neutral materials, making magnetic field-based navigation ideal for use between obstacles, at night, or in difficult weather conditions such as snow or fog. A large number of publications exist dealing with techniques for object navigation and localization. Depending on the application, they all use one or more distributed magnetic field sources that emit an alternating magnetic signal. This is usually sampled and measured by several sensors, which may be from different technologies. By combining these measurement data with the mathematical model of the source used, the position and orientation of the observed object are then determined. This can be divided into four different types of sources.

The first type is a simple rectangular or circular coil that typically has a large number of turns. It represents the basic element for the construction of more complex coil types and is often referred to as a dipole, since above a minimum distance from the winding, its magnetic field distribution can be described by the equations of a magnetic dipole. In various research projects [2–6], such as MILPS (magnetic indoor local positioning system), multiple coils mounted on walls are used to locate cell phones and robots in buildings or patients in hospitals. Other use cases include the navigation of medical instruments inside the human body [7] or the localization of an implantable transponder for intensity-modulated radiation therapy [8]. In addition, several projects have investigated how these coils can be used to locate trapped persons as quickly as possible [9–12]. These could be buried miners or avalanche victims, who can be found by the emitted signal from a small coil worn close to the body. Another interesting topic is the detection of buried landmines in old war zones, also known as humanitarian demining [13,14]. This involves the use of hand-held coils such as metal detectors. Finally, a new application has emerged with the advent of the electric vehicle age.

In the context of inductive charging, the receiver coil mounted on the underbody of the car must be positioned relatively precisely above the transmitter coil in the parking lot, as shown in Figure 1. Since this is very difficult for the driver to achieve, the MPPS (magnetic pulse positioning system) was developed in [15]. Here, the transmitting coil in the ground emits a low-frequency pulsed magnetic signal that allows precise localization by the vehicle in the sub-centimeter range.

Figure 1. Navigation using a one-axis coil [15].

The next two coil types consist of two or three base coils wound orthogonally to each other. An example of a three-axis magnetic field source is shown in Figure 2.

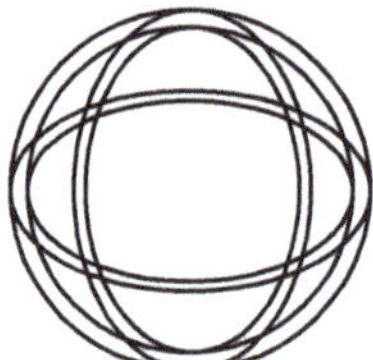

Figure 2. Principle of a three-axis coil.

For two-axis sources, the coils carry phase quadrature currents that produce a rotating magnetic dipole [1,16,17]. Together with phase detection algorithms and amplitude evaluation, the object position and also the orientation can be determined. For three-axis sources, all source axes are successively excited with currents of the same frequency and phase. In this way, each axis generates an equivalent dipole of arbitrary orientation [18–20]. The three sensor responses are linearly independent and provide sufficient information to determine both the sensor position and its orientation.

The last type of source to be mentioned is a long wire with an alternating signal. In [21–23], its magnetic field was used to allow an underwater vehicle to locate and track a buried underwater cable.

Until now, however, no attention has been paid to magnetic sources consisting of two long parallel conductors. They enable many new useful engineering applications, especially when it comes to the realization of unmanned aerial vehicles (UAVs). For example, UAVs could autonomously perform inventory in warehouses. For this purpose, parallel guides can be attached to the floor or ceiling for navigation between shelves. In addition, UAVs could provide an inspection of rail systems efficiently when a current flows through them.

Another interesting topic is the autonomous performance of inspection and assembly work around power line systems. This is exactly what was investigated in the Aerial-Core (https://aerial-core.eu/, accessed on 06.04.2021) project. One of the goals of the project was to develop a technology that will enable a UAV to perform (Figure 3) the placement of distance separators between power lines.

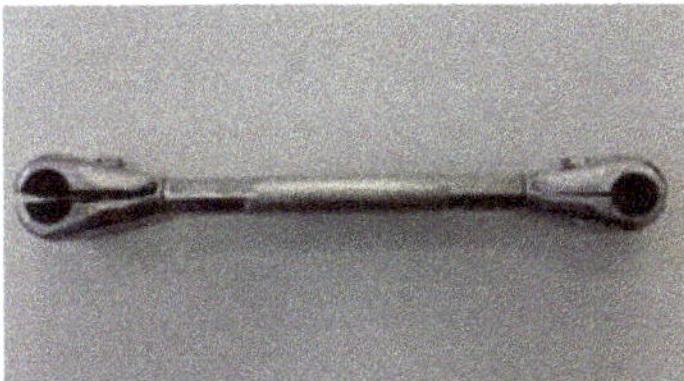

Figure 3. Distance separator for transmission lines.

This is usually performed by workers who are exposed to very high electric and magnetic fields. Moreover, such a task requires a highly skilled helicopter pilot to keep the helicopter close to the power lines while a worker performs the installation. Therefore, the main objective of this work is to derive the equations for determining the position and orientation of a UAV moving through the magnetic field of two parallel conductors, thus enabling precise navigation of a UAV in close proximity to the power lines. We assumed that a UAV is equipped with visual and other sensors that are used both for navigating farther away from the lines and for approaching the lines.

2. Orientation and Localization Equations

2.1. System Overview

Figure 4 shows two parallel lines, which together form one phase of a three-phase power transmission system and a UAV flying in close proximity. The transmission lines lie in the xy-plane of the global coordinate frame $\{G\}$ positioned exactly between the lines. For the purpose of sensing the magnetic field of the transmission lines, there are magnetometers mounted on the UAV, which are marked with yellow points.

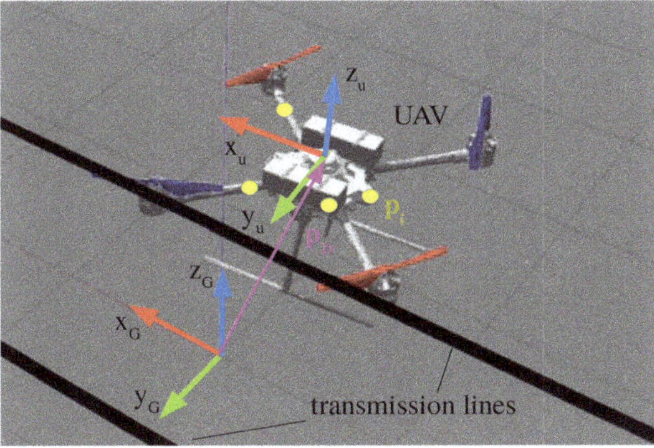

Figure 4. System overview. Global frame and UAV frame.

The transmission lines of a single phase carry sinusoidal currents of fixed frequency and phase, which can be used to our advantage. They generate an alternating magnetic field, which allows static and low-frequency magnetic interference fields such as the Earth's field to be canceled out. This is realized by a differential measurement of the amplitude of the magnetic signal. The calculation can be implemented in the time domain and is very efficient [24]. In addition, the system can be treated as a static field problem with constant currents. For simplicity, they are chosen as I in the following. The magnetic field thus generated can be calculated using the law of Biot–Savart for thin conductors. In general, transmission lines are not thin. However, as the distance from the lines increases, the magnetic flux density distribution approaches that of thin conductors relatively quickly:

$$\vec{B}(\vec{r}) = -\frac{\mu_0}{4\pi} I \int_{-\infty}^{\infty} \frac{(\vec{r} - \vec{r}') \times d\vec{s}'}{|\vec{r} - \vec{r}'|^3}. \tag{1}$$

The integration is executed only along the cables according to the arrangement in Figure 4, i.e., $d\vec{s}' = (d\vec{x}', 0, 0)^T$. Thus, from the cross-product in Equation (1), it immediately follows that:

$$B_x = 0, \forall (x, y, z)^T, \tag{2}$$

which means that there is no field component in the direction parallel to the cables. Solving the integral for one cable and using the superposition principle, the remaining field components B_y and B_z can be determined. The total magnetic field equation with the transmission lines at $\pm y_0$ is then:

$$\vec{B}(y, z) = C \begin{pmatrix} 0 \\ \frac{-z}{z^2 + (y + y_0)^2} + \frac{-z}{z^2 + (y - y_0)^2} \\ \frac{y + y_0}{z^2 + (y + y_0)^2} + \frac{y - y_0}{z^2 + (y - y_0)^2} \end{pmatrix}, \quad with \quad C = \frac{I \mu_0}{2\pi}, \tag{3}$$

where C is a prefactor depending on the transmission line current.

Figure 5 shows the signal amplitude $|\vec{B}(y, z)|$ around the transmission lines for $I = 700$ A and $y_0 = 0.25$ m. Figure 6 shows the locations of equal amplitude in the yz-plane expressed by isolines. The magnetic field contributions of the other two phases are neglected because the cables are typically at least 2 m apart.

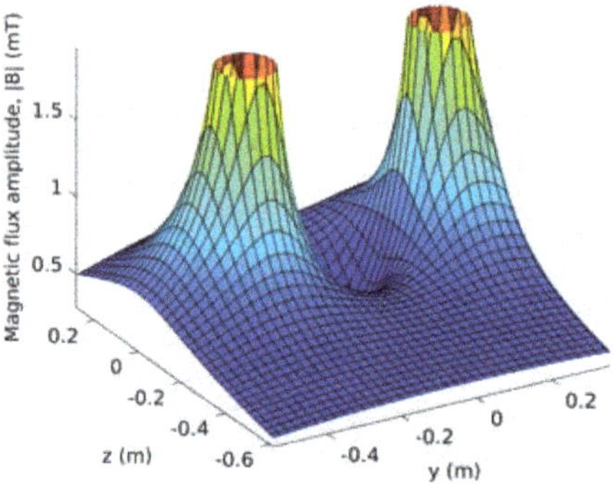

Figure 5. Signal power around the transmission lines.

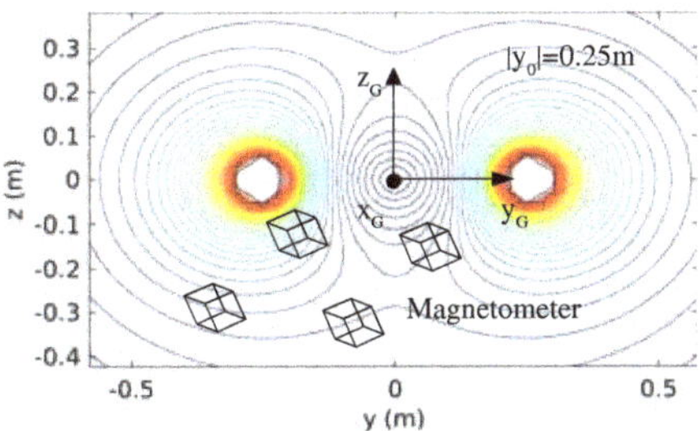

Figure 6. Isolines of the magnetic field describing equal power.

The magnetic field-based navigation of a UAV requires that the UAV can sense the magnetic field in all spatial directions. It uses this information to determine its relative position and orientation with respect to the fixed position and orientation of the global coordinate frame $\{G\}$. The flight capabilities allow the UAV to achieve an arbitrary orientation by rotating the UAV's body about the z- (blue), y- (green), and x-axes (red) of the local drone frame L_u for the respective yaw, pitch, and roll angles α, β, and γ. The origin position of the local frame L_u in $\{G\}$, denoted by $\vec{p}_D$, also represents the position of the UAV to be determined. The sensors have predefined positions $\vec{p}_{i0}$ in L_u, where the index i refers to a particular sensor. Moreover, each magnetometer has its own local coordinate frame L_i. Then, their respective positions $\vec{p}_i = (x_i, y_i, z_i)^T$ in $\{G\}$ can be expressed using the rotation matrix $R_G^{L_u}(\alpha, \beta, \gamma) = R_x(\gamma)\, R_y(\beta)\, R_z(\alpha)$:

$$\vec{p}_i = R_G^{L_u}(\alpha, \beta, \gamma)\, \vec{p}_{i0} + \vec{p}_D, \tag{4}$$

$$R_z(\alpha) = \begin{pmatrix} c_\alpha & s_\alpha & 0 \\ -s_\alpha & c_\alpha & 0 \\ 0 & 0 & 1 \end{pmatrix}, \quad R_y(\beta) = \begin{pmatrix} c_\beta & 0 & -s_\beta \\ 0 & 1 & 0 \\ s_\beta & 0 & c_\beta \end{pmatrix}, \quad R_x(\gamma) = \begin{pmatrix} 1 & 0 & 0 \\ 0 & c_\gamma & s_\gamma \\ 0 & -s_\gamma & c_\gamma \end{pmatrix}.$$

where s_x and c_x are abbreviations for $\sin x$ and $\cos x$ needed in the following. From Equation (3), it can be seen that the position of the UAV along the cables (x-direction) cannot be determined because the magnetic field does not provide sufficient information. Thus, only the projections of the magnetometers and the UAV frame origin in the yz-plane can be computed. The representation of the magnetic field $\vec{v}_i$ measured by the i-th sensor at position $\vec{p}_i$ expressed in $\{G\}$ then becomes:

$$\vec{B}(\vec{p}_i) = R_G^{L_u}(\alpha, \beta, \gamma)\, R_{L_u}^{L_i}\, \vec{v}_i, \tag{5}$$

where $R_{L_u}^{L_i}$ is the user-defined orientation of L_i in the UAV body frame L_u and $R_G^{L_u}$ is the orientation of L_u in the global frame $\{G\}$. If L_i is aligned with $\{G\}$, then it is $R_G^{L_u} R_{L_u}^{L_i} = I$, where I is the identity matrix. In this case, the sensor measures exactly the magnetic field given by Equation (3), i.e., $\vec{B}(\vec{p}_i) = \vec{v}_i$. The system of three equations in (5) establishes a relationship between the sensor measurement, sensor position, and UAV orientation. It is nonlinear and has five unknowns $(y_i, z_i, \alpha, \beta, \gamma)^T$, which are determined in the following sections.

2.2. Orientation Determination

In this section, the UAV angles α, β, and γ are derived. As will be shown below, the calculation is possible without knowledge of the field equation and the transmission line current. Three magnetometers were considered. For the following analysis, their corresponding equations were set up in matrix notation, where $0_{3\times3}$ is a 3×3 zero matrix:

$$
\begin{pmatrix} \vec{B}(\vec{p}_1) \\ \vec{B}(\vec{p}_2) \\ \vec{B}(\vec{p}_3) \end{pmatrix} = \begin{pmatrix} R_G^{L_u} R_{L_u}^{L_1} & 0_{3\times3} & 0_{3\times3} \\ 0_{3\times3} & R_G^{L_u} R_{L_u}^{L_2} & 0_{3\times3} \\ 0_{3\times3} & 0_{3\times3} & R_G^{L_u} R_{L_u}^{L_3} \end{pmatrix} \begin{pmatrix} \vec{v}_1 \\ \vec{v}_2 \\ \vec{v}_3 \end{pmatrix}.
\tag{6}
$$

It is obvious that this system of equations is solvable since the rank of the matrix is nine, which is equal to the number of unknowns. In general, in addition to the three angles, each row provides two unknowns y_i and z_i. Now, the UAV's yaw and pitch angles can be calculated directly from the two sensor measurements $\vec{v}_l$ and $\vec{v}_n$ and the zero field property (2). With this and the Cartesian unit vector $\vec{e}_x$, it follows from Equation (6) that:

$$
0 = (R_G^{L_u} R_{L_u}^{L_i} \vec{v}_i) \cdot \vec{e}_x, \qquad i = l, n.
\tag{7}
$$

Obviously, this equation does not depend on the sensor positions or the choice of their local frames. Thus, they can be located anywhere on the UAV and have any local frame. However, to avoid additional transformations or computational effort, the sensor frames are aligned with the UAV body frame. Then, the equation system (7) becomes:

$$
\begin{aligned}
0 &= & c_\beta c_\alpha v_{lx} & +c_\beta s_\alpha v_{ly} & -s_\beta v_{lz} \\
0 &= & c_\beta c_\alpha v_{nx} & +c_\beta s_\alpha v_{ny} & -s_\beta v_{nz}
\end{aligned}
\tag{8}
$$

After division by c_β and subsequent subtraction, $\tan \beta$ can be eliminated. Solving the remaining equation for α leads to:

$$
\alpha = atan\left(\frac{v_{lz}v_{nx} - v_{lx}v_{nz}}{v_{ly}v_{nz} - v_{lz}v_{ny}} \right) = atan\left(\frac{(\vec{v}_l \times \vec{v}_n) \cdot \vec{e}_y}{(\vec{v}_l \times \vec{v}_n) \cdot \vec{e}_x} \right)
\tag{9}
$$

and further with $A = tan\,\alpha$ to:

$$
\beta = atan\left(\frac{v_{nx} + v_{ny}A}{v_{nz}\sqrt{A^2 + 1}} \right) = atan\left(\frac{v_{lx} + v_{ly}A}{v_{lz}\sqrt{A^2 + 1}} \right).
\tag{10}
$$

The roll angle γ can also be computed by exploiting the zero field property (2). However, the rotation matrix $R_G^{L_u}$ does not in general contain γ in its x-row. Thus, a different sensor k must have a different local frame L_k in order for the zero to appear in the y- or z-row containing all three angles. This can be achieved by, for example, placing the sensor so that it is rotated relative to the UAV body frame L_u by an angle δ about the y-axis, i.e., it is:

$$
R_{L_u}^{L_k} = R_y(\delta).
\tag{11}
$$

Then, from Equation (5), it follows that after applying the coordinate transformation $A = R_G^{L_u} R_y^T(\delta) R_{L_u}^G$, the vector $\vec{B}(\vec{p}_k)$ is expressed in L_u:

$$
\hat{\vec{B}}(\vec{p}_k) = A\,\vec{B}(\vec{p}_k) = R_G^{L_u} R_y^T(\delta) R_{L_u}^G R_G^{L_u} R_y(\delta)\,\vec{v}_k,
$$

$$
\text{for } \delta = \tfrac{\pi}{2} \Rightarrow \begin{pmatrix} \hat{B}_x(\vec{p}_k) \\ \hat{B}_y(\vec{p}_k) \\ 0 \end{pmatrix} = R_G^{L_u}\,\vec{v}_k.
\tag{12}
$$

From Equation (12), it follows that the rotation of $\vec{B}(\vec{p}_k)$ for a given angle $\delta = \tfrac{\pi}{2}$ around the y-axis of L_u has forced the z-component of the resulting vector $\hat{\vec{B}}(\vec{p}_k)$ to zero:

$$
\hat{B}_z(\vec{p}_k) = 0 = (R_G^{L_u}\,\vec{v}_k) \cdot \vec{e}_z,
\tag{13}
$$

where $\vec{e}_z$ is the Cartesian unit vector. Then, in a first step, Equation (13) can be written as:

$$0 = (s_\gamma s_\alpha + c_\gamma c_\alpha s_\beta)v_{kx} + (c_\gamma s_\beta s_\alpha - c_\alpha s_\gamma)v_{ky} + c_\gamma c_\beta v_{kz}.$$

Next, the division by c_γ forms the tangent function t_γ, which can be isolated and finally solved for γ:

$$0 = (t_\gamma s_\alpha + c_\alpha s_\beta)v_{kx} + (s_\beta s_\alpha - c_\alpha t_\gamma)v_{ky} + c_\beta v_{kz}$$
$$\Rightarrow \gamma = atan\left(\frac{v_{ky}s_\beta s_\alpha + v_{kx}s_\beta c_\alpha + v_{kz}c_\beta}{v_{ky}c_\alpha - v_{kx}s_\alpha}\right). \tag{14}$$

The sensor can be placed anywhere on the UAV. However, the result only applies to the local frame definition in (11). Theoretically, arbitrary local frame definitions can be chosen, but in this case, there is no zero in any coordinate, and the whole system of Equation (12) must be solved since the sensor position $\vec{p}_k$ is unknown. However, the equations are complicated, and it is questionable whether an analytical solution exists. The presented formulas are valid if the UAV angles are within the definition range D, respectively $\alpha, \beta, \gamma \in D = [-\frac{\pi}{2}, \frac{\pi}{2}]$. The limits of the equations presented are discussed in Section 3.

2.3. Position Determination

In the previous section, it was shown that the UAV orientation and position computation can be decoupled into two independent problems. In doing so, it was shown how the UAV orientation can be computed using the zero field property and without knowledge of the transmission line currents or the magnetic field equation. This section now deals with the computation of the respective sensor positions $\vec{p}_i$, which are needed to finally determine the UAV position $\vec{p}_D$ via Equation (4).

With the decoupling presented, the complexity of the equation system (6) can be greatly reduced, since by knowing the angles, the right side can be computed and becomes a simple real-numbered vector. With $R_G^{L_u} R_{L_u}^{L_i} \vec{v}_i = \vec{V}_i$, the system becomes:

$$\begin{pmatrix} \vec{B}(\vec{p}_1) \\ \vec{B}(\vec{p}_2) \\ \vec{B}(\vec{p}_3) \end{pmatrix} = \begin{pmatrix} \vec{V}_1 \\ \vec{V}_2 \\ \vec{V}_3 \end{pmatrix}. \tag{15}$$

Obviously, the computation of the respective sensor positions is also decoupled. Thus, to obtain $\vec{p}_D$, it is sufficient to solve one of these rows in (15). Then, for a sensor i, the corresponding row is:

$$\vec{B}(\vec{p}_i) = \begin{pmatrix} B_y(y_i, z_i) \\ B_z(y_i, z_i) \end{pmatrix} = \begin{pmatrix} V_{iy} \\ V_{iz} \end{pmatrix} = \vec{V}_i, \tag{16}$$

which represents a system of two equations. The components B_y and B_z describe the magnetic field in two dimensions in $\{G\}$ and were obtained by calculating the Biot–Savart integral (1). This means, on the one hand, that the two equations in (16) are linearly independent and theoretically allow the determination of the unknowns y_i and z_i. On the other hand, the equations are nonlinear, and the question arises whether it is possible to solve (16) analytically and, if so, how. In this context, there are several approaches, all of which were investigated in the context of this work. Since the following considerations are equally valid for all sensors, the subscript i is omitted for simplicity.

The first row in (16) can be solved for one coordinate y or z, which is then inserted into the second row to compute the other coordinate; and the same in reverse. However, solving $B_y = V_y$ for z or $B_z = V_z$ for y in the general form leads to extremely large solutions $z(y, V_y)$ and $y(z, V_z)$, which contain root terms with eighth-order polynomials as arguments. The other two cases, i.e., solving $B_y = V_y$ for y or $B_z = V_z$ for z, lead to smaller solutions. For example, from the latter, the four solutions for z follow:

$$z(y, V_z) = \pm\sqrt{\pm D(y^2, y) + a(y^2, y)},$$

$$\text{with} \quad D = \frac{1}{V_z}\sqrt{4V_z^2 y^2 y_0^2 - 4CV_z y y_0^2 + C^2 y^2}, \tag{17}$$

$$a = -y_0^2 - y^2 + \frac{Cy}{V_z}.$$

The result $y(z, V_y)$ has the same structure, but is not used, as the easiest final equation can be found by inserting (17) into $B_y^2(y, z) = V_y^2$. With D from (17), it becomes:

$$\frac{V_z(Cy \pm D)^2(-V_z y_0^2 - V_z y^2 + Cy \pm D)}{y^2(-2V_z y_0^2 + Cy \pm D)^2} = V_y^2. \tag{18}$$

The equation can in principle be solved for y. However, no suitable substitution could be found, so that the root terms disappear, and at the same time, a polynomial of a smaller order than five is obtained. For example, after the substitution with $E = Cy \pm D$, unwanted y^2 expressions still remain in the equation. The square roots can also be eliminated by subsequent isolation and squaring. However, this leads to a high-order polynomial. In summary, it is difficult to impossible to derive an analytical solution only from the equation system (16). In this context, the idea is to consider the signal power as an additional equation. As will be shown in the next steps, its combination with (16) leads to a compact analytical solution, which is derived below: At a given position, the signal power can be calculated by taking the dot product of the corresponding sensor response $\vec{V}$ with itself. This must be equal to the dot product of $\vec{B}(y, z)$ with itself, leading to the following equation:

$$\vec{V} \cdot \vec{V} = |\vec{V}|^2 = \vec{B}(y, z) \cdot \vec{B}(y, z) \Rightarrow |\vec{V}|^2 = \frac{4C^2(z^2 + y^2)}{z^4 + y^4 + (2y_0^2 + 2y^2)z^2 + y_0^4 - 2y^2 y_0^2}. \tag{19}$$

The most simple equation is found by solving the formula for z^2 and inserting it into $B_z(y, z) = V_z$. For this, the two solutions for z^2 are obtained with $P = |\vec{V}|$:

$$z^2(y, P) = \pm 2\sqrt{y^2 y_0^2 - \frac{C^2 y_0^2}{P^2} + \frac{C^4}{P^4}} - y_0^2 - y^2 + \frac{2C^2}{P^2}. \tag{20}$$

Next, with the positive solution of (20), the equation $B_z(y, z^2(y, P)) = V_z$ can be written as:

$$\frac{-P^4 y y_0^2 + \hat{D}P^2 y + C^2 P^2 y}{-CP^2 y_0^2 + 2C\hat{D} + 2C^3} = V_z, \quad \text{with} \quad \hat{D} = \sqrt{(P^4 y^2 - C^2 P^2)y_0^2 + C^4}.$$

Further, $\hat{D}$ is isolated and squared, which leads to:

$$\hat{D}^2 = \frac{((P^4 y - CP^2 V_z)y_0^2 - C^2 P^2 y + 2C^3 V_z)^2}{(P^2 y - 2CV_z)^2}.$$

This eliminates the square root of $\hat{D}$. After re-substituting and transforming the equation, the position calculation is reduced to a root-finding problem of a fourth-order polynomial,

$$\sum_{n=0}^{4} a_n y^n = 0, \quad a_4 = P^8 y_0^2, \; a_3 = -4CP^6 V_z y_0^2,$$

$$a_2 = -P^8 y_0^4 + C^2 P^6 y_0^2 + 4C^2 P^4 V_z^2 y_0^2, \tag{21}$$

$$a_1 = 2CP^6 V_z y_0^4 - 2C^3 P^4 V_z y_0^2, \; a_0 = -C^2 P^4 V_z^2 y_0^4.$$

The general analytic root expressions of such a polynomial are extremely large. Therefore, it is important to make Equation (3) as simple as possible by placing the global coordinate frame exactly halfway between the transmission lines. In addition, the lines should be placed in the xy-plane parallel to the x-axis. Then, with the coefficients a_n given in (21), the four large expressions for y are greatly reduced and become:

$$y = \pm \frac{\sqrt{P^4 y_0^2 + 2C^2 V_z \sqrt{V_z^2 - P^2} + 2C^2 V_z^2 - C^2 P^2}}{2P^2} \pm \frac{\sqrt{P^4 y_0^2 - 2C^2 V_z \sqrt{V_z^2 - P^2} + 2C^2 V_z^2 - C^2 P^2}}{2P^2} + \frac{CV_z}{P^2} \tag{22}$$
$$= \frac{S}{2P^2} + \frac{CV_z}{P^2}$$

where S represents the two square root addends. The same result is obtained for the negative solution of Equation (20). Now, in the next step, four different cases can be distinguished, all of which lead to a certain class of solutions.

(I) $V_y, V_z \neq 0$:

Theorem 1. *Two of the y-solutions in (22) can immediately be excluded, because they are complex.*

Proof. $V_z \leq |\vec{V}| = P, \ \forall \vec{V} \in \mathbb{R}^3 \Rightarrow \sqrt{V_z^2 - P^2} \in \mathbb{C}$. With this, the above square root addends represented by S can be written as:

$$S = \pm\sqrt{a + jb} \pm \sqrt{a - jb},$$
$$a = P^4 y_0^2 + (2V_z^2 - P^2)C^2, \ b = 2C^2 V_z \sqrt{P^2 - V_z^2}. \tag{23}$$

Next, the parameters r and φ are introduced as:

$$r = \sqrt{a^2 + b^2}, \ \varphi = tan^{-1}\left(\frac{b}{a}\right) + \pi. \tag{24}$$

Now, if the addends have a different sign S, this results in:

$$S = \pm\sqrt{a + jb} \mp \sqrt{a - jb} = \pm\sqrt{r}\,e^{j\frac{\varphi}{2}} \mp \sqrt{r}\,e^{-j\frac{\varphi}{2}}$$
$$= \pm 2j\sqrt{r}\,\tfrac{1}{2j}(e^{j\frac{\varphi}{2}} - e^{-j\frac{\varphi}{2}}) = \pm 2j\sqrt{r}\,\sin\tfrac{\varphi}{2},$$

whereby the e-expressions are obtained using Euler's formula. $\square$

Applying this procedure for the case of the same sign, the real-numbered and more compact expressions for y can be derived. Therefore, in a first step, S is determined:

$$S = \pm\sqrt{a + jb} \pm \sqrt{a - jb} = \pm\sqrt{r}\,e^{j\frac{\varphi}{2}} \pm \sqrt{r}\,e^{-j\frac{\varphi}{2}}$$
$$= \pm 2\sqrt{r}\,\tfrac{1}{2}(e^{j\frac{\varphi}{2}} + e^{-j\frac{\varphi}{2}}) = \pm 2\sqrt{r}\,\cos\tfrac{\varphi}{2}.$$

The square root addends in Equation (22) represented by S now can be replaced by this result, leading to:

$$y_{1,2} = \frac{S}{2P^2} + \frac{CV_z}{P^2} = \frac{1}{P^2}\left(\pm\sqrt{r}\,\cos\tfrac{\varphi}{2} + CV_z\right), \tag{25}$$

with r and φ according to (24).

Theorem 2. *There exist two solutions $|y_1| \neq |y_2|$. One each is in $y < 0$ and $y > 0$, respectively.*

Proof. The hypothesis is true if the absolute value of the square root addends in Equation (22) is always greater than $C|V_z|P^{-2}$.

$$\frac{|\pm(\sqrt{a + jb} + \sqrt{a - jb})|}{2P^2} > \frac{C|V_z|}{P^2} \ \Rightarrow \ |(\sqrt{a + jb} + \sqrt{a - jb})|^2 > (2C|V_z|)^2$$
$$\Rightarrow q_l = \sqrt{a + jb}\sqrt{a - jb} > 2C^2 V_z^2 - a = q_r.$$

Further, with a, b from (23), it follows for the left and right side:

$$q_l = P^8 y_0^4 + (4C^2 P^4 V_z^2 - 2C^2 P^6) y_0^2 + C^4 P^4,$$

$$q_r = P^8 y_0^4 - 2C^2 P^6 y_0^2 + C^4 P^4,$$

$$\Rightarrow q_l - q_r = 4C^2 P^4 V_z^2 y_0^2 > 0.$$

$\square$

(II) $V_y = 0 \Rightarrow P = |V_z|$: In this case, the excluded complex solutions must also be considered, since they become real. Both give the same result y_3. The two real solutions $y_{1,2}$ can again be obtained from (25) or from the general Equation (22).

$$y_{1,2} = \tfrac{1}{V_z^2}(\pm\sqrt{V_z^4 y_0^2 + C^2 V_z^2} + CV_z), \quad y_3 = \tfrac{C}{V_z} \tag{26}$$

$|y_1| \neq |y_2| \neq |y_3|$. One of the solutions lies in the opposite half plane.

(III) $V_z = 0 \Rightarrow P = |V_y|$: The solutions are obtained analogous to Case (II). $|y_1| = |y_2|$.

$$y_{1,2} = \pm\frac{\sqrt{V_y^4 y_0^2 - C^2 V_y^2}}{V_y^2}, \quad y_3 = 0 \tag{27}$$

(IV) $V_y = V_z = 0$: If this case occurs, the sensor lies exactly in the origin of the global frame.

Once the corresponding y is determined, in a final step, z can be calculated from Equation (20):

$$z = \pm\sqrt{\pm 2\sqrt{(y^2 - \frac{C^2}{P^2})y_0^2 + \frac{C^4}{P^4}} - y_0^2 - y^2 + \frac{2C^2}{P^2}}, \tag{28}$$

offering four possible z-solutions, all of which can be real-numbered. There are also situations where, for a given y, a subset of the z-solutions in (28) can become complex. Therefore, their use must be treated with caution, especially when the formulas are implemented on a microcontroller of a real UAV. A safer way is to check the arguments of the square roots. The argument of the outer square root can be written as:

$$A_o = \pm g + h, \quad h = -y_0^2 - y^2 + \frac{2C^2}{P^2}, \tag{29}$$

where g is the inner square root expression. For all solutions $y_{1,2,3}$ obtained, $g \geq 0$ holds, since they were obtained using the z^2-expression (20). Thus, the argument of the inner square root does not need to be checked. In contrast, A_o can become negative in general and should be checked to satisfy that the condition $A_o > 0$ is fulfilled before applying the square root.

2.4. Selecting the Right Solution

Equation (28) yields four z-solutions for each y-solution of Cases (I)–(IV). Thus, depending on the case, there are up to twelve possible positions in the yz-plane for a given sensor. For navigation, it is necessary to know which of these points is the correct one. Finding it is the task of this subsection.

Now, in the first approach, the number of possible positions can be narrowed down by excluding Cases (II)–(IV). The reason for this is that they can only occur in very few locations and for a very short time during navigation, which extremely reduces their probability of occurrence. For example, the solutions of Case (II) all lie on the z-axis. In Case (III), one solution lies on the y-axis and the other two lie on a circle with a diameter equal to the distance between the transmission lines, i.e., $2y_0$. This can be shown by computing the radius $\sqrt{y_{1,2}^2 + |z^2(y_{1,2})|}$, where $y_{1,2}$ comes from Equation (27) and z^2 comes from (20). The case with the lowest probability of occurrence is Case (IV), where the sensor is located directly at the origin. The locations of Cases (II)–(IV) are summarized in Figure 7 and shown as black and red lines. Since they are not relevant for navigation and only increase the implementation effort, they can be ignored. Thus, since only Case (I) is considered, eight possible positions for a single sensor remain, from which the correct one must still be selected.

In the next steps, it will be seen that six can be excluded. To do this, some constraints must be met. As explained in Section 2.1, the sinusoidal transmission line current can be considered as an equivalent constant current I, i.e., it never changes its direction. Furthermore, the UAV orientation must not exceed the definition range D from Section 2.2 with $\alpha, \beta, \gamma \in D = [-\frac{\pi}{2}, \frac{\pi}{2}]$. This would otherwise lead to ambiguities due to the symmetry of the magnetic field. The details concerning ambiguities are the subject of Section 3. Now, the idea is to investigate which of the eight remaining positions lead to a match with the sensor measurement $\vec{V}$, i.e., which positions satisfy Equation (16).

Theorem 3. *Given a sensor k that measures $\vec{V}_k = (V_{ky}, V_{kz})^T$ and has two y-solutions according to (25) in Case (I), for each of the two values, there are four associated z-solutions according to Equation (28). However, for each y, there is only one z-solution that leads to a match with $\vec{V}_k$, i.e., there are only two possible locations in the yz-plane where the sensor can lie.*

Proof. Let y_k be one of the two y-solutions. Then, it generates the four z-solutions, which in simplified notation are $z_n = \pm\sqrt{\pm g + h}$, $g, h \neq 0$. z_n has two outer and two inner signs. Further, the B_z-component in Equation (3) has a z^2-addend in the nominator as the only z-dependency, so it can be written as $B_z(y_k, z^2)$. Now, if for $z_l = \pm\sqrt{g + h}$, it follows that:

$$B_z(y_k, z_l^2) = B_z(y_k, g + h) = V_{kz},$$

then with $g + h \neq -g + h$:

$$V_{kz} = B_z(y_k, g + h) \neq B_z(y_k, -g + h),$$

because $B_z(y_k, z^2)$ is strictly monotonous. This means, that only one of the inner signs of z_n leads to a match with V_{kz}. This fact is independent of the outer sign. Once the inner sign is known, similarly, the B_y-component in Equation (3) can be analyzed. Therefore, let $z_m^2 = g + h$ lead to a match with V_{kz}. Now, if $B_y(y_k, z_m) = V_{ky}$, then $B_y(y_k, -z_m)$ is:

$$C\left(\frac{-(-z_m)}{z_m^2 + (y_k + y_0)^2} + \frac{-(-z_m)}{z_m^2 + (y_k - y_0)^2}\right) = -V_{ky}. \tag{30}$$

As shown in the last step, only one of the outer signs of z_n can be the correct one, leading to a match with V_{ky}. $\square$

In summary, there are only two possible locations for a single sensor, so that Equation (16) is satisfied. With the constraints defined above, another fact follows from the proof. It is needed for the consideration of the ambiguities in Section 3.

Corollary 1. *The sign of V_y of the sensor response $\vec{V}$ with respect to the global frame $\{G\}$ depends only on the half plane $z > 0$ or $z < 0$ in which the sensor is located.*

Consequently, both solutions are always in the same half plane. By Theorem 2, one lies in $y < 0$ and the other in $y > 0$. An example is given in Figure 7, which shows the solutions as bold dots.

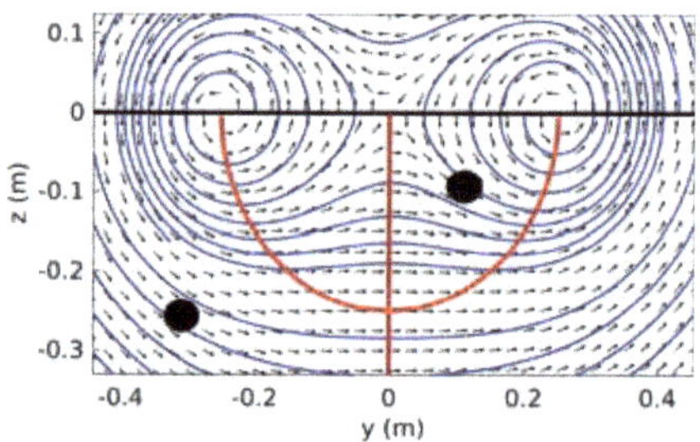

Figure 7. Two remaining positions of a sensor.

Since both points must be in areas with the same direction of the magnetic field lines to satisfy Equation (16), one is outside the circle and one is inside. Of the two remaining sensor positions that match $\vec{V}$, one must still be excluded. By considering only a single sensor, it is impossible to determine the correct one. However, in the following, multiple sensors are considered instead, which provide much more information through which a whole set of correct sensor positions can be selected.

The idea is as follows. On the one hand, the user-defined positions $\vec{p}_{i0}$ of N sensors form an original polygon in the yz-plane of the UAV frame L_u. It has a fixed geometric structure S_u like a fingerprint. On the other hand, the two positions $\vec{p}_{i1,2}$ computed for each sensor can now be combined into 2^N polygons in the global frame $\{G\}$. Finally, the one that matches the structure S_u is selected. Therefore, before comparison, each polygon must be transformed back to L_u using Equation (4). However, this is not possible since the UAV position $\vec{p}_D$ must already be known for this. One way around this problem is to choose a different reference point. For example, all polygons are shifted so that their upper left corner appears in the origin of $\{G\}$ before comparison. What is important here is that the original polygon is first transformed to $\{G\}$. Its new geometric structure S_G can be computed using the known UAV orientation and $\vec{p}_{i0}$ via Equation (4) with $\vec{p}_D = \vec{0}$. Then, its upper left corner is moved to the origin. The other polygons obtained from the sensor measurements are already representations with respect to $\{G\}$, so their upper left corner only needs to be shifted to the origin. The shift is allowed since it does not affect the polygon geometry. Finally, the comparison can be done using the least- squares estimation. Here, the deviation of the corresponding polygon nodes is squared and added up. The polygon with the smallest sum best fits S_G and is selected. From each of its nodes, the UAV position $\vec{p}_D$ can then finally be computed using Equation (4). It is recommended to place the sensors in the yz-plane of L_u so that they span an area as large as possible. This allows for better polygon detection, since during the aerial manipulation, the roll and pitch angle of the UAV will typically have a rather small misalignment with respect to the global frame $\{G\}$. The polygon detection method proposed here was successfully tested. The implementation details and test results are presented in Section 4.

3. Ambiguities

3.1. Ambiguities Due to Field Symmetry

An ambiguity exists if all of the UAV's sensors show the same output in at least two different locations. Under the condition of equivalent static transmission line current, several cases can be identified, which are shown in Figure 8.

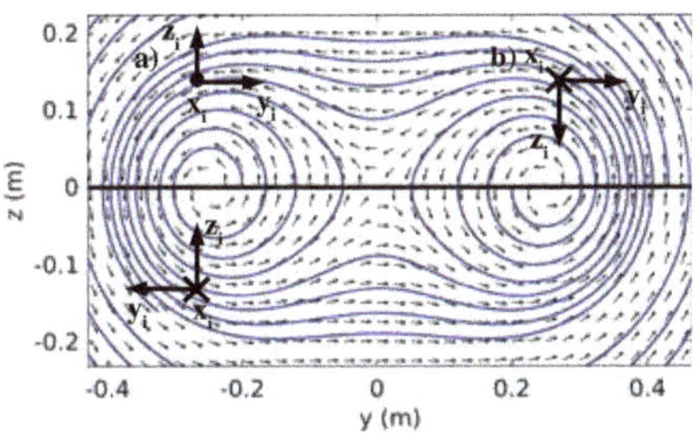

Figure 8. Ambiguities due to field symmetry.

In the lower left, the real position of the local frame L_i of the i-th sensor is shown. The UAV can now change its flight direction and additionally be above the transmission line in Case (a). The sensor output would be the same. In Case (b), the whole UAV is rotated by angle π around the x-axis. Here, since the magnetic field is mapped to itself, the sensor will again measure the same. In this case, the UAV is flying on its back.

3.2. Ambiguities Due to Signal Symmetry

As introduced in Section 2.1, transmission lines carry an alternating current. In addition to improved filtering capabilities, this also introduces a number of ambiguities in navigation, as shown below. The magnitude and sign of the three field components measured by a three-axis magnetometer can be determined in the time domain from the samples. For this purpose, a method was proposed in [24], which is beyond the scope of the present paper. It is taken as given for the following considerations. For example, a sensor with the sample values depicted in Figure 9 would yield the measured value $\vec{v} = (1\,mT, -200\,\mu T, -500\,\mu T)^T$.

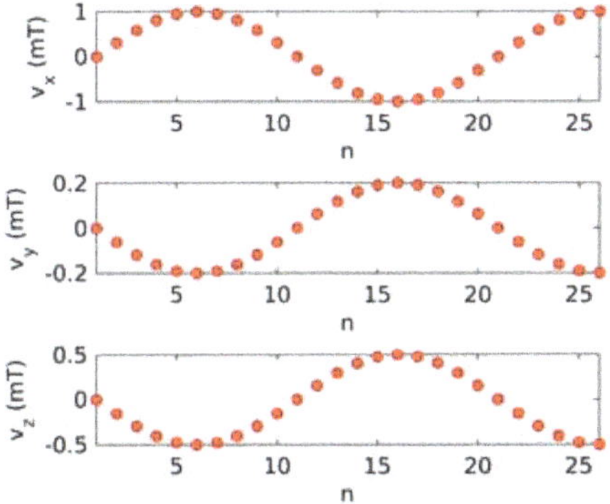

Figure 9. Sample values of a magnetometer.

However, a half period later, the signals in Figure 9 appear shifted, and the signs of the individual field components change; i.e., the sensor now measures $-\vec{v}$. Since technically, only signal amplitudes are used, it is as if the equivalent static current I changes its direction every half period. A direct consequence of this fact follows from Corollary 1. Under the defined constraints in Section 2.4, it allows detecting whether the UAV is flying under or above the transmission lines, since the sign of the measured y-components V_{iy} in the global frame depends only on which half plane $z > 0$ or $z < 0$ the sensors are located. However, since the direction of I changes cyclically, this is no longer possible without additional detection methods, since V_{iy} also changes sign. Such a detection could be realized, for example, by comparing the signal energy of several sensors. This could exploit the fact that the closer the sensor is to the cables, the higher the energy is. However, this is beyond the scope of this paper. The fact that it is impossible to uniquely determine the sensor position can even be seen directly from Equation (25). It contains the prefactor C, which, as shown in (17), depends on the current; thus, it can be $C > 0$ or $C < 0$, leading to different solutions. Furthermore, the whole approach proposed in Theorem 3 to narrow down the solutions no longer works.

3.3. Resolving the Ambiguities

If no additional electronic means are to be used, the ambiguities can be resolved as suggested in [15], where an electric vehicle approaches the inductive charging coil of a parking lot. In the case of using only two magnetometers, it never knows whether the coil is in front or behind it. The vehicle therefore assumes that the coil is always in front of it or that the driver is always moving ahead of the parking space. In the same way, concrete constraints for the navigation of a UAV must be defined and passed to it as a priori knowledge. These are:

1. The equivalent static current I always flows away from the UAV towards that end of the transmission line to which the UAV is looking; i.e., $I > 0 \Rightarrow C > 0$.
2. During the navigation, the UAV always remains under the transmission lines, i.e., in $z < 0$.
3. The UAV angles never exceed the definition range D, so that $\alpha, \beta, \gamma \in D = [-\frac{\pi}{2}, \frac{\pi}{2}]$. This means that the UAV never makes large turns and never flies on its back.

Furthermore, a selection criterion for the measured values must be established. This is because, as mentioned earlier, the sign of the measurement $\vec{v}_i$ of the sensor i changes with each half period. It must now be decided which is the correct one. The selection criterion can be derived from the above constraints. With $C > 0$ and $z < 0$ from Conditions (1) and (2), it follows from the right-hand rule that the magnetic field always has a positive y-component, i.e., $B_y(y,z) > 0$. This can also be shown directly with Equation (3). Moreover, according to Theorem 3, this circumstance never changes as long as the UAV remains in the half plane $z < 0$ and as long as Constraint (3) is not broken. Using Equation (5) and the Cartesian unit vector $\vec{e}_y$ leads to the necessary selection criterion:

$$(R\,\vec{v}_i) \cdot \vec{e}_y > 0. \tag{31}$$

If $\vec{v}_i$ does not satisfy (31), $-\vec{v}_i$ does, and vice versa. Only one sign can be a match. Direct comparison of the signs of $\vec{v}_{iy}$ and $B_y(y_i, z_i)$ is not allowed, since from $B_y(y_i, z_i) > 0$, $\vec{v}_{iy} > 0$ does not necessarily follow. The last point to be clarified is the orientation of the UAV. Since the rotation matrix in (31) is used, it must be ensured that the angles can be computed independently of the direction of I.

Theorem 4. *Given two sensors i and k with equally aligned local frames and fixed locations $\vec{p}_i \neq \vec{p}_k$, as long as their measurement values $\vec{v}_i$ and $\vec{v}_k$ are obtained from the same signal period, the yaw angle α and pitch angle β are independent of the direction of the static current I.*

Proof. Referenced to the same time interval containing a whole signal period, the sensors measure $\vec{v}_i, \vec{v}_k \neq \vec{0}$. Half a signal period later, when both measurements are calculated from the same new time interval and the sensors have not moved, the sign of the measurements is reversed. Then, the cross-product in Equation (9) for the yaw angle becomes $-\vec{v}_i \times -\vec{v}_k$. Now, by applying the rule $(r\vec{a}) \times \vec{b} = \vec{a} \times (r\vec{b}) = r(\vec{a} \times \vec{b})$, it follows that:

$$-\vec{v}_i \times -\vec{v}_k = (-1)^2 (\vec{v}_i \times \vec{v}_k) = \vec{v}_i \times \vec{v}_k.$$

Since the direction of the static current obviously has no effect on the cross-product, the result for the yaw angle also remains unchanged. Flipping the signs of the field components in Equation (10) for the pitch angle leads to:

$$\frac{-v_{kx} - v_{ky}A}{-v_{kz}\sqrt{A^2+1}} = \frac{v_{kx} + v_{ky}A}{v_{kz}\sqrt{A^2+1}}, \quad A = tan\,\alpha.$$

The direction of the static current obviously also has no influence on the pitch angle. $\quad\square$

Equation (14) for the roll angle γ uses the measurement from only one sensor as the input. Thus, it is also independent of the sign of I. In summary, the sign of I does not affect the orientation as long as the measurements are referenced to the same time interval of the corresponding signal.

4. Simulation Results

In this section, the presented equations for determining the UAV position and orientation are tested in the context of a software-in-the-loop (SIL) simulation.

4.1. SIL Design and Test Procedure

The SIL simulation was implemented in MATLAB and cyclically ran through a specified flight trajectory created with the Gazebo tool. Each waypoint of the trajectory is represented by a vector $(y_D, z_D, \hat{\alpha}, \hat{\beta}, \hat{\gamma})^T$, which describes the position and orientation of the local UAV frame L_u with respect to the global frame $\{G\}$. From this, the SIL computes the positions of the virtual sensors i and the corresponding magnetic measurements $\vec{v}_i$ with respect to the local sensor frame L_i in each run. For this purpose, Equations (4) and (5) were used, where the rotation matrix was used according to the convention in Gazebo, i.e., $\hat{R}_G^{L_u} = R_x^T(\hat{\gamma}) R_y^T(\hat{\beta}) R_z^T(\hat{\alpha})$. This corresponds to the transposed version of the definition used in this paper, i.e., $\hat{R}_G^{L_u} = (R_G^{L_u})^T$. In order for the angles α, β, and γ obtained in the test run to be compared later with the Gazebo angles $\hat{\alpha}$, $\hat{\beta}$, and $\hat{\gamma}$, they must first be converted at the end of each run. This is done automatically by the SIL. To calculate the measured values $\vec{v}_i$, the SIL assumes that there is a transmission line system with the arrangement described in Section 2.1; i.e., the cables are in the xy-plane at $-y_0$ and y_0. The SIL then passes the computed $\vec{v}_i$ as the input values to the test procedure, which verifies the entire method described in this publication. The test procedure uses the constraints defined in Section 3.3 for navigation as a priori knowledge. It is called cyclically from the SIL until all waypoints on the flight trajectory have been passed and is divided into the following steps:

1. Select the correct sign for the measurement values $\vec{v}_i$ according to the selection criterion (31) (the SIL skips this step because it always calculates with the equivalent static current).
2. Calculate the UAV angles using Equations (9), (10) and (14) for all possible sensor combinations and calculate the average. The results are $\bar{\alpha}$, $\bar{\beta}$, and $\bar{\gamma}$.
3. Filter out the outliers by applying the moving median filter to the last 10 calculated values of each angle.
4. Calculate the measurement values $\vec{V}_i$ referred to the global frame $\{G\}$ via $\vec{V}_i = R_G^{L_u}(\bar{\alpha}, \bar{\beta}, \bar{\gamma}) \, R_{L_u}^{L_i} \vec{v}_i$ from Equation (6). Calculate the signal powers $P_i = |\vec{V}_i|$.
5. Calculate the y-coordinates y_{i1} and y_{i2} using Equation (25).
6. Select the correct signs in Equation (28) using the Theorem 3 in order to reduce the amount of possible z-solutions and calculate the z-coordinates z_{i1} and z_{i2}.
7. Calculate the nodes $\vec{p}_{i,e}$ of the original polygon referred to $\{G\}$. They are calculated from the user-defined sensor positions $\vec{p}_{i0}$ in L_u and the orientation of the UAV in $\{G\}$, i.e., $\vec{p}_{i,e} = R_G^{L_u}(\bar{\alpha}, \bar{\beta}, \bar{\gamma})\vec{p}_{i0}$. Shift the resulting polygon so that its upper left corner appears in the origin of $\{G\}$.
8. Calculate from the determined sensor positions $\vec{p}_{i1,2}$ all possible polygons and shift them so that their upper left corner appears in the origin of $\{G\}$. The nodes of the k-th polygon are represented by $\vec{p}_{i,k}$.
9. Compare all polygons with the original one. Therefore, first calculate the sum U_k of the least-squared errors of the corresponding polygon nodes with:

$$U_k = \sum_{i=1}^{N} (\vec{p}_{i,e} - \vec{p}_{i,k})^T (\vec{p}_{i,e} - \vec{p}_{i,k}), \tag{32}$$

where N is the number of nodes, respectively sensors. Choose the polygons of the two smallest sums. Finally, choose one out of these two whose perimeter most closely matches that of the original polygon.

10. Calculate the UAV position from the found polygon for each node using Equation (4), i.e.,

$$\vec{p}_{D,i} = \vec{p}_{i,k} - R_G^{L_u}(\bar{\alpha}, \bar{\beta}, \bar{\gamma})\,\vec{p}_{i0}.$$

Take the average of all $\vec{p}_{D,i}$ as the UAV position $\vec{p}_D$.

11. Filter out the outliers by applying the moving median filter to the last 10 calculated positions.
12. Convert the angles to Gazebo angles.

If the entire trajectory has been run through by the SIL, the specified and actual position are output graphically. The same goes for the angles. The software offers several configuration parameters. These are:

1. Strength of the current I
2. Position of the transmission lines $\pm y_0$
3. Number of sensors
4. Sensor positions in L_u
5. The SIL can simulate white noise that is superimposed on the magnetic measurements $\vec{v}_i$. For this, the standard deviation σ can be set.
6. The flight trajectory can be shifted in the yz-plane in order to test the algorithms, both close to and further away from the transmission lines.

4.2. Test Parameters

The test cases and the associated results are presented below. If not otherwise stated, the following configuration is used:

- Transmission lines: $y_0 = 0.25\,\text{m}$, $I = 700\,\text{A}$
- Two sensors aligned with L_u. Their positions in L_u are:

$$\vec{p}_1 = (0.05m, 0.2m, -0.05)^T, \quad \vec{p}_2 = (0.05m, -0.2m, -0.05)^T.$$

- Two sensors rotated by $90°$. Their positions in L_u are:

$$\vec{p}_3 = (0.05m, 0.2m, -0.25)^T, \quad \vec{p}_4 = (0.05m, -0.2m, -0.25)^T.$$

In the previous publication [25], it was shown that the electric motor of a small 1:5 scale electric toy car can produce stray magnetic fields of several μT; especially when the vehicle accelerates. The reason for this is that inadequate magnetic shielding measures are often taken in technical installations. The supply lines of electronic components with high power requirements can therefore emit strong magnetic interference fields. Since the UAV was also equipped with several electric motors, the test cases were performed with different standard deviations σ of the white noise. A value of $1\,\mu T$ was assumed for the case of good shielding and $8\,\mu T$ for poorly shielded electric motors. These were realistic values based on the measurements in [25] and took into account that the UAV may have multiple electric motors.

4.3. Test Results

Figure 10 shows the localization results using the derived equations and the proposed polygon detection method.

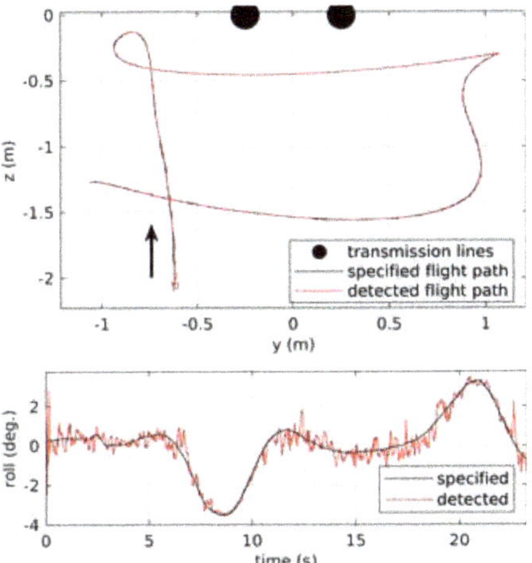

Figure 10. Localization quality, good shielding: $\sigma = 1\,\mu$T.

The flight started at the arrow in the indicated direction, with the UAV flying one lap under the transmission lines. As was to be expected, good shielding allowed more precise navigation. The quality of localization was obviously very high with a magnetic noise standard deviation of $\sigma = 1\,\mu$T. The mean deviation from the specified flight path was 21 mm, with a maximum error of 54 mm. The standard deviation of the roll angle error was $0.4°$, and its mean was very close to zero, around $0.025°$. Only the roll angle is shown here, as it had the largest values since the UAV moved almost exclusively in the yz-plane in the simulation.

Figure 11 shows the localization results for the case where poorly shielded electrical drives in the UAV produce magnetic noise with a standard deviation of $\sigma = 8\,\mu$T.

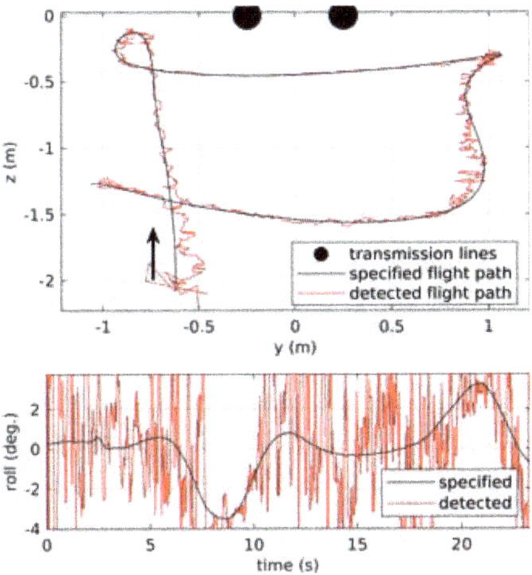

Figure 11. Localization quality, poor shielding: $\sigma = 8\,\mu$T.

Despite the high magnetic interference, the navigation worked very well near the transmission lines due to the high signal-to-noise ratio (SNR). The deviations increased with greater distance from the transmission lines. Figure 12 shows the deviation from the specified flight path and the specified roll angle for each waypoint.

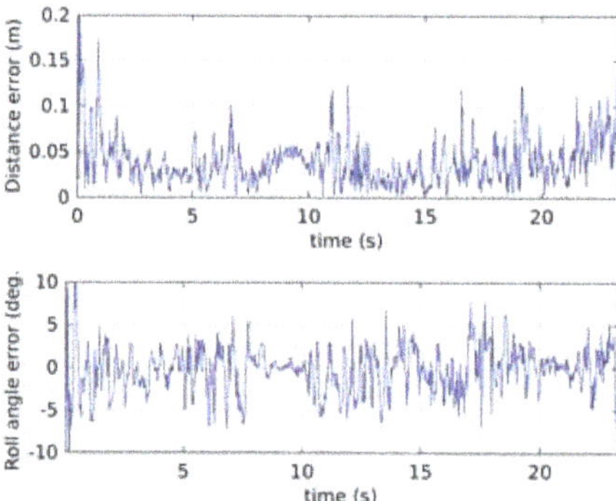

Figure 12. Path and roll angle errors, poor shielding: $\sigma = 8\,\mu T$.

The mean deviation from the specified flight path was 41 mm, where the maximum error was 221 mm. The mean deviation from the specified roll angle was $-0.22°$, where the maximum error was $12.3°$. The standard deviation of the roll angle error was $2.96°$.

Figure 13 shows the localization results for the same noise standard deviation of $\sigma = 8\,\mu T$ when only three sensors were used, i.e., Sensor 4 at $\vec{p}_4$ was excluded.

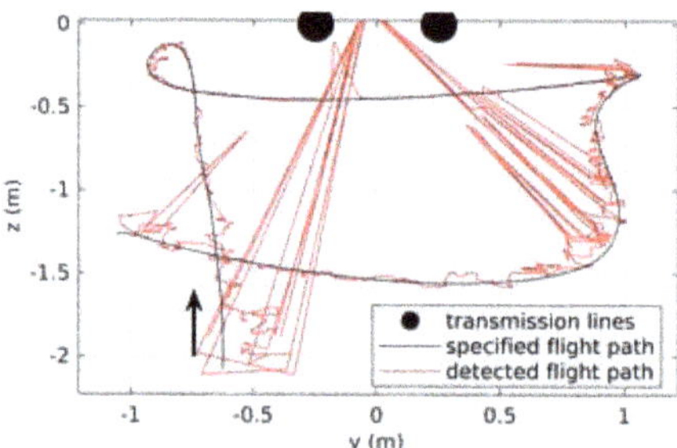

Figure 13. Localization quality, three sensors, $\sigma = 8\,\mu T$.

The mean deviation from the specified flight path was 91 mm. In areas far from the transmission lines, as in this case, very strong outliers could occur occasionally, so that the maximum error was now about 2260 mm. The outliers occurred very frequently compared to the case with four sensors, as demonstrated by repeated experiments. These were not due to an incorrect calculation of the UAV position, but to an incorrect detection of the polygons. These were extremely distorted by the strong magnetic interference and could therefore hardly be distinguished from each other. This was to be expected, because the more sensors were used, the higher the noise immunity was. In practice, the UAV would carry out aerial manipulations in the vicinity of the transmission lines and would therefore be in an area with a high SNR. Therefore, it may also be possible to use only three sensors, as shown in Figure 14. The UAV started its flight near the transmission lines and then flew between the lines. Only when it then moved away did strong outliers occur. Repeated tests always led to the same result.

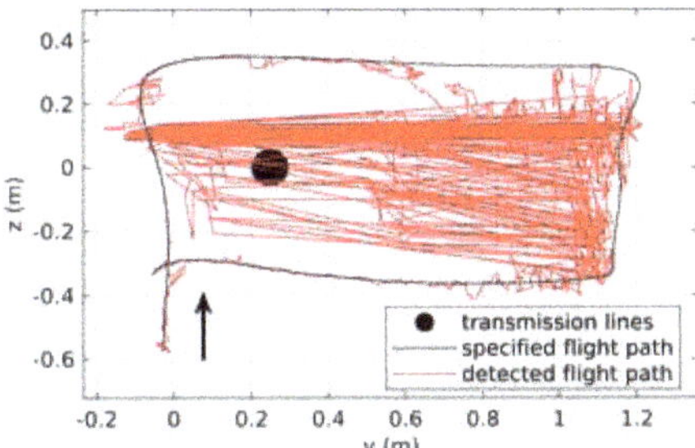

Figure 14. Localization quality close to the transmission lines and far away, three sensors, $\sigma = 8\,\mu\text{T}$.

In the tests, a current of 700 A was assumed. It should be mentioned that the quality of the localization depended on the current strength in the transmission lines. This in turn depended on the per capita energy consumption of the residents, which may vary from city to city. Thus, for completeness, Figure 15 shows additionally the same case as Figure 10 with a low transmission line current of 100 A.

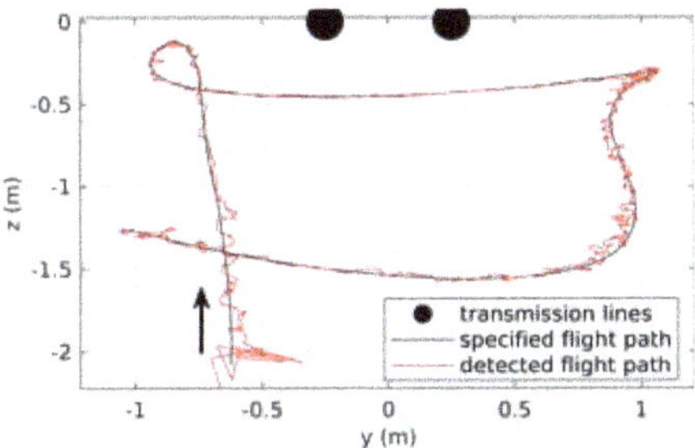

Figure 15. Localization quality in the case of low transmission line current of 100 A, four sensors, $\sigma = 1\,\mu\text{T}$.

5. Discussion

As mentioned in Section 3.2, in a real transmission line system, it is difficult to detect whether the UAV is located below or above the cables without additional detection mechanisms. This is only possible in Figure 14 because the SIL calculated with a static equivalent current and was only used in the context of the tests.

In this paper, transmission lines were modeled as perfectly straight cables, which may not be the case in real-world circumstances. Transmission lines, for example, hang downward, thus forming more of a curved line. Even though the radius of this curve is very large, the influence on navigation accuracy should be investigated in a further work. In UAV indoor navigation applications, this problem does not occur because the cables are shorter and can be routed to fit the straight line model.

Furthermore, the magnetic field contributions of the other two phases of the three-phase system were neglected, since they are typically at least 2 m away from the phase under consideration. This should also be investigated in more detail in a further study in the context of transmission line navigation.

Finally, the magnetometers should be as far away as possible from ferromagnetic materials that may be in the UAV; otherwise, they may affect the sensor readings.

6. Conclusions

In this work, it was demonstrated that UAV navigation is possible in the magnetic field of two long parallel conductors. In this context, equations were derived to be able to calculate the UAV orientation analytically. It was shown that only three magnetometers are needed for this purpose and that neither the current, nor the magnetic field equations need to be known. Furthermore, it was shown that combining the magnetic field equation with the magnetic energy equation leads to analytical solutions for the calculation of the sensor position. This results in 16 possible positions for a single sensor, from which the correct one must be determined. For this purpose, theorems were derived, which can be used to exclude 14 of the 16. The last two remaining positions are not distinguishable. However, it was shown that considering multiple sensors allows a whole set of correct positions to be selected. Here, the positions were combined into a number of polygons in the yz-plane, from which the one most similar to the original polygon was selected. This can be determined from the known orientation and the user-defined sensor positions in the local UAV frame. This ultimately led to a unique determination of the correct sensor positions and hence the UAV position. All equations, as well as the polygon selection method were successfully verified for different noise.

Author Contributions: Investigation, D.M.; methodology, D.M.; software, D.M.; validation, D.M.; writing—original draft preparation, D.M.; writing—review and editing, Z.K., S.B.; visualization, D.M.; supervision, S.B., Z.K.; project administration, S.B. All authors read and agreed to the published version of the manuscript.

Funding: The project AERIAL COgnitive integrated multi-task Robotic system with Extended operation range and safety (Aerial-Core) received funding by the European Union's Horizon 2020 research and innovation program under grant agreement No. 871479.

Institutional Review Board Statement: Not applicable.

Informed Consent Statement: Not applicable.

Data Availability Statement: Not applicable.

Acknowledgments: I would like to thank my colleague Filip Zorić who generated the test flight data in Gazebo and thus supported the tests.

Conflicts of Interest: The authors declare no conflict of interest.

Abbreviations

The following abbreviations are used in this manuscript:

| SNR | signal-to-noise ratio |
| UAV | unmanned aerial vehicle |

References

1. Kalmus, H.P. A New Guiding and Tracking System. *Ire Trans. Aerosp. Navig. Electron.* **1962**, *ANE-9*, 7–10. [CrossRef]
2. Prigge, E.A.; How, J.P. Signal architecture for a distributed magnetic local positioning system. *IEEE Sensors J.* **2004**, *4*, 864–873. [CrossRef]
3. Blankenbach, J.; Norrdine, A. Position estimation using artificial generated magnetic fields. In Proceedings of the IPIN International Conference on Indoor Positioning and Indoor Navigation, Zurich, Switzerland, 15–17 September 2010; pp. 1–5.
4. Blankenbach, J.; Norrdine, A.; Hellmers, H. A robust and precise 3D indoor positioning system for harsh environments. In Proceedings of the 2012 International Conference on Indoor Positioning and Indoor Navigation (IPIN), Sydney, Australia, 13–15 November 2012; pp. 1–8. [CrossRef]
5. Hellmers, H.; Eichhorn, A.; Norrdine, A.; Blankenbach, J. Indoor localisation for wheeled platforms based on IMU and artificially generated magnetic field. In Proceedings of the 2014 Ubiquitous Positioning Indoor Navigation and Location Based Service (UPINLBS), Corpus Christi, TX, USA, 20–21 November 2014; pp. 255–264. [CrossRef]
6. Pasku, V.; Angelis, A.D.; Dionigi, M.; Angelis, G.D.; Moschitta, A.; Carbone, P. A Positioning System Based on Low-Frequency Magnetic Fields. *IEEE Trans. Ind. Electron.* **2016**, *63*, 2457–2468. [CrossRef]
7. Plotkin, A.; Paperno, E. 3-D magnetic tracking of a single subminiature coil with a large 2-D array of uniaxial transmitters. *IEEE Trans. Magn.* **2003**, *39*, 3295–3297. [CrossRef]
8. Loke, W.; Choi, T.; Maleki, T.; Papiez, L.; Ziaie, B.; Jung, B. Magnetic Tracking System for Radiation Therapy. *IEEE Trans. Biomed. Circuits Syst.* **2010**, *4*, 223–231. [CrossRef] [PubMed]
9. Olsen, R.G.; Farstad, A.J. Electromagnetic Direction Finding Experiments for Location of Trapped Miners. *IEEE Trans. Geosci. Electron.* **1973**, *11*, 178–185. [CrossRef]
10. Wait, J. Electromagnetic Induction Technique for Locating a Buried Source. *IEEE Trans. Geosci. Electron.* **1971**, *9*, 95–98. [CrossRef]

11. Caffey, T.W.H.; Romero, L. Locating a Buried Magnetic Dipole. *IEEE Trans. Geosci. Remote. Sens.* **1982**, *GE-20*, 188–192. [CrossRef]
12. Pinies, P.; Tardos, J.D. Fast localization of avalanche victims using sum of Gaussians. In Proceedings of the 2006 IEEE International Conference on Robotics and Automation (ICRA), Orlando, FL, USA, 15–19 May 2006; pp. 3989–3994. [CrossRef]
13. Vasić, D.; Ambruš, D.; Bilas, V. Analytical modelling of soil effects on electromagnetic induction sensor for humanitarian demining. *J. Phys. Conf. Ser.* **2013**, *450*, 012052. [CrossRef]
14. Ambruš, D.; Vasić, D.; Bilas, V. Innovating on top of I&M fundamentals for safer humanitarian demining. *IEEE Instrum. Meas. Mag.* **2020**, *23*, 35–41. [CrossRef]
15. Martinović, D. Precise Vehicle Alignment by Eliminating the Distortion of the Pulsed Magnetic Flux Density Distribution of a Charging Coil. Doctoral Thesis, University of Stuttgart, Stuttgart, Germany, 2017.
16. Paperno, E.; Sasada, I.; Leonovich, E. A new method for magnetic position and orientation tracking. *IEEE Trans. Magn.* **2001**, *37*, 1938–1940. [CrossRef]
17. Song, S.; Hu, C.; Li, B.; Li, X.; Meng, M.Q. An Electromagnetic Localization and Orientation Method Based on Rotating Magnetic Dipole. *IEEE Trans. Magn.* **2013**, *49*, 1274–1277. [CrossRef]
18. Raab, F.H.; Blood, E.B.; Steiner, T.O.; Jones, H.R. Magnetic Position and Orientation Tracking System. *IEEE Trans. Aerosp. Electron. Syst.* **1979**, *AES-15*, 709–718. [CrossRef]
19. Raab, F.H. Quasi-Static Magnetic-Field Technique for Determining Position And Orientation. *IEEE Trans. Geosci. Remote. Sens.* **1981**, *GE-19*, 235–243. [CrossRef]
20. Jinglan, T.; Shuang, S.; Xiaojing, W.; Tingfang, Y.; Chao, H.; Meng, M.Q. An improved method and algorithm for electromagnetic localization. In Proceedings of the 2011 IEEE International Conference on Information and Automation, Shenzhen, China, 6–8 June 2011; pp. 406–411. [CrossRef]
21. Ito, Y.; Kato, N.; Kojima, J.; Takagi, S.; Asakawa, K.; Shirasaki, Y. Cable tracking for autonomous underwater vehicle. In Proceedings of the IEEE Symposium on Autonomous Underwater Vehicle Technology (AUV'94), Cambridge, MA, USA, 19–20 July 1994; pp. 218–224. [CrossRef]
22. Xiang, X.; Yu, C.; Niu, Z.; Zhang, Q. Subsea Cable Tracking by Autonomous Underwater Vehicle with Magnetic Sensing Guidance. *Sensors* **2016**, *16*. [CrossRef] [PubMed]
23. Yu, C.; Xiang, X.; Lapierre, L.; Zhang, Q. Robust Magnetic Tracking of Subsea Cable by AUV in the Presence of Sensor Noise and Ocean Currents. *IEEE J. Ocean. Eng.* **2018**, *43*, 311–322. [CrossRef]
24. Martinović, D.; Binz, C.; Reuss, H.C. Magnetic Field based Localization of the Charging Coil using Trilateration. In *Autoreg 2015—VDI-Berichte 2233*; VDI Verlag GmbH: Düsseldorf, Germany, 2015; pp. 129–140.
25. Hisung, M.; Martinović, D.; Reuss, H.C. Feasibility study on the basis of a prototype for automated vehicle positioning in inductive charging process. In Proceedings of the 19. Internationales Stuttgarter Symposium; Springer Fachmedien: Wiesbaden, Germany, 2019; pp. 667–678. [CrossRef]

Article

Safe Local Aerial Manipulation for the Installation of Devices on Power Lines: AERIAL-CORE First Year Results and Designs

Jonathan Cacace [1,2,*], Santos M. Orozco-Soto [2], Alejandro Suarez [3], Alvaro Caballero [3], Matko Orsag [4], Stjepan Bogdan [4], Goran Vasiljevic [4], Emad Ebeid [5], Jose Alberto Acosta Rodriguez [6] and Anibal Ollero [3]

[1] Department of Electrical Engineering and Information Technology, University of Naples Federico II, Via Claudio 21, 80125 Naples, Italy
[2] CREATE Consortium, Via Claudio 21, 80125 Naples, Italy; santos.orozco.cav@gmail.com
[3] GRVC Robotics Labs, University of Seville, 41092 Sevilla, Spain; asuarezfm@us.es (A.S.); alvarocaballero@us.es (A.C.); aollero@us.es (A.O.)
[4] Department of Control and Computer Engineering, University of Zagreb, 10000 Zagreb, Croatia; matko.orsag@fer.hr (M.O.); stjepan.bogdan@fer.hr (S.B.); goran.vasiljevic@fer.hr (G.V.)
[5] SDU UAS Center, University of Southern Denmark, 5230 Odense, Denmark; esme@mmmi.sdu.dk
[6] E-distribución, 41007 Sevilla, Spain; josealberto.acosta@enel.com
* Correspondence: jonathan.cacace@unina.it

Abstract: The power grid is an essential infrastructure in any country, comprising thousands of kilometers of power lines that require periodic inspection and maintenance, carried out nowadays by human operators in risky conditions. To increase safety and reduce time and cost with respect to conventional solutions involving manned helicopters and heavy vehicles, the AERIAL-CORE project proposes the development of aerial robots capable of performing aerial manipulation operations to assist human operators in power lines inspection and maintenance, allowing the installation of devices, such as bird flight diverters or electrical spacers, and the fast delivery and retrieval of tools. This manuscript describes the goals and functionalities to be developed for safe local aerial manipulation, presenting the preliminary designs and experimental results obtained in the first year of the project.

Keywords: aerial manipulation; inspection and maintenance; power lines

Citation: Cacace, J.; Orozco-Soto, S.M.; Suarez, A.; Caballero, A.; Orsag, M.; Bogdan, S.; Vasiljevic, G.; Ebeid, E.; Rodriguez, J.A.A.; Ollero, A. Safe Local Aerial Manipulation for the Installation of Devices on Power Lines: AERIAL-CORE First Year Results and Designs. *Appl. Sci.* **2021**, *11*, 6220. https://doi.org/10.3390/app11136220

Academic Editor: Oscar Reinoso García

Received: 8 June 2021
Accepted: 2 July 2021
Published: 5 July 2021

Publisher's Note: MDPI stays neutral with regard to jurisdictional claims in published maps and institutional affiliations.

1. Introduction

The power grid is one of the most extended and critical infrastructures in any country, comprising thousands of kilometers of power lines that require periodic inspection and maintenance (I&M) to ensure continuous service, despite the diverse factors that may affect correct operation. Adverse climatic conditions (i.e., rain, wind, snow), vegetation growth, and the impact on the bird species in the environment are some typical problems faced by companies responsible for its management and maintenance, involving a significant economic investment due to the vast extension of this infrastructure. Not only that, but the high altitude (15–50 m) and high voltage (15–500 kV) of the power lines make them a highly risky workspace for human operators, who must follow strict safety protocols [1] and use special clothes [2], which reduce the effective operation time and the feeling of comfort in the realization of manipulation tasks. The installation of devices, such as bird flight diverters or electrical spacers, such as those shown in Figure 1, is an illustrative example of an operation that requires the transportation and deployment of the worker on power lines. As indicated in [1], different methods and tools can be considered for positioning the personnel on the points of interest, including non-conductive ropes, bucket trucks, or manned helicopters.

The development of aerial manipulation robots in the last decade [3–5] has been motivated by the ability of these platforms to easily and quickly reach high altitude work spaces and conduct diverse operations involving physical interaction with the environment, demonstrating their application in contact-based inspection in chemical plants [6] and bridges [7], the installation of sensor devices [8,9], valve turning [10] and other torsional operations [11]. In this sense, the use of aerial manipulation robots are particularly useful to install devices, such as the ones shown in Figure 1, avoiding risk for human operators while reducing the time required to reach the points of interest. However, some new challenges arise. First, the electrical interaction between the high voltage power line and the aerial robot, considered a floating conductor [12], should be investigated to avoid possible malfunctions due to electrostatic discharge [13] or magnetic noise. Second, the risk of collision of the aerial robot with the transmission line conductors, especially outdoors, where the wind gusts may be significant, makes convenient the use of long-reach [8] or cable-suspended [14,15] configurations to increase the separation distance between the manipulator and the aerial platform. Third, the manipulator should provide a sufficient level of dexterity and force to install the mentioned devices, taking into account that these are intended for human operators.

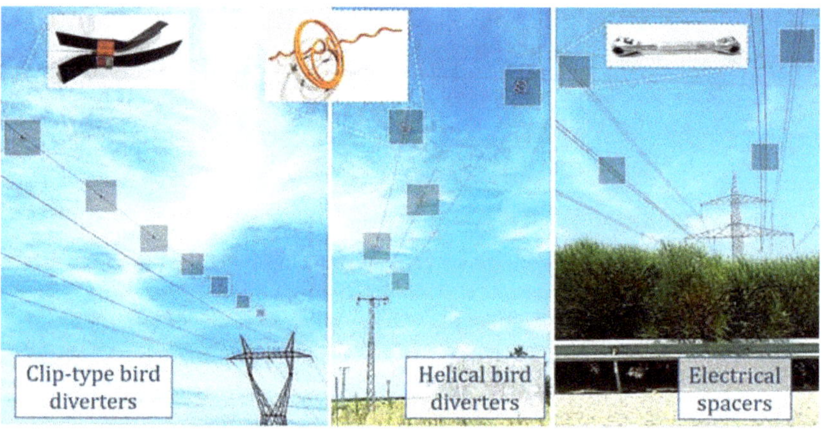

Figure 1. Three devices typically installed on the power lines.

This paper proposes the application of aerial manipulation robots for the installation of bird flight diverters and electrical spacers on power lines, presenting the preliminary designs and results derived from the first year of the AERIAL-CORE (AERIAL COgnitive integrated multi-task Robotic system with Extended operation range and safety) H2020 project [16]. The paper describes the technological solutions under development, including prototypes of arms and end effectors, methods for safe and accurate aerial manipulation, and augmented human–machine interfaces. The paper analyzes first the requirements and functionalities of the system from an illustrative scenario, identifying later the effects of the high voltage/current of the power line on the aerial robot. The paper aims to provide a wider perspective of the problem and the possible solutions, considering the different approaches followed by the authors.

The rest of the paper is organized as follows. Section 2 introduces the current procedures in the inspection and maintenance of power lines with a focus on the installation of devices based on the requirements and functionalities. Section 3 analyzes the electromagnetic effects of the power lines on the aerial manipulation robot. Section 4 presents the preliminary designs of arms and end effectors for the installation of the devices, whereas Section 5 is devoted to the methods for safe aerial manipulation. Section 6 describes the human–aerial robot interaction, comprising augmented cognitive human–machine interfaces intended to facilitate the aerial manipulation operation, and the use of long-reach aerial manipulators for fast and safe tool delivery to the operators. Finally, the conclusion and future works are summarized.

2. Procedures in the Inspection and Maintenance of Power Lines

In this section, an overview of the inspection and maintenance procedures is provided along with a set of system functionalities to be implemented by the aerial manipulation robots.

2.1. Power Lines Inspection

The inspection and maintenance of power line assets are of great importance to ensure that the infrastructure can deliver electricity to millions of inhabitants, while maximizing the operational life of the lines before replacement is required. The task of power line inspection is undertaken through various approaches [17]. The most common approach is to use a helicopter to transport a photographer within proximity of the line, taking high-resolution photos for further analysis. Unmanned Aerial Vehicles (UAVs) are also used by technicians to inspect faults and areas of importance. In both cases, pilots and technicians require special training and permissions to complete the inspection. Even though UAVs are faster, more accurate, and cost-effective, the working range for UAVs is limited to the pilot's visual line of sight. To overcome this limitation, onboard sensors and navigation algorithms are embedded into the UAVs to enable the system autonomy to inspect cables, beyond visual line of sight (BVLOS) [18,19].

2.2. Installation of Devices on Power Lines

The aerial manipulators developed in the AERIAL-CORE project (see Section 4) are intended to conduct the installation of the three devices illustrated in Figure 1: the clip-type bird diverter, the helical bird diverter, and the electrical spacer. Although the particular devices employed in each country or region may vary according to the local regulation or available manufacturers, the mechanisms employed to attach these to the power lines are similar between the different models. Nowadays, the installation operation is carried out by human workers employing elevating work platforms or manned helicopters to reach the points of interest. Some solutions with rolling robots [20,21] and multirotors [22] have been already demonstrated. To identify the most suitable solution according to the features of each device, Table 1 summarizes the associated operations and requirements for the corresponding manipulator. The clip-type bird diverter is installed, exerting a relatively high pushing force (around 200 N) over the device once the line is introduced in the slot up to the circular hole that crosses the plastic body of the device. The helical bird diverter requires dexterous bimanual manipulation capabilities, given the geometry of the device and the complex installation procedure [23]. In this case, a human-like and human-size dual-arm system results in a particularly suitable solution since it allows to replicate in an intuitive way the movements of the operators. As occurs with the clip-type bird diverter, the installation of the electrical spacer requires the design of a specific manipulator that holds the main body of the device along with the two clamps and the screws, which should be also screwed automatically with a screwdriver mechanism. The manipulator is attached to the multirotor platform, exploiting the yaw control to achieve an adequate orientation when it is introduced between the two cables.

Table 1. Procedure for the installation of the different devices.

Device	Operation	Manipulator	Main Requirement
Clip-Type	Insertion/pushing	Linear actuator	High force (>100 N)
Helical	Grasping and rolling	Dual arm	Dexterity, human-size
Electrical spacer	Screwing	Screwdriver	Support screw/nut

2.3. Requirements and Functionalities

In order to perform the installation of devices on the power line using aerial robotic manipulators, the following main system requirements have been identified:

1. Installation of clip-type bird diverters, helical bird diverters, and electrical spacers. These are the target devices on which the solution design is focused.
2. Landing and takeoff from the power line. This allows an increment in the accuracy of the manipulation since the robot can operate with its base fixed on the cables.
3. Locomotion along power lines using the cables. In this way, the installation of several devices along a segment of the power line can be simplified, avoiding continuous landings and take-offs and improving energy efficiency.
4. Compatibility with the live-line operation. Cut-offs of the energy transmission during I&M tasks on power lines entail high costs due to the supply losses and should be planned to mitigate their effects. For this reason, these cut-offs should be avoided.
5. Autonomous installation of devices. The installation of devices, such as clip-type bird diverters, can become repetitive and relatively easy. Consequently, it should be automated to speed up the process and to reduce the operator workload.
6. Teleoperation capabilities for manipulation. In contrast to the previous requirement, the helical bird diverters require complex interaction with the power line that can be difficult to automate. In this case, teleoperation methods can be applied to achieve better performance by introducing the human in the loop.
7. Cognitive safe aerial manipulation. During the approaching phase to the power line and when operating around it, potential obstacles, such as trees, electric towers, or cables can be present and the associated collision and entrapment risks should be detected and mitigated.

According to the previous requirements, the developed solution should incorporate the functionalities listed below:

1. Task-oriented end effectors. The use of end effectors specially designed for the manipulation of a specific device leads to a more effective operation.
2. Dual-arm manipulation capabilities. Devices such as helical bird diverters are relatively large and difficult to manipulate due to their elasticity, so a dual-arm system is more adequate to perform their installation.
3. Manipulation with force interaction. The installation of devices involves the application of high forces to attach them in a reliable way to the power line. Since this can be critical, particularly if the robot is flying, the aerial manipulator and its control strategy need to be designed carefully.
4. Aerial manipulation with extended locomotion. The aerial manipulator can integrate a rolling base that allows movement along the power line once the robot has landed on it. In this way, flight segments can be substituted by rolling segments on the line, whose energy consumption is lower.
5. High-voltage-tolerant design. The electric and electromagnetic fields associated with the high voltage and current of the power line can disturb the robot electronics. Consequently, the robot design needs to deal with this difficulty.
6. Planning system for the installation of devices on power lines. The efficient installation of many devices in terms of time and energy requires the use of mission plans computed according to the particular characteristics of both the robot and the environment.
7. Autonomous navigation and manipulation. In order to autonomously install devices on power lines, the aerial manipulator needs to detect the power line, estimate its relative position, and apply the proper hybrid position/force control techniques.
8. Augmented cognitive human–machine interface. The aerial manipulator will incorporate a module that allows an operator to easily operate it from a ground control station in order to perform and supervise the installation of complex devices.
9. Obstacle detection and avoidance. Thanks to perception sensors installed onboard the aerial manipulator, the different obstacles in the operation area will be identified, and the robot trajectory will be locally modified to avoid them.

3. Effects of Power Lines on Aerial Manipulators

Power transmission lines generate magnetic and electric fields in their environment. The magnetic field strength depends on the current flowing through the conductor, while the electric field strength depends on the voltage. Both fields can affect the aerial manipulator operating near the power line.

3.1. Effects of the Magnetic Field

Power line can be seen as a thin straight long conductor, which generates the magnetic field proportional to the current flowing through it:

$$B = \frac{\mu_0 I}{2\pi r} \tag{1}$$

where I is the current flowing through the conductor, B is the magnetic field strength, r is the distance from the conductor and μ_0 is the permeability of vacuum.

The magnetic field isomagnetic lines of a single conductor are circular (Figure 2), centered on the conductor, and their direction is determined by a right-hand rule. With two or more conductors, the magnetic field becomes more complex.

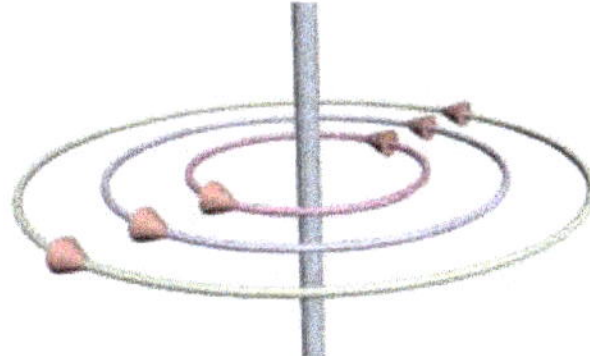

Figure 2. Magnetic field around the long straight conductor.

The typical current flowing through the power line conductor is an alternating current with a frequency of 50 or 60 Hz and with the strength of several hundred amperes. It generates a strong alternating magnetic field with the same frequency as the current. In such a setup, at a distance of less than 1 m from the power line, the magnetic field strength is stronger than the earth's magnetic field, rendering the information from the drone's on-board compass useless.

On the other hand, since the idea is to perform operations on the power line, the magnetic field generated by the current flowing through the conductors can be used for precise localization with respect to the power line. We explored this approach in [24], where the position of the drone with respect to the two-wire power line was calculated, using measurements from magnetometers mounted on the drone. Here, we present an analytical solution derived from the nonlinear magnetic field equations that promises low computational time. To determine the orientation of the drone with respect to the power lines, only three magnetometers are needed, and neither the current nor the magnetic field equations need to be known. It was shown that an analytical solution for calculating the drone position can be determined by combining the magnetic field equation with the magnetic energy equation.

3.2. Electrostatic Discharge

Preliminary flight tests were conducted in the context of the AERIAL-CORE project [25] in order to identify possible malfunctions of an aerial manipulation robot interacting with a live power line, taking into account that the installation procedures described in Section 2.2 involve the potential risk of crash and entrapment due to the proximity with the cables. The experiments revealed that the multirotor platform can be severely affected by the electrostatic discharge (ESD) due to the high voltage difference between the power line and the robot, whose charge increases as it approaches to the power line due to the capacitive coupling [12]. The ESD event is triggered when the electric field between the two bodies (the power line and the aerial robot) exceeds the breakdown strength of the air, causing a charge transfer between them that equalizes the electric potential. This ESD current is introduced through the robotic manipulator and propagated through the wires and conductors of the different on-board electronic components, reaching the autopilot and electronic speed controllers (ESC). Although these components were not damaged in the realization of the tests, the ESD caused the crash of the platform, due to the perturbation introduced in the PWM (pulse width modulation) signal taken as input by the ESC of the propellers. This fault is particularly critical in platforms that integrate ESCs with active brakes, which is a common feature in many commercial multirotors, causing a sudden stop of the rotors when the PWM signal presents irregular pulses. In order to overcome this problem, two solutions were investigated in [25]: shielding the aerial manipulator to prevent the entry of the ESD, and insulating electrically the manipulator from the aerial platform to prevent its propagation.

4. Arms and End Effectors for Aerial Manipulation

4.1. Aerial Manipulation Platforms

According to the requirements and functionalities identified in Section 2.3, four different aerial robotic manipulators are developed to perform the installation of the devices shown in Figure 1. These are the Main Local-Manipulation Platform (MLMP), the Deployable Manipulator with Carriage (DM-C), the Linear-Actuator Platform (LAP), and the Dual-Arm Platform (DAP). Figure 3 depicts these platforms.

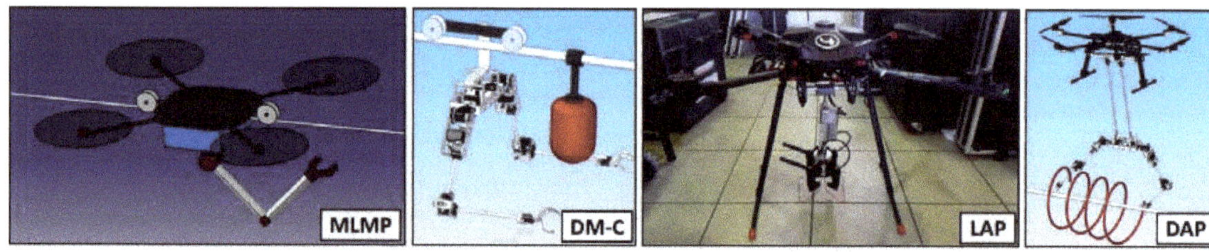

Figure 3. Aerial manipulation platforms: Main Local-Manipulation Platform (MLMP), Deployable Manipulator with Carriage (DM-C), Linear-Actuator Platform (LAP), and Dual-Arm Platform (DAP).

The MLMP platform is a multi-purpose aerial manipulator conceptually designed for the efficient installation of devices after landing on the power line and moving along it, using a rolling system inspired by [26]. This approach combines the benefits from rolling [20] and aerial solutions [22] into a compact system. To reach the stability required to operate on the cables, the mass is distributed in such a way that the center of mass is located below the cables. This aerial manipulator is endowed with a robotic arm with interchangeable end effectors that can be used for the teleoperated installation of clip-type bird diverters and electrical spacers. Nevertheless, opposite to other specialized solutions, such as [22], its multi-purpose design makes the MLMP platform suitable for other I&M tasks on power lines.

The DM-C platform was conceived as an alternative to the MLMP. The main idea is to keep similar capabilities while allowing the decoupling of the aerial platform to the manipulator and the rolling system. Thus, the rolling manipulator can be deployed on the power line for the installation of devices. Simultaneously, the aerial platform can be used to provide devices from a supply point and to assist the rolling manipulator in aerial transitions between different segments of the power line. With this novel approach, the robot complexity increases but the installation of many devices can be addressed faster and with more efficient use of the whole system.

The LAP platform is devoted to the fast and autonomous installation of clip-type bird diverters on flight. In contrast to the few state-of-the-art solutions available for the installation of bird diverters while flying, the LAP platform allows the application of high forces during installation, without the need to use a heavy aerial vehicle. Hence, the installation mechanism consists of a high force (up to 90 kg) linear actuator that exerts a pushing force directly on the device, which is held by a clamp mechanism that transmits the reaction forces to the power line, isolating in this way the aerial platform. This manipulator is attached to an aerial platform on its bottom part through a passive spherical joint that avoids the propagation of wrench disturbances from the mechanism to the aerial platform during the realization of the operation on flight.

The DAP platform is intended to install helical bird diverters for the first time on flight, using aerial robots. Due to the complexity in the manipulation of these large and flexible devices, this robot makes use of a teleoperated dual arm in the long-reach configuration [8,27], replicating the dexterous manipulation capabilities of human operators. Similarly to the LAP platform, the long-reach manipulator is attached to the aerial platform through a passive revolute joint that reduces the disturbance transmission. Moreover, the long-reach feature offers two main advantages. Firstly, it increases the safety distance between the manipulation area and the robot propellers, which reduces collision risks. Secondly, it acts as an electrical insulator for the aerial platform because the link is made of PVC. Thus, after placing the main system electronics on the aerial platform, high-voltage physical interaction with the power line is possible.

4.2. Design of End Effectors for Holding/Grabbing and Manipulation

The devices shown in Figure 1 are commercially available and intended for installation on the power lines by human workers. To overcome this, new types of end effectors need to be designed to enable the autonomous installation of these devices. Typically, in this scenario, there are two main functions of end effectors: holding/carrying the devices to be installed, and fixing them to the power line. The installation procedure depends on each particular device, as mentioned in Section 2.2. This can be done by applying sufficient force (clip-type), twisting the device around the power line (helical), or tightening the screw on the device (electrical spacer). After the device is fixed to the power line, the end effector must be released and detached from the installed device.

The two main functions of end effectors can be achieved by two individual end effectors, e.g., one end effector holds and carries the device to be installed, and the other end effector takes care of fixing it to the power line. This approach requires two different manipulators: one for each end effector, and a certain coordination between them. This increases the weight of the whole system and requires more precise manipulation. The advantage of this approach is that the individual end effectors are mechanically simpler and can be used for different purposes, e.g., one gripper type can be used to hold a clip-type bird diverter as well as a helical-type bird diverter.

The other approach integrates both functions on the same end effector, designed specifically for the device in question. These types of end effectors are more complex and are suited only for operation with one device type. The advantage of this approach is that only one end effector is needed to install (remove) the device to (from) the power line, using only one manipulator, which results in the need for less manipulation precision. The example of such an end effector for the installation of electrical spacers is shown in Figure 4, whereas Figure 5 shows the installation of the clip-type bird diverter.

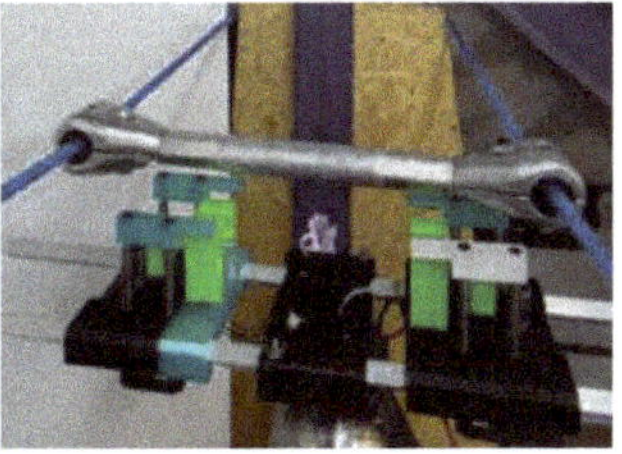

Figure 4. End effector for electrical spacers installation.

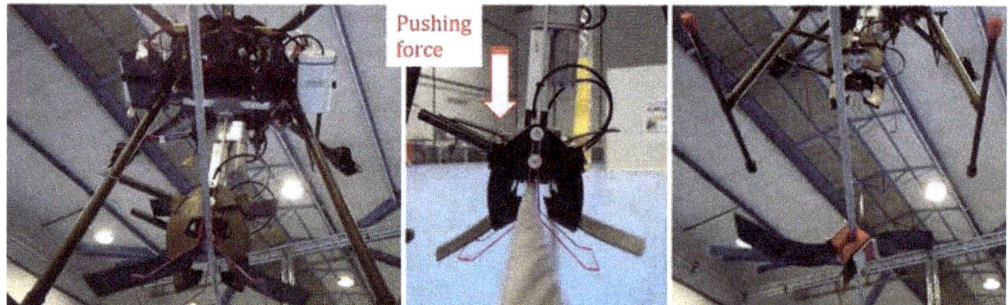

Figure 5. Installation of clip-type bird diverter with linear actuator and clamp mechanism.

4.3. Dual Arm Manipulation

As motivated in Section 2.2, the installation of the helical bird diverter will be conducted by a human-like dual arm system, considering the long-reach configuration [8] for the integration in the aerial platform to improve safety during the realization of the operation on flight. The manipulator employed in the preliminary tests shown in Figure 6 is the LiCAS A1 [28], a lightweight and compliant anthropomorphic dual-arm system, built with smart servos and a very low weight frame structure that protects the actuators against axial–radial loads and tilting moments and integrates a compact spring-lever transmission mechanism in all the joints to provide mechanical compliance [29]. The arms provide 4 DOF (degrees of freedom) for end effector positioning—three at the shoulder and one at the elbow—with the kinematic configuration described in [29] and depicted in Figure 6. The manipulator consists of a shoulder structure that supports the left/right arms in the usual upper arm/forearm link configuration. The main mechanical features of the arms are summarized in Table 2.

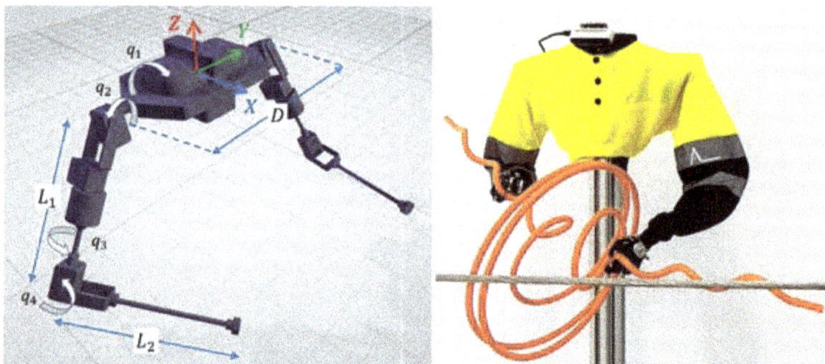

Figure 6. Kinematic model of the LiCAS A1 representing the joint angles and link lengths (**left**) and manipulation of helical bird diverter with LiCAS A1 anthropomorphic dual arm (**right**).

Table 2. Main mechanical features of the LiCAS A1 dual-arm manipulator.

Weight (kg)	2.2	Payload (at elbow) (kg)	0.7
Forearm/Upper arm links length (mm)	250 250	Separation between arms (mm)	360
Max joint speed (deg/s)	300	Joint stiffness (Nm/rad)	5–20

The compliant mechanism integrated in the joints is a particular implementation of the series elastic actuators presented in [30,31], allowing the estimation and control of the torques and forces from the deflection of the manipulator, defined as the position deviation referred at the joints or at the end effector with respect to an equivalent stiff-joint manipulator [8,29,32]. Regardless of the sensor employed to measure the deflection, either a magnetic encoder (joint deflection) or a camera head (Cartesian deflection), the control problem can be formulated in the joint or Cartesian space, relying on the forward and inverse kinematic models of the manipulator. It is necessary to remark that most servo actuators do not provide torque feedback or control. At most, they can be controlled directly in PWM mode, although the friction of the gearbox, with reduction ratios in the order of 1:200, reduces significantly the accuracy in the estimation of the transmitted torque. The performance of a torque/force controller based on joint deflection measurement was evaluated on flight in [32], using a 3-DOF compliant dual arm, proposing the use of one of the arms (grabbing arm) for estimating the position of the aerial manipulator relative to a fixed point, a power line in this case, while the other arm performs the operation, exploiting the passive/active compliance of the manipulator.

5. Safe and Accurate Aerial Manipulation

5.1. Hybrid Position/Force Control of Aerial Manipulators

The force interaction of aerial manipulators can be accomplished, using a hybrid position/force controller; the motion part is helpful while approaching the force application point, and thereafter the force control is activated for the interaction task. Consider Cartesian-space dynamics of an aerial manipulator given by the following:

$$H(\mathbf{X})\ddot{\mathbf{X}} + B(\mathbf{X}, \dot{\mathbf{X}}) + F_e = F \tag{2}$$

where $\mathbf{X} \in \mathbb{R}^6$ is the Cartesian-space position coordinates vector, $H(\mathbf{X}) \in \mathbb{R}^{6\times6}$ is the inertial forces matrix, $b(\mathbf{X}, \dot{\mathbf{X}}) \in \mathbb{R}^6$ is a bias vector containing centripetal, Coriolis and gravity effects, $F_e \in \mathbb{R}^6$ is the external forces vector due to interaction with the environment and $F \in \mathbb{R}^6$ is the control wrench vector. The control target is to regulate the interaction force between the aerial manipulator and the environment. The following hybrid position/force controller is proposed [33]:

$$F = B(\mathbf{X}, \dot{\mathbf{X}}) + F_e + K_f S(F^d - F_e) + (I - S)K_p(\mathbf{X}^d - \mathbf{X}) + K_d(\dot{\mathbf{X}}^d - \dot{\mathbf{X}}) \tag{3}$$

where $\mathbf{X}^d, \dot{\mathbf{X}}^d \in \mathbb{R}^6$, are the desired position and velocity vectors, $F^d \in \mathbb{R}^6$ is the desired wrench, K_f, K_p and K_d are diagonal 6×6 constant positive definite matrices, I is the 6×6 identity and $S \in \mathbb{R}^{6\times6}$ is a ones and zeros matrix that is useful to switch between position and force control depending on the task. The selected approach is the base of many other controllers found in the literature, which can also be enhanced, according to the requirements of the tasks performed by the aerial manipulators.

Case Study: Hybrid Position/Force Control of a 2-Link Aerial Manipulator

In this example, the control of a planar aerial manipulator is presented. The experiment consists of applying the wrench $F^d = (20\text{ N}, 0\text{ N}, 0\text{ N}, 0\text{ Nm}, 0\text{ Nm}, 0\text{ Nm})$ at the point $\mathbf{X}^d = (0.2\text{ m}, 0\text{ m}, 1.7\text{ m}, 0°, 0°, 0°)$, with respect to the world reference frame, and considering the initial position of the end effector at $\mathbf{X}(0) = (0\text{ m}, 0\text{ m}, 1.5\text{ m}, 0°, 0°, 0°)$. The controller used is (3), with a switching condition of $||\mathbf{X}^d - \mathbf{X}|| < 0.02\text{ m}$. In Figure 7, the controlled position of the end effector at $(r^d, z^d) \subset \mathbf{X}^d$ is presented, where $r^d = ||(x^d, y^d)||$. Additionally, the forces applied by the end effector along the horizontal plane and vertical axis are also shown. Notice that the force errors and the z position error are 0; the $r-$plane error is not zero, due to the control switch, at the pointed time, when the corresponding condition was reached. The Figure 8 shows the trajectory of the aerial manipulator applying the force at the corresponding point.

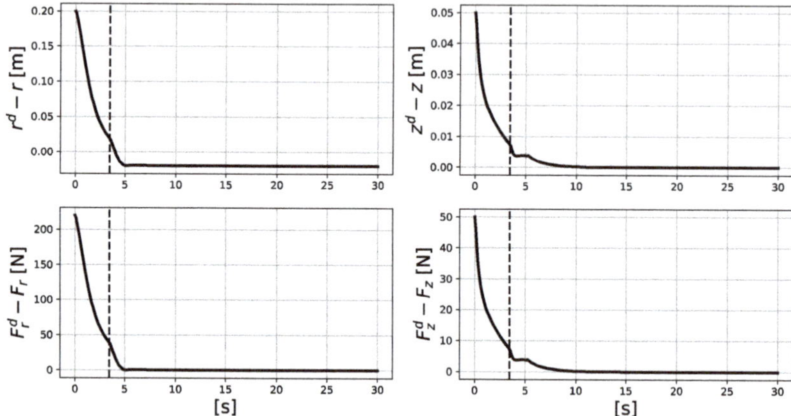

Figure 7. Error signals for the controlled end effector of the planar aerial manipulator. The dashed vertical line points out the time when the position controller is switched to force controller.

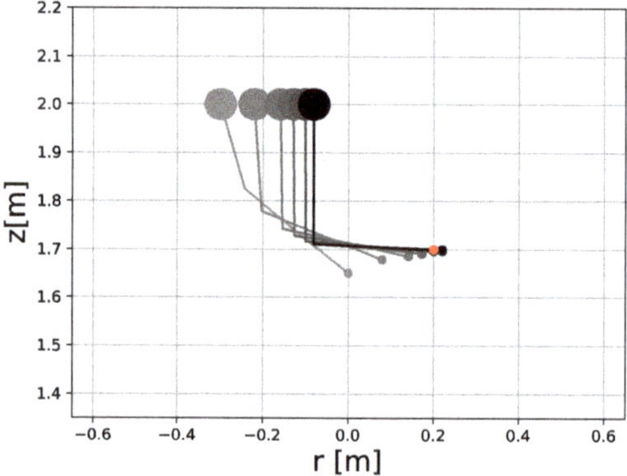

Figure 8. Cartesian-space trajectory of the planar aerial manipulator to the target force application point.

5.2. Control Methods for Aerial Manipulation

Different methods exist to control aerial manipulators. Some of them consider decentralized approaches, developing control strategies to actuate the floating base and the manipulation tool as two separate systems [34,35]. Differently, aerial manipulators can also be considered as highly-coupled nonlinear systems. In this context, the dynamic model parameters are usually unknown or difficult to identify; in addition, wind gusts, the motion of the aerial manipulator, and its interaction with the environment produce external disturbances for the aerial floating base. Here, the control problem consists of stabilizing the UAV at a desired pose while the manipulator completes a task, minimizing the effects of external disturbances and the inaccurate representation of its dynamics [36,37]. In the following, some control methods to accomplish aerial manipulation tasks are discussed.

5.2.1. Dynamics and Actuation Properties of Multirotor UAVs

Control strategies for aerial manipulators strictly depend on their controllable space. In this work, we mainly focus on rotary-wings UAVs. The generalized dynamics of these platforms can be expressed as follows: [38]:

$$\begin{bmatrix} m\mathbf{I}^{3\times3} & \mathbf{0}^{3\times3} \\ \mathbf{0}^{3\times3} & \mathbf{J} \end{bmatrix} \begin{bmatrix} \ddot{\mathbf{p}} \\ \dot{\omega} \end{bmatrix} = -\begin{bmatrix} mg\mathbf{e}_3 \\ \omega \times \mathbf{J}\omega \end{bmatrix} + \begin{bmatrix} {}^{W}\mathbf{R_B} & \mathbf{0}^{3\times3} \\ \mathbf{0}^{3\times3} & \mathbf{I}^{3\times3} \end{bmatrix} u \tag{4}$$

where m is the mass of the UAV, $\mathbf{J} \in \mathbb{R}^{3\times3}$ is the inertia tensor of the airframe, $\mathbf{p} \in \mathbb{R}^3$ is the position of the UAV with respect to the world frame, $\omega \in \mathbb{R}^3$ is the angular velocity of the UAV, $^W\mathbf{R_B} \in \mathcal{SO}(3)$ is the orientation of the airframe with respect to the world frame, $\mathbf{0}^{3\times3}$ is the zeros matrix of the corresponding dimensions, $\mathbf{I}^{3\times3}$ is the identity matrix of the corresponding dimensions, g is the gravity acceleration, $\mathbf{e}_i$ is the canonical basis vector of $i = 1, 2, 3$ and $u \in \mathbb{R}^6$ is the control vector in the generalized coordinates. For UAVs with passively or actively tilting rotors, the control vector is statically mapped from the rotors thrust, using the following:

$$u = \mathbf{F}(\alpha, \varsigma, L, \sigma) f_r \tag{5}$$

where $f_r \in \mathbb{R}^{2j}$ is the vector containing the vectorized thrusts supplied by the j rotors [39], and $\mathbf{F}(\alpha, \varsigma, L, \sigma) \in \mathbb{R}^{6\times2j}$ is the allocation matrix that depends on the following:

- The vector containing the tilting angles of the j rotors about their respective x-axis $\alpha = \left[\alpha_1, \alpha_2, ..., \alpha_j\right]$.
- The offset angle depending on the x_B axis of the UAV (for example "X" or "+" quadrotor configurations) ς.
- The length of the arms of the UAV L.
- The relation between the rotor torque constant k_τ and the rotor force constant k_f $\sigma = k_\tau / k_f$.

For fully actuated UAVs, i.e., those that are capable of moving in 6D with its number of actuators, there exists the allocation pseudoinverse $\mathbf{F}(\alpha, \varsigma, L, \sigma)^\dagger$ such that $f_r = \mathbf{F}(\alpha, \varsigma, L, \sigma)^\dagger u : \mathbb{R}^6 \to \mathbb{R}^{2j}$. Therefore, the velocity of each rotor can be computed as follows [40]:

$$\omega_{r_i} = \sqrt{\frac{\sqrt{f_{r_i} + f_{r_{j+i}}}}{k_f}}, \ i = 1, 2, ..., j \tag{6}$$

The tilting angles to exert the desired thrust are given by the following [40]:

$$\alpha_i = atan2(f_{j+i}, |f_i|), \ i = 1, 2, ..., j \tag{7}$$

For the case when the rotors of the UAV are not able to tilt, i.e., $\alpha = 0 \forall t$, the UAV becomes underactuated since the thrust of the rotors can be only exerted, in a first stage, to control the orientation and the z axis of the airframe; hence, a cascade control is required to lead the UAV to a desired Cartesian position.

5.2.2. On Robust UAV Stabilization

In order to deal with both internal and external disturbances of aerial manipulators, appropriate control techniques must be implemented. First, consider the generalized multirotor UAV dynamics (4) in state-space representation, as follows:

$$\frac{d}{dt}\begin{bmatrix}\mathbf{x}\\\dot{\mathbf{x}}\end{bmatrix} = \begin{bmatrix}\dot{\mathbf{x}}\\-\mathbf{M}^{-1}b + \mathbf{M}^{-1}\mathbf{R}u + \mathcal{D}(t)\end{bmatrix} \tag{8}$$

where $\mathbf{x} = \begin{bmatrix}\mathbf{p}^T & \phi & \theta & \psi\end{bmatrix}$, $\mathbf{M} = \begin{bmatrix}m\mathbf{I}^{3\times3} & \mathbf{0}^{3\times3}\\\mathbf{0}^{3\times3} & \mathbf{J}\end{bmatrix}$, $\mathbf{R} = \begin{bmatrix}^W\mathbf{R_B} & \mathbf{0}^{3\times3}\\\mathbf{0}^{3\times3} & \mathbf{I}^{3\times3}\end{bmatrix}$, $b \in \mathbb{R}^6 : b = \begin{bmatrix}mg\mathbf{e}_3\\\omega \times \mathbf{J}\omega\end{bmatrix}$ and $\mathcal{D}(t) \in \mathbb{R}^6$ is the external disturbances vector function, usually unknown but bounded and Lipschitz continuous, or at least with a removable discontinuity. In this context, there are some approaches that use higher order dynamics and the isolation of nonlinearities and disturbances [41]. To illustrate how robust control works on a UAV system, consider it as a 6-dimensional motion unit mass such as the following [42]:

$$\frac{d}{dt}\begin{bmatrix}\mathbf{x}\\\dot{\mathbf{x}}\end{bmatrix} = \begin{bmatrix}\dot{\mathbf{x}}\\u + \mathcal{F}(\mathbf{x}, \dot{\mathbf{x}}, t)\end{bmatrix}, \tag{9}$$

where $0 < |\mathcal{F}(\mathbf{x}, \dot{\mathbf{x}}, t)| \leq \mathcal{F}_{max}$ is a bounded disturbance term, including both the UAV dynamics and external disturbances. The target is to lead the UAV (9) to a displaced equilibrium $\mathbf{x}^d$. Below are presented some techniques to achieve such a target.

5.2.3. Active Disturbance Rejection Control

The main idea of this strategy is to estimate online $\mathcal{F}(\mathbf{x}, \dot{\mathbf{x}}, t)$ in some accurate manner [43]; thereafter, it is used as part of the feedback control action. The total disturbance can be estimated with a high-order observer such as the following:

$$
\begin{bmatrix} \dot{\hat{\mathbf{x}}}_1 \\ \dot{\hat{\mathbf{x}}}_2 \\ \dot{\hat{\mathbf{x}}}_3 \end{bmatrix} = \begin{bmatrix} \hat{\mathbf{x}}_2 + \varepsilon \Phi_1 \left(\dfrac{\mathbf{x}_1 - \hat{\mathbf{x}}_1}{\varepsilon^2} \right) \\ \hat{\mathbf{x}}_3 + \Phi_2 \left(\dfrac{\mathbf{x}_1 - \hat{\mathbf{x}}_1}{\varepsilon^2} \right) + u \\ \dfrac{1}{\varepsilon} \Phi_3 \left(\dfrac{\mathbf{x}_1 - \hat{\mathbf{x}}_1}{\varepsilon^2} \right) \end{bmatrix},
\tag{10}
$$

where $\mathbf{x}_1 = \mathbf{x}$ is the measured pose of the UAV, $\hat{\mathbf{x}}_i \in \mathbb{R}^6 \; i = 1, 2$, are the estimated states of the UAV, $\hat{\mathbf{x}}_3 \in \mathbb{R}^6$ is an estimated augmented state vector that contain the estimated lumped disturbance $\hat{\mathbf{x}}_3 \triangleq \hat{\mathcal{F}}(\mathbf{x}, \dot{\mathbf{x}}, t)$, $\Phi_i : \mathbb{R}^6 \to \mathbb{R}^6$ are functions that guarantee $|\mathbf{x}_1 - \hat{\mathbf{x}}_1| \to 0$ as $t \to \infty$ and ε is a positive scalar that the smaller it is, the faster (10) converges.

Thus, the feedback control that yields (9) despite disturbances and uncertainty affecting the UAV is the following:

$$
u = \mathbf{R}^{-1} \tilde{\mathbf{M}} (u_0 - \hat{\mathbf{x}}_3),
\tag{11}
$$

where $\tilde{\mathbf{M}}$ is an approximation of $\mathbf{M}$. Note that, if the mass or the inertia tensor of the UAV are not accurately known, an approximation of such values are suitable to be proposed. If they are completely unknown, a unit value is also a good alternative; the estimation of the total disturbance is even capable of compensating them.

5.2.4. Variable Structure Control

The target is to lead the UAV (9) to a displaced equilibrium $\mathbf{x}^d$. Therefore, the following changes of variables are proposed [42]:

$$
\mathbf{z} = \mathbf{x}^d - \mathbf{x}
\tag{12}
$$

$$
\dot{\mathbf{z}} = \dot{\mathbf{x}}^d - \dot{\mathbf{x}}
\tag{13}
$$

$$
\sigma = \mathbf{C}\mathbf{z} + \dot{\mathbf{z}}
\tag{14}
$$

where $\sigma \in \mathbb{R}^6$ is called the *sliding surface* and $\mathbf{C} \in \mathbb{R}^{6\times 6}$ is a constant diagonal matrix. Now the control target is to reach the origin (0,0). In this context, the following candidate Lyapunov function $V(\sigma) = \frac{1}{2}\sigma^T \sigma$ is proposed; its derivative is given by the following [42]:

$$
\dot{V}(\sigma) = \sigma^T \dot{\sigma} = \sigma^T \left(\mathbf{C}\dot{\mathbf{z}} + \ddot{\mathbf{x}}^d - u - \mathcal{F}(\mathbf{x}, \dot{\mathbf{x}}, t) \right)
\tag{15}
$$

Using the following controller with acceleration feedforward,

$$
u = \mathbf{C}\dot{\mathbf{z}} + \ddot{\mathbf{x}}^d - K_s \text{sign}(\sigma)
\tag{16}
$$

$\dot{V}$ can be rewritten as follows:

$$
\dot{V}(\sigma) \leq |\sigma|(\mathcal{F}_{max} - K_s)
\tag{17}
$$

where it can be appreciated that $(\mathbf{z}, \dot{\mathbf{z}}) \to (0,0)$ as long as $K_s > \mathcal{F}_{max}$. The proposed controller (16) forces the motion of $(\mathbf{z}, \dot{\mathbf{z}})$ to reach the surface and to remain on it despite the disturbance $\mathcal{F}(\mathbf{x}, \dot{\mathbf{x}}, t)$, by means of finite-amplitude high-frequency switching along (14) called *chattering*, which is important to avoid in systems such as DC motors or aircrafts [42] because it might be dangerous for the actuators. Different approaches have been proposed for *chattering* attenuation [42,44], which are suitable to be implemented for UAVs, specifically higher order SMC, since they conserve robustness and do not necessarily require the derivative of **z**. In this context, the u_0 must be properly regulated (9).

5.2.5. Case Study: Full-Pose Regulation of Passively Tilted Hexarotor

In this example, a passively tilted hexarotor is controlled using a SMC. The proposed controller is as follows:

$$u = K_p(\mathbf{z}) + K_d(\dot{\mathbf{z}}) + K_s \text{sign}(\sigma) \tag{18}$$

where K_p, K_d and K_s are diagonal matrices with positive constants. The target position is $x^d = (1\ \text{m}, 1\ \text{m}, 2\ \text{m}, 0°, 0°, 10°)$. In Figure 9 the position errors are presented. Notice that all error signals converge to 0; therefore, the full pose of the UAV is regulated.

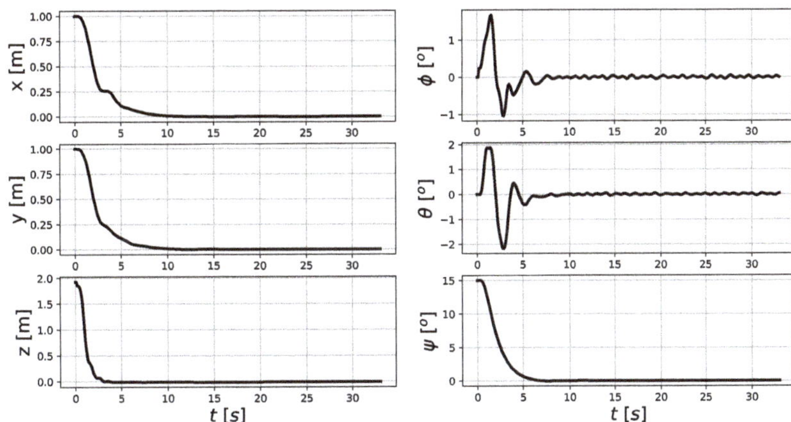

Figure 9. Error signals for the controlled states of the passively tilted hexarotor.

5.3. Planning Methods for Aerial Manipulation

Devices such as bird diverters have to be installed every 5–10 m. Since the length of each segment of a power line is typically above 150 m, this implies the installation of a large number of devices. Moreover, taking into account the limited payload capacity of the aerial platform, it is expected that the aerial robot has to move frequently to a supply point where the devices are stored. Consequently, it is proposed the application of planning methods that generate the most efficient trajectories in terms of energy consumption for the aerial manipulator in order to maximize the number of installed devices. To the best of the authors' knowledge, this paper proposes for the first time the use of planning methods for the efficient installation of devices on power lines.

Power lines are linear structures that admit a graph-based representation, where the nodes model the electric towers and the edges model the segments of the power line. These power lines are usually located in open environments with sparse obstacles, such as trees or buildings, but where weather factors, especially wind, play an important role when operating with aerial robots. Taking into account the features above, the planning problem is suitable to be addressed as a vehicle routing problem in which the obstacles are neglected. Later, during the robot operation, these obstacles can be avoided using a collision-avoidance system working at a local level.

Focusing on the formulation of the vehicle routing problem for the aerial manipulation platforms described in Section 4.1, the general approach proposed in this paper is based on covering all the segments of the power line while minimizing the energy consumption. For this, simplified but realistic models of energy consumption are required to define edge costs. These models should consider the hybrid locomotion capabilities of some of the presented aerial manipulators. Therefore, the graph representation of the power line is composed of two kinds of edges. Firstly, rolling edges are associated with the power line. Secondly, flying edges are associated with either aerial transitions between segments/points of the power line or connections of the power line with the supply point. Moreover, the consideration of effects, such as the wind, transforms the graph into a directed one since its edges have asymmetric costs. Finally, constraints such as the maximum robot payload or the battery capacity allow closing the definition of the problem.

Figure 10 shows some preliminary planning results for the installation of clip-type bird diverters on the represented power line, using both MLMP and the DM-C platforms. The problem was solved using linear programming after imposing the following requirements: (a) all the power lines should be covered rolling once, (b) minimizing the total energy consumption, (c) taking into account the maximum robot payload (10 devices for the MLMP and no limits for the DM-C because the aerial platform provides the devices), and (d) visiting the supply point to load the devices when it is needed. The number of diverters to be installed in each segment of the power line is indicated inside the circles in the figure. For these results, no wind was considered and the cost of rolling per unit of distance was fixed to be lower than the cost of flying. As can be seen in the figure, the proposed planning method computes efficient routes that are compatible with the imposed constraints.

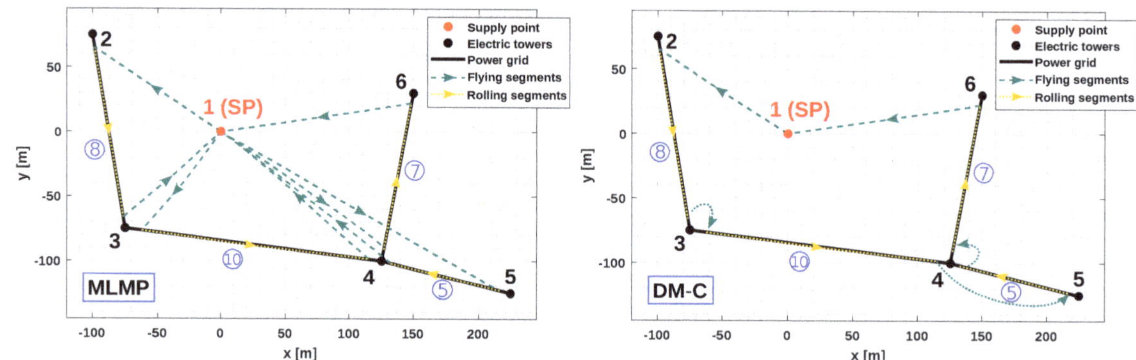

Figure 10. Preliminary planning results for the installation of clip-type bird diverters, using the MLMP platform (**left**) and the DM-C platform (**right**).

6. Human–Aerial Robot Interaction

6.1. Augmented Cognitive Human–Machine Interfaces

The remote control of the aerial manipulator is carried out by the newly developed human–machine interface (HMI). The idea of the new augmented HMI is to create a method by which an operator can control the drone, not through the classic control that uses only hands and fingers, but through the full range of human body movement as well as voice commands. On the other hand, it is planned to provide the user with enhanced feedback that includes not only visual, but also acoustic and haptic information about the state of the vehicle. In addition to functionality, special attention is paid to the ergonomics of the newly developed HMI, which is personalized for each operator [45].

The user gives commands by the movement of the body (see Figure 11), which are detected based on measurements from IMUs (Inertial Measurement Units) placed at specific points on the body. This approach is the extension of the FlyJacket system presented in [46]. Another type of command output is based on the smart gloves worn by the user, which have the ability to measure the pose of the hand and the flexion of each finger. Finally, the voice recognition system is used for issuing direct commands to the drone.

The main feedback to the operator is visual information displayed on the smart glasses, allowing the user to see telemetry and camera information from the drone as well as the drone itself through the transparent lenses. Another feedback channel is based on the acoustic information, which can be used to inform the operator about hazards, the distance to the obstacle or to the ground, but also to follow a previously defined path or trajectory [47]. Tactile feedback is provided by the vibration motors placed at certain points on the operator's body. A particular motor may start vibrating when the drone is in contact with the manipulated object, but it may also vibrate depending on whether the drone is performing a left–right or up–down movement [48].

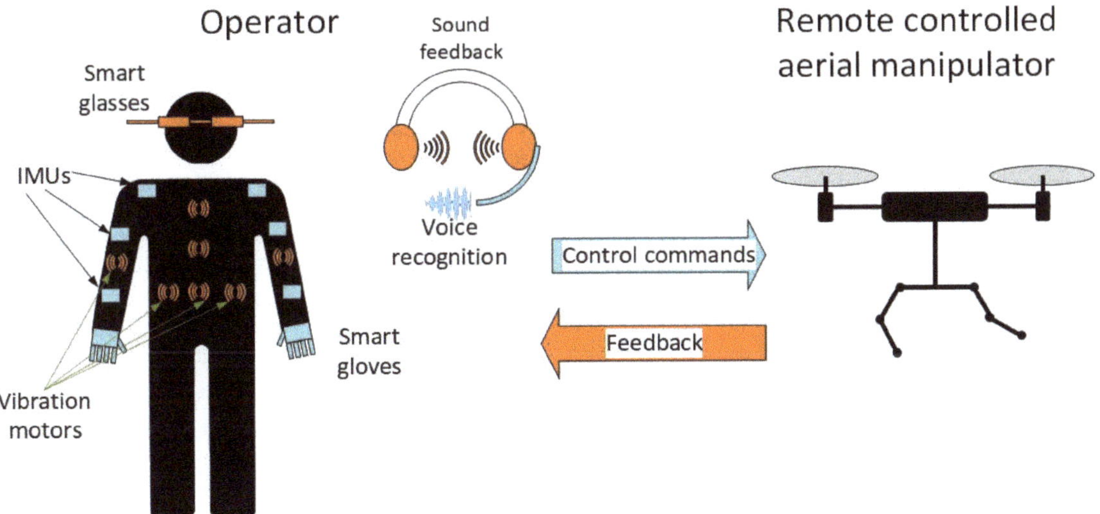

Figure 11. Augmented human–machine interface.

6.2. Aerial Co-Worker for Fast Tool Delivery

One application in which the use of aerial manipulators can be particularly useful for the operators working on the power lines is the delivery of tools, such as screwdrivers and pliers, or devices, such as clip-type bird diverters. This is motivated by taking into account the time and effort required to climb to the power lines, and the possibility to lose accidentally part of the equipment during operations. As occurs with other human–robot collaborative tasks [49,50], the close interaction of the aerial robot with the human operator must be safe, especially taking into account the risk due to the high altitude and the difficulty in standing upright in this kind of workspace. Since the proximity of the propellers to the worker will result in an uncomfortable feeling and a sense of danger [51], it is preferable that the handover operation [8,52] is carried out in such a way that the manipulator is separated from the aerial platform. Thus, the proposed solution consists of employing an aerial manipulator in long-reach [8,27] or cable-suspended configurations [14,15], using a cable (around 1.5 m in length) to attach the manipulator to the multirotor base. Figure 12 illustrates the use of a dual-arm aerial manipulator in a long-reach configuration for the delivery of tools (a screwdriver and some pliers) to a human operator working on a line.

Four reference frames are considered in the definition of the operation: the Earth fixed frame $\{E\}$ (inertial), the multirotor body frame $\{E\}$, the manipulator frame $\{0\}$, and the operators head/helmet frame $\{H\}$. These are illustrated in Figure 12. In the navigation phase, the aerial robot takes off from the supply point and navigates to the proximity of the power line employing GNSS (Global Navigation Satellite System) sensors, such as GPS, to estimate the position relative to the Earth fixed frame. In the approaching phase, the aerial robot is positioned relative to the operators helmet, using, for example, visual markers on it for its detection and localization [53]. Two constraints should be imposed in the realization of the handover operation: (1) keep a minimum separation distance between the multirotor base and the operator's helmet for safety, and (2) keep the end effector of the manipulator within the reach of the operator's hand to ensure a comfortable interaction with the aerial robot.

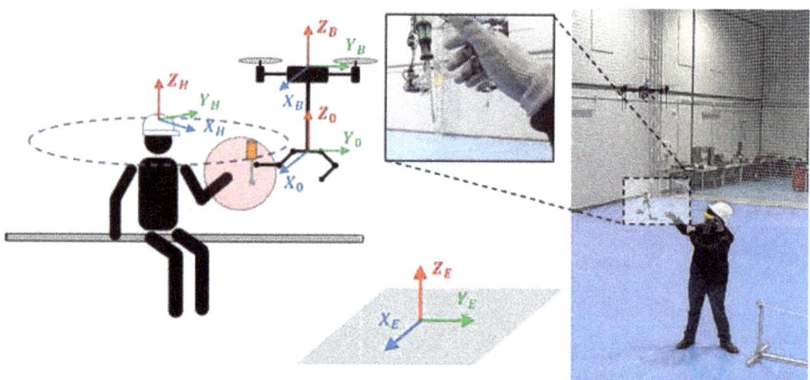

Figure 12. Reference frames in the aerial delivery application (**left**). Aerial tool delivery to a human operator with a dual-arm aerial manipulator in a long-reach configuration (**right**).

7. Conclusions

This paper presented the first-year results of the AERIAL-CORE project, focusing on aerial manipulation operations. The AERIAL-CORE project is framed in the Horizon 2020 European research and innovation program and aims to develop new aerial robots able to autonomously perform or assist human operators in the inspection and maintenance of power grid lines. The goal of this work is to introduce the aerial manipulation problem in the context of power line maintenance, describing the general procedure for the installation of bird flight diverters and electrical spacers to outline the requirements in the design of the aerial manipulators. First prototypes of aerial manipulation platforms and end effectors satisfying such requirements were presented along with low-level control techniques and planning methods proposed to achieve safe operation on flight and energy efficiency, taking into account the vast extension of this kind of infrastructure. The paper also covered the interaction of the aerial manipulator with the human operators through the design of appropriate human–machine interfaces, as well as the application of aerial co-workers in aerial tool delivery tasks. Future directions of this work involve the integration of manipulation, control, perception, planning and navigation functionalities in the proposed platforms, and their experimental validation in a real power-line scenario.

Author Contributions: Conceptualization, investigation and original draft presentation, J.C., S.M.O.-S., A.S., A.C., M.O., S.B., G.V., E.E., J.A.A.R., supervision J.C., A.S., and A.O., funding acquisition A.O. Writing–review & editing, J.C. All authors have read and agreed to thepublished version of the manuscript.

Funding: The research leading to these results has been supported by the AERIAL-CORE project (Horizon 2020 Grant Agreement No. 871479) and by the 2018 FPI Program of the Spanish Ministerio de Ciencia, Innovación y Universidades.

Conflicts of Interest: The authors declare no conflict of interest.

References

1. IEEE Guide for Maintenance Methods on Energized Power-Lines. 2009. Available online: https://ieeexplore.ieee.org/abstract/document/29044 (accessed on 2 July 2021). [CrossRef]
2. Hotte, P.; Gela, G.; Mitchell, J.; Lyons, P. Electrical performance of conductive suits. *IEEE Trans. Power Deliv.* **1997**, *12*, 1193–1201. [CrossRef]
3. Ruggiero, F.; Lippiello, V.; Ollero, A. Aerial manipulation: A literature review. *IEEE Robot. Autom. Lett.* **2018**, *3*, 1957–1964. [CrossRef]
4. Orsag, M.; Korpela, C.; Oh, P.; Bogdan, S.; Ollero, A. *Aerial Manipulation*; Springer: Berlin/Heidelberg, Germany, 2018.
5. Meng, X.; He, Y.; Han, J. Survey on aerial manipulator: System, modeling, and control. *Robotica* **2020**, *38*, 1288–1317. [CrossRef]

6. Trujillo, M.Á.; Martínez-de Dios, J.R.; Martín, C.; Viguria, A.; Ollero, A. Novel aerial manipulator for accurate and robust industrial ndt contact inspection: A new tool for the oil and gas inspection industry. *Sensors* **2019**, *19*, 1305. [CrossRef] [PubMed]

7. Ikeda, T.; Yasui, S.; Fujihara, M.; Ohara, K.; Ashizawa, S.; Ichikawa, A.; Okino, A.; Oomichi, T.; Fukuda, T. Wall contact by octo-rotor UAV with one DoF manipulator for bridge inspection. In Proceedings of the 2017 IEEE/RSJ International Conference on Intelligent Robots and Systems (IROS), Vancouver, BC, Canada, 24–28 September 2017; pp. 5122–5127.

8. Suarez, A.; Real, F.; Vega, V.M.; Heredia, G.; Rodriguez-Castaño, A.; Ollero, A. Compliant bimanual aerial manipulation: Standard and long reach configurations. *IEEE Access* **2020**, *8*, 88844–88865. [CrossRef]

9. Hamaza, S.; Georgilas, I.; Fernandez, M.; Sanchez, P.; Richardson, T.; Heredia, G.; Ollero, A. Sensor installation and retrieval operations using an unmanned aerial manipulator. *IEEE Robot. Autom. Lett.* **2019**, *4*, 2793–2800. [CrossRef]

10. Orsag, M.; Korpela, C.; Bogdan, S.; Oh, P. Valve turning using a dual-arm aerial manipulator. In Proceedings of the 2014 International Conference on Unmanned Aircraft Systems (ICUAS), Orlando, FL, USA, 27–30 May 2014; pp. 836–841.

11. Shimahara, S.; Leewiwatwong, S.; Ladig, R.; Shimonomura, K. Aerial torsional manipulation employing multi-rotor flying robot. In Proceedings of the 2016 IEEE/RSJ International Conference on Intelligent Robots and Systems (IROS), Daejeon, Korea, 9–14 October 2016; pp. 1595–1600.

12. Greason, W.D. Generalized model of electrostatic discharge (ESD) for bodies in approach: analyses of multiple discharges and speed of approach. *J. Electrost.* **2002**, *54*, 23–37. [CrossRef]

13. Wan, S.; Bian, X.; Chen, L.; Yu, D.; Wang, L.; Guan, Z. Electrostatic discharge effect on safe distance determination for 500 kV ac power line's helicopter inspection. *J. Electrost.* **2013**, *71*, 778–780. [CrossRef]

14. Sarkisov, Y.S.; Kim, M.J.; Bicego, D.; Tsetserukou, D.; Ott, C.; Franchi, A.; Kondak, K. Development of sam: Cable-suspended aerial manipulator. In Proceedings of the 2019 International Conference on Robotics and Automation (ICRA), Montreal, QC, Canada, 20–24 May 2019; pp. 5323–5329.

15. Miyazaki, R.; Paul, H.; Kominami, T.; Shimonomura, K. Wire-Suspended Device Control Based on Wireless Communication with Multirotor for Long Reach-Aerial Manipulation. *IEEE Access* **2020**, *8*, 172096–172104. [CrossRef]

16. AERIAL-CORE Project Home Page. Available online: https://aerial-core.eu/ (accessed on 9 May 2021).

17. Yang, L.; Fan, J.; Liu, Y.; Li, E.; Peng, J.; Liang, Z. A Review on State-of-the-Art Power Line Inspection Techniques. *IEEE Trans. Instrum. Meas.* **2020**, *69*, 9350–9365. [CrossRef]

18. Schofield, O.B.; Lorenzen, K.H.; Ebeid, E. Cloud to Cable: A Drone Framework for Autonomous Power line Inspection. In Proceedings of the 2020 23rd Euromicro Conference on Digital System Design (DSD), Kranj, Slovenia, 26–28 August 2020; pp. 503–509. [CrossRef]

19. Iversen, N.; Schofield, O.B.; Ebeid, E. LOCATOR—Lightweight and Low-Cost Autonomous Drone System for Overhead Cable Detection and Soft Grasping. In Proceedings of the 2020 IEEE International Symposium on Safety, Security, and Rescue Robotics (SSRR), Abu Dhabi, United Arab Emirates, 4–6 November 2020; pp. 205–212. [CrossRef]

20. LineRanger: A Revolution in Transmission Line Robotics. Available online: https://www.youtube.com/watch?v=OItActG9S6U" (accessed on 25 April 2021).

21. FulcrumAir's Heavy Lift UAV and Line Fly installing Bird Flight Diverters on a 144kV Power Line. Available online: https://www.youtube.com/watch?v=FxpnxHnVGPw (accessed on 25 April 2021).

22. UAV Installation Services—Bird Diverters. Available online: https://www.youtube.com/watch?v=PfkRt2W7NAk" (accessed on 25 April 2021).

23. Swan-Fight Diverters. Available online: https://www.preformed.com/za/energy/distribution/wildlife-protection/swan-flight-diverter (accessed on 25 April 2021).

24. Martinović, D.; Bogdan, S.; Kovačić, Z. Mathematical Considerations for Unmanned Aerial Vehicle Navigation in the Magnetic Field of Two Parallel Transmission Lines. *Appl. Sci.* **2021**, *11*, 3323. [CrossRef]

25. Suarez, A.; Salmoral, R.; Zarco-Periñan, P.J.; Ollero, A. Experimental Evaluation of Aerial Manipulation Robot in Contact with 15 kV Power Line: Shielded and Long Reach Configurations. *IEEE Access* **2021**. [CrossRef]

26. Suarez, A.; Caballero, A.; Garofano, A.; Sanchez-Cuevas, P.J.; Heredia, G.; Ollero, A. Aerial Manipulator With Rolling Base for Inspection of Pipe Arrays. *IEEE Access* **2020**, *8*, 162516–162532. [CrossRef]

27. Caballero, A.; Suarez, A.; Real, F.; Vega, V.M.; Bejar, M.; Rodriguez-Castaño, A.; Ollero, A. First Experimental Results on Motion Planning for Transportation in Aerial Long-Reach Manipulators with Two Arms. In Proceedings of the 2018 IEEE/RSJ International Conference on Intelligent Robots and Systems (IROS), Madrid, Spain, 1–5 October 2018; pp. 8471–8477. [CrossRef]

28. LiCAS Robotic Arms Home Page. Available online: https://licas-robotic-arms.com/ (accessed on 25 April 2021).

29. Suarez, A.; Heredia, G.; Ollero, A. Design of an anthropomorphic, compliant, and lightweight dual arm for aerial manipulation. *IEEE Access* **2018**, *6*, 29173–29189. [CrossRef]

30. Pratt, G.A.; Williamson, M.M. Series elastic actuators. In Proceedings of the 1995 IEEE/RSJ International Conference on Intelligent Robots and Systems, Human Robot Interaction and Cooperative Robots, Pittsburgh, PA, USA, 5–9 August 1995; Volume 1, pp. 399–406.

31. Oh, S.; Kong, K. High-precision robust force control of a series elastic actuator. *IEEE/ASME Trans. Mechatron.* **2016**, *22*, 71–80. [CrossRef]

32. Suarez, A.; Sanchez-Cuevas, P.J.; Heredia, G.; Ollero, A. Aerial Physical Interaction in Grabbing Conditions with Lightweight and Compliant Dual Arms. *Appl. Sci.* **2020**, *10*, 8927. [CrossRef]

33. Raibert, M.H.; Craig, J.J. Hybrid position/force control of manipulators. *ASME J. Dyn. Syst. Meas. Control* **1981**, *103*, 126–133. [CrossRef]

34. Lippiello, V.; Cacace, J.; Santamaria-Navarro, A.; Andrade-Cetto, J.; Trujillo, M.A.; Esteves, Y.R.R.; Viguria, A. Hybrid Visual Servoing With Hierarchical Task Composition for Aerial Manipulation. *IEEE Robot. Autom. Lett.* **2016**, *1*, 259–266. [CrossRef]

35. Buonocore, L.R.; Cacace, J.; Lippiello, V. Hybrid visual servoing for aerial grasping with hierarchical task-priority control. In Proceedings of the 2015 23rd Mediterranean Conference on Control and Automation (MED), Torremolinos, Spain, 16–19 June 2015; pp. 617–623. [CrossRef]

36. Ruggiero, F.; Cacace, J.; Sadeghian, H.; Lippiello, V. Impedance control of VToL UAVs with a momentum-based external generalized forces estimator. In Proceedings of the 2014 IEEE International Conference on Robotics and Automation (ICRA), Hong Kong, China, 31 May–5 June 2014; pp. 2093–2099. [CrossRef]

37. Ruggiero, F.; Cacace, J.; Sadeghian, H.; Lippiello, V. Passivity-based control of VToL UAVs with a momentum-based estimator of external wrench and unmodeled dynamics. *Robot. Auton. Syst.* **2015**, *72*, 139–151. [CrossRef]

38. Michieletto, G.; Ryll, M.; Franchi, A. Fundamental actuation properties of multirotors: Force—Moment decoupling and fail—Safe robustness. *IEEE Trans. Robot.* **2018**, *34*, 702–715. [CrossRef]

39. Invernizzi, D.; Lovera, M. Geometric tracking control of a quadcopter tiltrotor UAV. *IFAC-PapersOnLine* **2017**, *50*, 11565–11570. [CrossRef]

40. Invernizzi, D.; Giurato, M.; Gattazzo, P.; Lovera, M. Full pose tracking for a tilt-arm quadrotor UAV. In Proceedings of the 2018 IEEE Conference on Control Technology and Applications (CCTA), Copenhagen, Denmark, 21–24 August 2018; pp. 159–164.

41. Dullerud, G.E.; Paganini, F. *A Course in Robust Control Theory: A Convex Approach*; Springer: Berlin/Heidelberg, Germany, 2013; Volume 36.

42. Shtessel, Y.; Edwards, C.; Fridman, L.; Levant, A. *Sliding Mode Control and Observation*; Springer: Berlin/Heidelberg, Germany, 2014; Volume 10.

43. Sira-Ramírez, H.; Luviano-Juárez, A.; Ramírez-Neria, M.; Zurita-Bustamante, E.W. *Active Disturbance Rejection Control of Dynamic Systems: A Flatness Based Approach*; Butterworth-Heinemann : Oxford, UK, 2018.

44. Shtessel, Y.B.; Moreno, J.A.; Plestan, F.; Fridman, L.M.; Poznyak, A.S. Super-twisting adaptive sliding mode control: A Lyapunov design. In Proceedings of the 49th IEEE Conference on Decision and control (CDC), Atlanta, GA, USA, 15–17 December 2010; pp. 5109–5113.

45. Macchini, M.; Schiano, F.; Floreano, D. Personalized Telerobotics by Fast Machine Learning of Body-Machine Interfaces. *IEEE Robot. Autom. Lett.* **2020**, *5*, 179–186. [CrossRef]

46. Rognon, C.; Mintchev, S.; Dellagnola, F.; Cherpillod, A.; Atienza, D.; Floreano, D. FlyJacket: An Upper Body Soft Exoskeleton for Immersive Drone Control. *IEEE Robot. Autom. Lett.* **2018**, *3*, 2362–2369. [CrossRef]

47. Vasilijevic, A.; Jambrosic, K.; Vukic, Z. Teleoperated path following and trajectory tracking of unmanned vehicles using spatial auditory guidance system. *Appl. Acoust.* **2018**, *129*, 72–85. [CrossRef]

48. Orsag, M.; Haus, T.; Tolić, D.; Ivanović, A.; Car, M.; Palunko, I.; Bogdan, S. Human-in-the-loop control of multi-agent aerial systems. In Proceedings of the 2016 European Control Conference, ECC 2016, Aalborg, Denmark, 29 June–1 July 2016; pp. 2139–2145. [CrossRef]

49. Bauer, A.; Wollherr, D.; Buss, M. Human–robot collaboration: A survey. *Int. J. Hum. Robot.* **2008**, *5*, 47–66. [CrossRef]

50. Edsinger, A.; Kemp, C.C. Human-robot interaction for cooperative manipulation: Handing objects to one another. In Proceedings of the RO-MAN 2007—The 16th IEEE International Symposium on Robot and Human Interactive Communication, Jeju Island, Korea, 26–29 August 2007; pp. 1167–1172.

51. Pounds, P.E.; Deer, W. The safety rotor—An electromechanical rotor safety system for drones. *IEEE Robot. Autom. Lett.* **2018**, *3*, 2561–2568. [CrossRef]

52. Eguíluz, A.G.; Rañó, I.; Coleman, S.A.; McGinnity, T.M. Reliable robotic handovers through tactile sensing. *Auton. Robot.* **2019**, *43*, 1623–1637.

53. Sani, M.F.; Karimian, G. Automatic navigation and landing of an indoor AR. drone quadrotor using ArUco marker and inertial sensors. In Proceedings of the 2017 International Conference on Computer and Drone Applications (IConDA), Kuching, Malaysia, 9–11 November 2017; pp. 102–107.

applied sciences

Article

Installation of Clip-Type Bird Flight Diverters on High-Voltage Power Lines with Aerial Manipulation Robot: Prototype and Testbed Experimentation

Angel Rodriguez-Castaño, Saeed Rafee Nekoo *, Honorio Romero, Rafael Salmoral, José Ángel Acosta and Anibal Ollero

GRVC Robotics Laboratory, Departamento de Ingeniería de Sistemas y Automática, Escuela Técnica Superior de Ingeniería, Universidad de Sevilla, 41004 Seville, Spain; castano@us.es (A.R.-C.); hromero_h@hotmail.com (H.R.); rafaslasla@gmail.com (R.S.); jaar@us.es (J.Á.A.); aollero@us.es (A.O.)
* Correspondence: saerafee@yahoo.com

Abstract: This work presents the application of an aerial manipulation robot for the semi-autonomous installation of clip-type bird flight diverters on overhead power line cables. A custom-made prototype is designed, developed, and experimentally validated. The proposed solution aims to reduce the cost and risk of current procedures carried out by human operators deployed on suspended carts, lifts, or manned helicopters. The system consists of an unmanned aerial vehicle (UAV) equipped with a custom-made tool. This tool allows the high force required for the diverter installation to be generated; however, it is isolated from the aerial robot through a passive joint. Thus, the aerial robot stability is not compromised during the installation. This paper thoroughly describes the designed prototype and the control system for semi-autonomous operation. Flight experiments conducted in an illustrative scenario validate the performance of the system; the tests were carried out in an indoor testbed using a power line cable mock-up.

Keywords: arial manipulation; inspection and maintenance; multirotor systems; high-voltage power lines; clip-type bird flight diverters

Citation: Rodriguez-Castaño, A.; Nekoo, S.R.; Romero, H.; Salmoral, R.; Acosta, J.Á.; Ollero, A. Installation of Clip-Type Bird Flight Diverters on High-Voltage Power Lines with Aerial Manipulation Robot: Prototype and Testbed Experimentation. *Appl. Sci.* **2021**, *11*, 7427. https://doi.org/10.3390/app11167427

Academic Editor: Alessandro Di Nuovo

Received: 2 July 2021
Accepted: 11 August 2021
Published: 12 August 2021

Publisher's Note: MDPI stays neutral with regard to jurisdictional claims in published maps and institutional affiliations.

1. Introduction

Overhead high-voltage power lines are built to transmit electricity between the source (dams, wind-turbine power stations, photovoltaic panel power generators, fossil fuel power stations, etc.) and regional/local distribution stations. The medium-voltage power line transmits electricity with less than 33 kV, and high-voltage lines are in the range of 110–750 kV [1]. The usual voltage of the high-voltage power lines is 220 or 380 kV, isolated far from the ground on pylons in two branches of cables in triangle arrangements [2]. The length of the power lines is so extensive, for example, in Spain that it was reported to be 44,372 km in 2019 [3]. This makes the inspection and maintenance (IM) topic so crucial in terms of economy, technology, time, personnel, and monitoring. The conventional methods for IM of power lines could be listed as human operators, vision inspection by helicopters, deployment of tools by elevators, etc. This current work proposes automatic and unmanned IM by unmanned aerial vehicles (UAVs) to reduce the cost and increase the safety of the task, defined through the framework of the current project "AERIAL COgnitive integrated multi-task Robotic system with Extended operation range and safety (AERIAL-CORE)" [4].

The use of UAVs is becoming more frequent in IM-based projects, such as the visual inspection of local distribution power networks [5], photovoltaic fields [6], vision-based positioning of an aerial platform on transmission lines [7], cracks in wind turbines [8], and structural health monitoring [9]. Kim and Ham studied wind-induced damage analysis by a vision inspection system based on received information by UAVs [5]. The leaning/damage of the utility poles generated power loss in local distribution networks. The use of UAVs and vision inspection reduced the cost of the operation and avoided different observations by human operators. Bizzarri et al. presented the application of UAVs in the inspection of photovoltaic panels [6]. One field of solar energy production possesses a huge number of photovoltaic panels; automation of an IM system presents a systematic method for regular checks. It was concluded that the automation for less than 50 MW plants was not cost-effective. Menéndez et al. researched a positioning system using vision on top of the overhead power transmission lines by UAVs [7]. Reddy et al. investigated the cracks and damage inspection in wind turbine blades using artificial intelligence-based image analytics [8]. UAVs and the idea of deep learning were used to train a neural network model for the classification of the cracks and faults in the blades. Accuracy was found to be 94.94% for binary fault classification and 91% for multiple class fault classification. Myeong et al. presented structural health monitoring using a wall-climbing drone prototype [9]. The mentioned research works were engaged with vision inspection and monitoring without intervention [5–9].

Physical intervention IM by UAVs could be regarded as a further complicated step in unmanned aerial vehicles used in this field. Physical interaction with the environment generates force reactions to the UAV and causes deviation of the system from a stable hover. Some of the applications of UAVs in IM with interaction with the environment are ultrasonic contact inspection [10], pipe inspection in refineries [11], aerial manipulator system [12], bridge inspection [13], tank thickness measurement [14], and the UAV with a parallel manipulator [15]. Kocer et al. presented contact-based autonomous inspection using UAVs equipped with ultrasonic sensors [10]. A soft compensation mechanism was designed to handle the oscillations in hovering while the UAV was inspecting the target. Ultrasonic inspection needed continuous contact in certain force ranges; to provide this condition compactly and feasibly, an optimization algorithm was used, so-called nonlinear moving horizon estimation. Ollero et al. employed UAVs with different configurations for contact inspection of pipes and infrastructures in refineries [11]. Customized add-ons and dual-arm compliant manipulators were used to provide the possibility of contact; the configuration of the UAVs was also different, with a simple hexarotor design and a tilted rotor structure to impose six-degree-of-freedom control. Suarez et al. studied a lightweight compliant arm with a compliant finger for aerial manipulation and inspection [12]. Joint compliance was proposed in the manipulator design to facilitate force interaction with the environment. Sanchez-Cuevas et al. used a tilted hexacopter design for UAV and an ultrasonic sensor placed on an add-on for bridge inspection [13]. Danko et al. employed a parallel manipulator for holding the sensor in UAV contact inspection [15].

UAVs are also very helpful in IM overhead transmission lines since the pylons are hard to reach and the length of transmission lines is so significant. The IM topic in this field is also diverse, but the majority of cases are devoted to visual inspection and monitoring. Larrauri et al. presented a UAV-based automatic system for inspection of overhead transmission power lines [16]. The objective was to collect consecutive images together with telemetry data sent from the autopilot to identify areas of vegetation, trees, and buildings close to transmission lines. The usage of an automatic mechanism or robot for the installation of devices (mounted on the cables with a rolling mechanism) was also reported, though the deployment was done by a helicopter or manually since the payload of the UAVs was not sufficient [17]. He et al. investigated the visual inspection of power lines using a multirotor UAV and image processing [18]. The range of the image capturing was calculated, and the efficiency of the automatic inspection was calculated to be 58–62% concerning manual conventional inspections. Teng et al. used a mini-UAV and LiDAR sensor for inspection and survey of the terrain under pylons [19]. Zhang et al. presented the automatic extraction of high-voltage power transmission objects from UAV LiDAR point clouds [20]. Visual inspection is common for power lines, though the maintenance part requires touching the cables; there has been some effort to grasp the cable [21]. This could be critical since even coming near the cables exposes the UAV to a high magnitude magnetic field and noise [22].

Working near the cables and pylons also has advantages, such as the wireless charging of batteries [23,24]. It was shown that direct contact with power lines caused a malfunction in the electronic speed controller and resulted in the UAV falling [25]. It was recommended to insulate the add-ons of the UAV to avoid electrical interference; here, in this work, we use the same approach and insulate the linear actuator with non-conductive material.

Among all the IM tasks carried out by energy distribution companies, in this work, we focus on bird flight diverters. These devices are the most widespread measure to reduce bird collisions with these infrastructures and therefore mitigate their impact on wildlife. Thus, most of the available studies reporting on these diverters are devoted to either confirming their effectiveness in the reduction of avian mortality at power lines, as in [26,27], or even in night collisions as in [28], or analyzing the impact of malfunctioning, as power line companies do not replace them often due to high cost, as in [29].

Importantly, unlike in vision inspection works, bird flight diverters need interaction with cables for their installation. Thus, the short circuit problem, shielding, isolation of the UAV, etc., must be considered. This research presents a prototype design and development of the system along with experimentation in an indoor testbed. This is the first and necessary step before a validation test on real power lines and pylons due to safety reasons.

Problem statement: Bird flight diverters, commonly known in the industry as *bird flappers, bird diverters, power line markers*, or simply *diverters*, are devices developed to increase overhead power line visibility for birds and reduce their risk of collision. They are also often installed on overhead cables near airfields to reduce the possibility of an aircraft or drone collision, and some of them use marking lights or strobe lights to improve visibility at night.

In this paper, a kind of clip-type diverter commonly used by Endesa® (the main Spanish electricity company) has been employed (see Figure 1). These diverters have two components: a supporting clip and two attached flaps. In this case, the clip is made of a $50 \times 50 \times 60$ mm red plastic part with a 12 mm hole in the middle and a guiding slot a little bit smaller than the diameter of the cables considered (about 10 mm). Two $300 \times 55 \times 3$ mm black rubber flaps are attached to each side of the clip. This kind of diverter is very robust and durable due to the lack of moving parts. It is installed by pushing the clip onto the power line cable until it is plugged in, as shown in Figure 1b. Several tests have been performed to obtain the required pushing force for the installation: 440 N (45 kg). Once it is installed, the diverter can rotate around the cable due to winds or bird strikes.

(a)

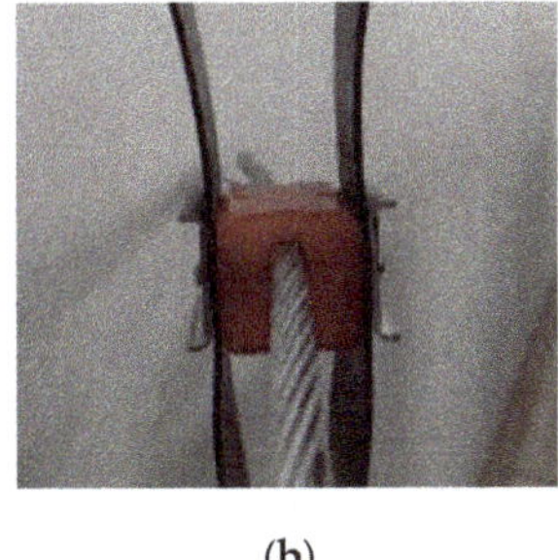

(b)

Figure 1. (a) Clip-type bird flight diverter developed by Wigeva currently installed on a power line cable; (b) detailed view of the diverter's clip part (red) with the cable inserted.

Most common procedures to install bird flight diverters on power lines require human intervention, using lifts, helicopters, suspended carts, or climbing the pylons. These operations are quite risky, particularly when the forces are high, or the required working time is long. Moreover, diverters are typically installed at 5 to 10 m intervals along the power line cable to be effective, so any time reduction of this operation would significantly contribute to decreasing the costs and risks.

The main goal of this work is the automation of the installation process of clip-type diverters. For this purpose, a prototype based on an unmanned aerial platform with a custom-made tool has been developed; it allows a user to install the diverters on the cable without human intervention. The tool has to generate enough force to push the diverter on the cable without endangering the aerial platform stability. The system is designed to isolate the high force needed to install the diverter (440 N) from the aerial robot, and so the UAV flying conditions are minimally perturbed. The required force is generated inside the tool, and a net-zero external force is exerted on the UAV.

The main contributions of this work are:

- Design and manufacturing of a novel prototype for the installation of a kind of clip-type bird flight diverter;
- Development of a semi-autonomous method for bird diverter installation on overhead power line cables using an unmanned aerial robot;
- Experimental validation of the prototype in an indoor testbed, closing the control loop with a vision-based Opti-Track system;
- Experimental validation and performance evaluation of the complete installation task.

The paper is organized as follows. Section 2 presents the prototype designed and developed. Section 2.1 describes the aerial platform and Section 2.2 the installation tool for clip-type bird diverters. Test results are reported in Section 3. Finally, the discussion and conclusions are presented in Section 4.

2. Aerial Robot Design and Development

This section describes the design and development of an aerial robot that can install bird clip-type diverters on high voltage power lines. The aerial robot is based on a custom-made multi-copter with a diverter installation tool attached using a pivot joint as shown in Figure 2. The multi-copter flies above the power line and approaches it slowly until the tool touches the cable. Then, it automatically installs the clip-type diverter while the multi-copter is hovering. Once the bird diverter is placed, the aerial platform flies out to the home location. The pivot joint provides some flexibility; it allows some lateral movement of the multi-copter while hovering above the power line during the attachment procedure. The next subsections describe the aerial platform, the diverter installation tool, and the procedure in detail.

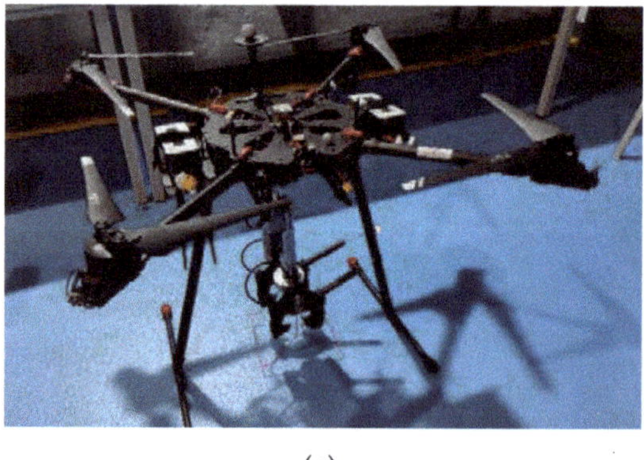
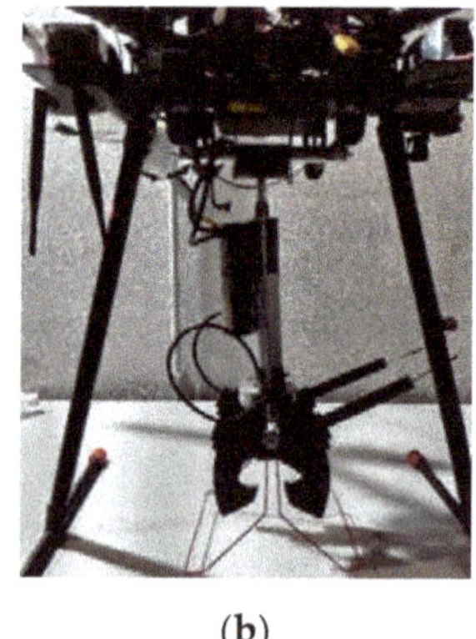

(a)   (b)

Figure 2. (**a**) The aerial platform of the diverter installation process and (**b**) the attachment of the clip installation tool through a pivot joint.

2.1. Aerial Platform

The aerial platform is a custom-made quadrotor (see Figure 2) that weighs 6.5 kg including batteries and avionics. It can carry up to a 3.5 kg payload which is enough for the diverter installation tool (1.7 kg) and one clip-type diverter (0.3 kg). There is a spare payload (1.5 kg) to add more sensors or carry heavier diverters if needed. The distance between its motors is 975 mm, and the height of the quadrotor is 500 mm. The hardware specification is as follows:

- Propellers: four ultra-carbon fiber 21-inch radius propellers are used to generate enough thrust and load capacity in a plus "+" shape configuration.
- Motors: brushless direct-current (BLDC) motors, 6010, with Kv = 130 (rpm/V), have been used, where Kv is the motor velocity constant. They provide 5.1 kg maximum pulling force per rotor, with a maximum of 770 W power. It has a cooling air circulation system, convenient for long-term operation. The weight of each motor is 270 g. The suggested battery is 12 S 45 V. Thus, the no-load speed by 12 V batteries is 1560 rpm.
- Drivers: a 1240 S electronic speed controller is used as the driver of the BLDC motors. It is waterproof and has silica thermal pads and heat sinks for maximum heat transfer and dissipation. It works with a maximum of 52.2 V and up to 25 A continuous current and 40 A peak.
- *Main* processor: an Intel NUC-i7, with 16 GB RAM, plays the role of the main processor of the UAV. It has USB 3.0, Ethernet, HDMI, Mini PCI Express ports, and low consumption that allows it to work with a 65 W power supply.
- – Autopilot: the UAV uses PIXHAWK Cube 2.1 autopilot. It has a 32-bit STM32F427 Cortex-M4F core with FPU, 168 MHz/252 MIPS, 256 KB RAM, 2 MB Flash (fully accessible), 32-bit STM32F103 failsafe co-processor, and 14 PWM/Servo outputs (8 with failsafe and manual override, 6 auxiliaries, high-power compatible).

2.2. Diverter Installation Tool

The tool has been designed to meet the following requirements:

- It should exert at least a 685 N (70 kg) force on the bird diverter: 440 N (45 kg) force to insert the clip plus 245 N (25 kg) force as a safety margin.
- The force exerted on the UAV during the installation should be minimal.
- It should be operated easily.

The next subsections describe the mechanical components and how they are mounted, the control electronics, and the insertion procedure.

2.2.1. Tool Design and Mechanical Components

The tool developed has three main components as shown in Figure 3: the electric linear actuator, the clamps, and the pulling cables.

The electric linear actuator is a Firgelli FA-PO-240-12-4. This actuator provides a 200 lb (90 kg) dynamic force with a 4-inch (10 cm) stroke and a 7.5 mm/s speed. It is equipped with a built-in 10 KΩ potentiometer to provide position feedback, and it is operated using a 12VDC input. The actuator housing is fixed to an aluminum disc that supports the clamps as shown in Figures 3 and 4. The rod end of the linear actuator holds the pusher. This is a square-shaped metallic part with two 6 mm pins that support the diverter and two ends of the pulling cables (see Figure 4).

Two plastic clamps are attached to the aluminum disc as shown in Figure 4. Each clamp is made of two jaws, a pivot joint, an innerspring between the jaws (K2 in Figure 4), and an outer spring (K1 in Figures 3 and 4). The outer spring stiffness (K1) is higher than the innerspring stiffness (K2) by design, and their equilibrium length keeps the clamp open. Moreover, a cable guide (red part in Figure 3) allows for centering the power line cable in the tool and supporting the diverter's flaps to avoid obstructions during the insertion process.

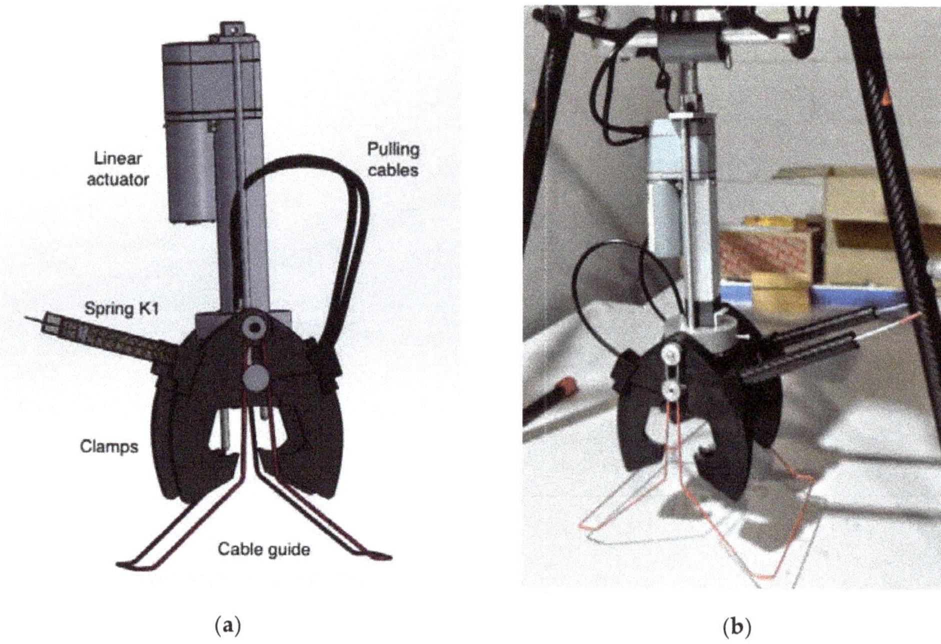

Figure 3. (**a**) Design of the diverter installation tool and (**b**) the current prototype developed.

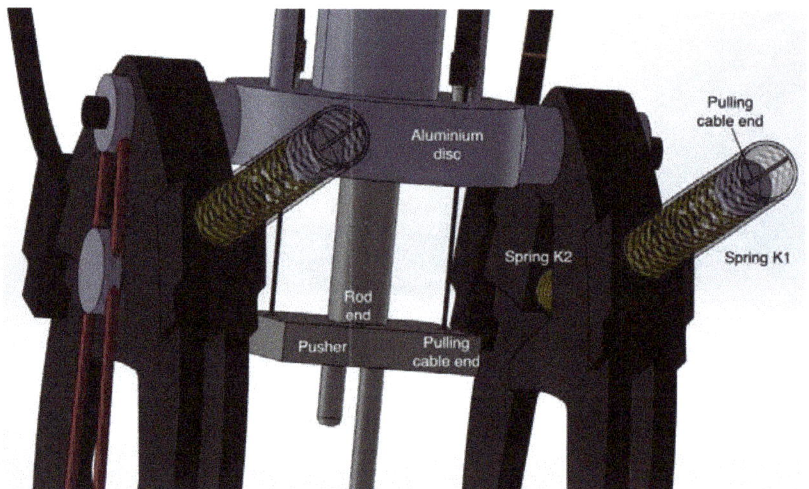

Figure 4. Detailed view of the attachment of the clamps to the linear actuator.

One pulling cable is needed to close or open each clamp moving the jaws. One end of the cable is attached to the rod end of the linear actuator, then the cable goes through the clamp innerspring K2 (see Figure 4), and the other end is attached to the outer spring K1 (see Figures 3 and 4). When the linear actuator stretches out, the cable is pulled so the clamp closes (compressing both springs). When the linear actuator retracts, the springs recover their equilibrium length, so the clamp opens.

2.2.2. Control Electronics

The control electronics are based on an Arduino Nano board as shown in Figure 5. The control board receives commands from the operator through a radio-control (RC) receiver. Then, the Arduino board regulates the linear actuator stroke with the feedback of the embedded potentiometer. The actuator is commanded through an H-bridge driver, and the whole system is powered using a 12 S LiPo battery, with 7000 mAh at 44.4 V. The details of the voltages and the connections are illustrated in Figure 6.

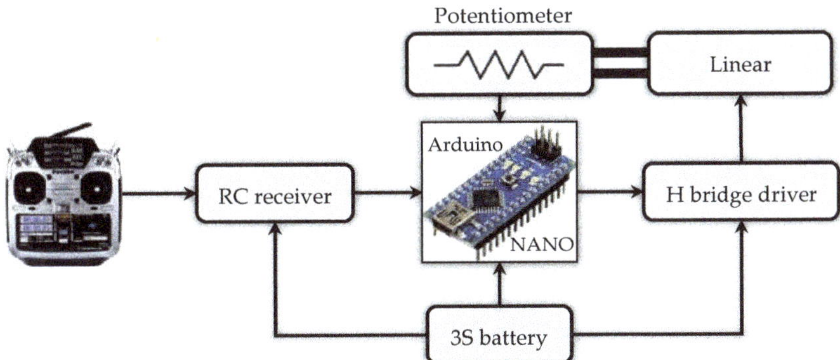

Figure 5. Block diagram of the control electronics.

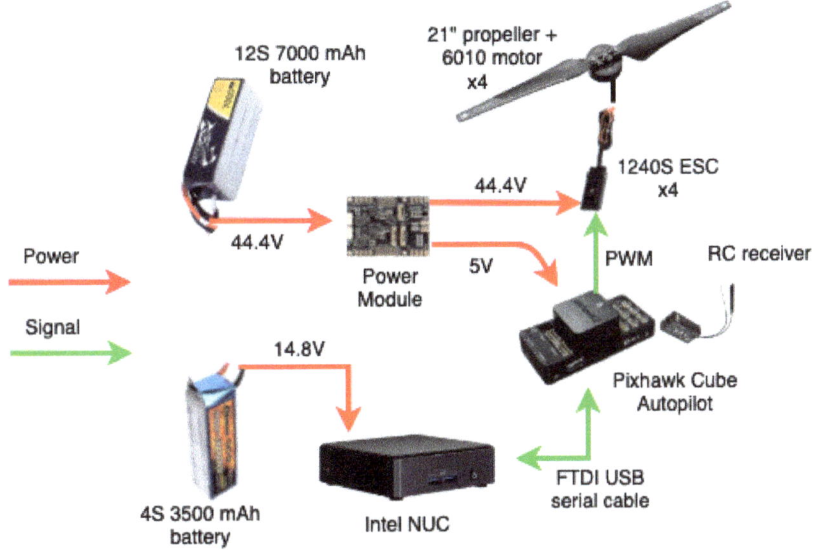

Figure 6. The details of the electronics and voltages of the units.

2.2.3. Insertion Procedure

The bird diverter has to be pre-charged in the installation tool before the process starts. For this purpose, the red clip part has to be inserted into the 6 mm pins of the pusher described in Section 2.2.1. Then, the aerial robot is ready to fly towards the power line cable and to approach the installation tool to the cable as shown in Figure 7a. The cable guides (in red) help to center the cable in the tool while the robot is descending (see Figure 7b). The clamps are still open at that moment. When the power line cable is centered in the tool, the insertion process is triggered. The linear actuator stretches out closing the clamps and pushing the clip on the cable as shown in Figure 7c. The force exerted by the linear actuator on the cable is canceled by the clamp's force, and therefore no force is applied to the aerial robot. In addition, the pivot joint between the installation tool and the aerial robot helps to avoid perturbations that could unbalance the UAV flight. Then the linear actuator retracts, disengaging the clip part (see Figure 7d) and opening the clamps. Finally, the aerial robot can move up and fly back to charge another diverter.

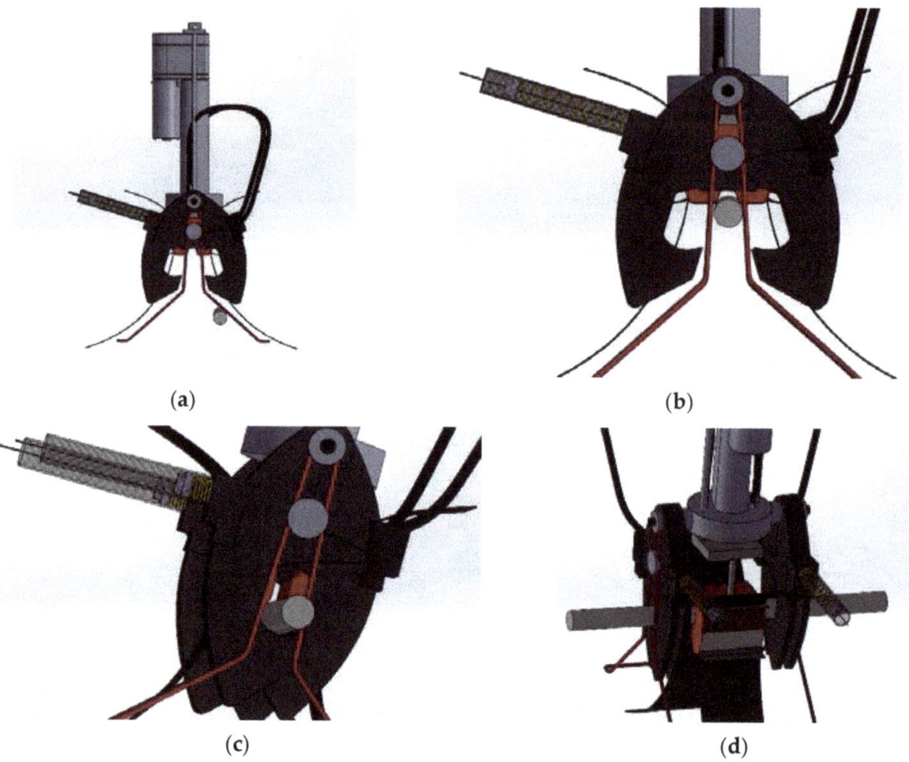

Figure 7. (a) Automatic installation process. The aerial robot carries the installation tool with a pre-charged diverter and approaches the power line cable. (b) The cable is centered in the tool while the aerial robot descends. (c) When the cable is in place, the linear actuator closes the clamps and pushes the diverter on the cable. (d) Finally, the linear actuator retracts, detaching the diverter and opening the clamps before flying out.

3. Experimental Results

The system has been validated in an indoor testbed using a mock-up as shown in Figure 8. The mock-up is a 2 m long cable suspended at 1 m above the ground. Four positions are defined, namely WP#, and prerecorded using an Opti-Track motion capture system:

— WP1 is defined at a 6.5 m distance from the insertion point on the cable and a 2.5 m height.
— WP2 is defined above the insertion point at a 2 m height (1 m above the cable).
— WP3 is defined above the insertion point at a 1.3 m height (30 cm above the cable).
— WP4 is defined as the insertion point on the cable.

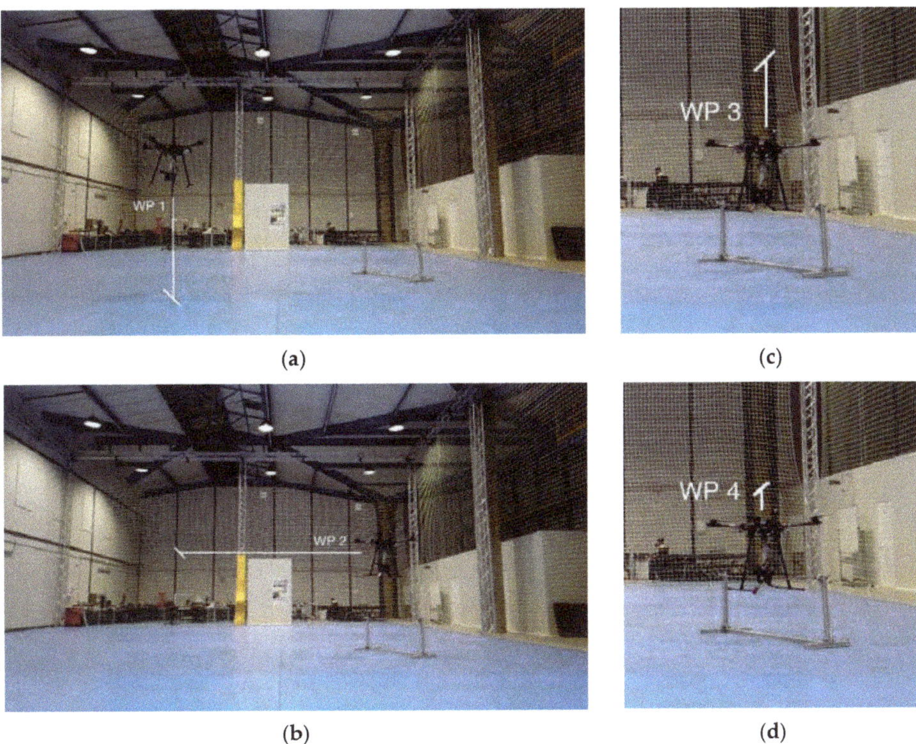

Figure 8. (**a**) Experimental setup for validation. The aerial robot takes off to WP1. (**b**) Then, it stabilizes on WP2, 1 m above the cable. (**c**) The robot slowly descends to WP3, 30 cm above the insertion point, and (**d**) it finally reaches WP4, the insertion point.

The Opti-Track is an ultra-precise camera tracking system for providing visual feedback to the controller. The current testbed dimension is $20 \times 15 \times 7$ m with high accuracy (1 mm) including 28 cameras. The high data rate translation and orientation feedback are read by the drone's controller and facilitate the semi-autonomous operation.

The multi-copter autopilot controls the position using the feedback provided by the Opti-Track system. The experiments, shown in the video provided as Supplementary Material, consisted of the following phases:

1. The pilot sends a "Goto WP1" command to the aerial robot, and it takes off to WP1.
2. After reaching WP1, the pilot sends a "Goto WP2" command, so the robot moves toward the cable, and the pilot waits until it is stabilized in that position.
3. Then, the pilot sends consecutive "Goto WP3" and "Goto WP4" commands, and the aerial robot slowly descends to the insertion point (WP4). In this position, the cable is centered in the tool and inside the clamps.
4. After WP4 has been reached, the pilot triggers the installation command, and the diverter installation tool starts its operation, automatically closing the clamps and inserting the clip on the cable.
5. When the pilot visually confirms that the bird diverter has been deployed, he/she sends a fly-up command to the aerial robot, which is finally landed.

The trajectory performed during one of the trials is shown in Figures 9–14. The Cartesian coordinates are shown in Figures 9–11, and the orientation variables are plotted in Figures 12–14. The 3D trajectory of the system is also presented in Figure 15. The oscillations in the y-axis seem bigger than the other x and z axes; however, they are in the same order. Since the motion in the y-axis is almost zero, the oscillation seems bigger than the other two. The UAV takes off and goes to WP1 from $t = 0$ s to $t = 14$ s. Then, it navigates to WP2, which is reached at $t = 27$ s. As shown in Figure 14, during this time slot, the yaw is changed to align the UAV with the cable. The pilot sends the "Goto WP3" command at $t = 27$ s and the "Goto WP4" command at $t = 33$ s. The insertion operation is then triggered at $t = 37$ s. After installing the bird diverter, the UAV flies away at $t = 40$ s.

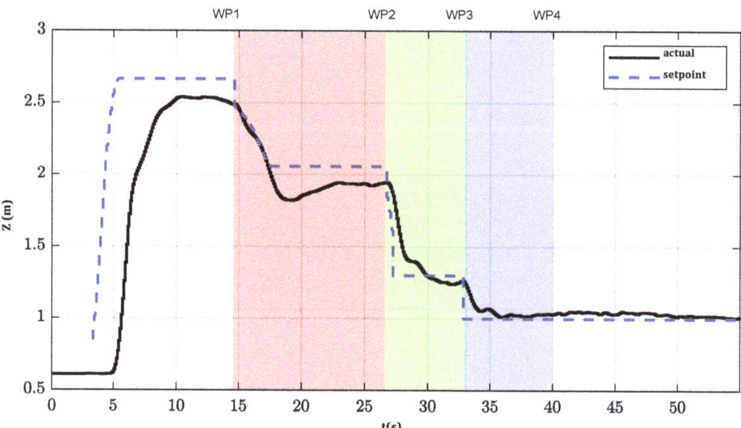

Figure 9. Z-axis motion during one of the trials.

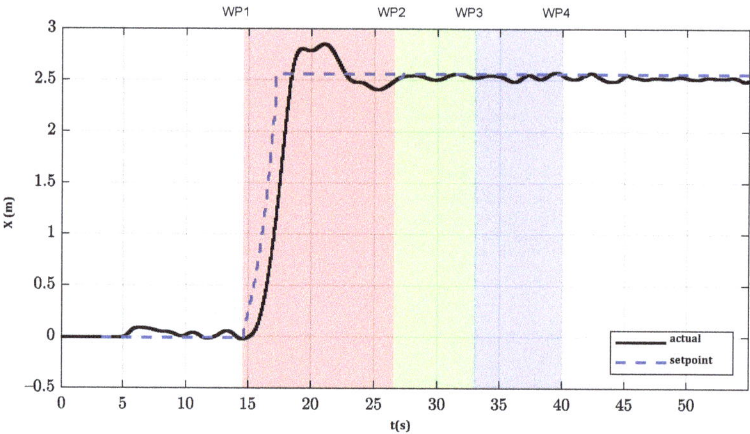

Figure 10. X-axis motion during one of the trials.

The yaw control would need more fine-tuning (Figure 14); however, it was intended to demonstrate that the system works even with misalignment between the tool and the power line. This has been experimentally evaluated; theoretically, the tool design (clamps and cable guide size) and the pivot joint should allow a maximum 8 cm lateral error (if it is perfectly aligned) or a maximum 20 degree misalignment (if the lateral error is zero). The UAV does not try to follow a continuous trajectory, though it tries to reach the predefined waypoints. The UAV reaches each waypoint with an error lower than 4 cm (see Figure 15).

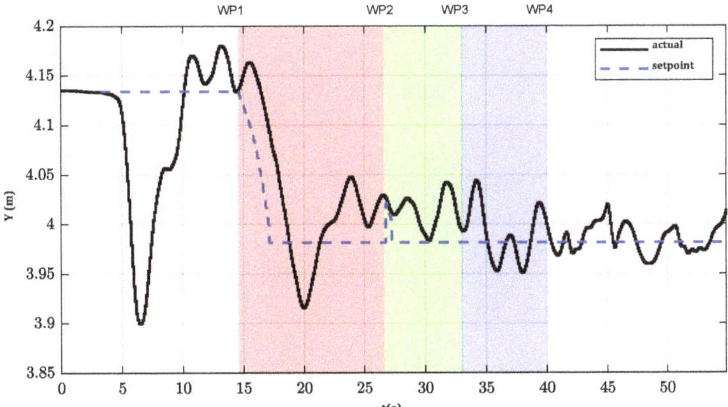

Figure 11. Y-axis motion during one of the trials.

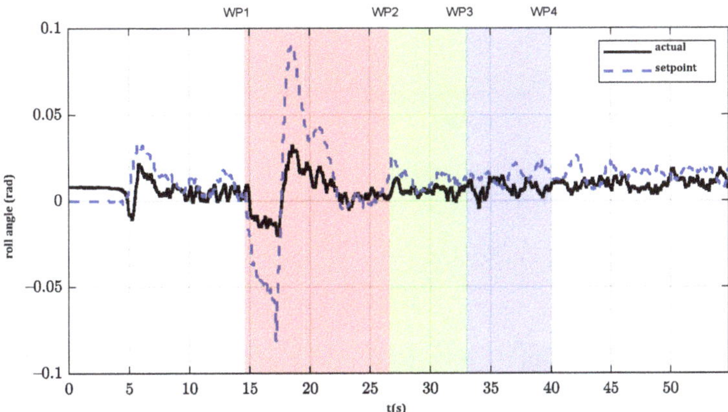

Figure 12. The roll angle of the UAV in clip installation.

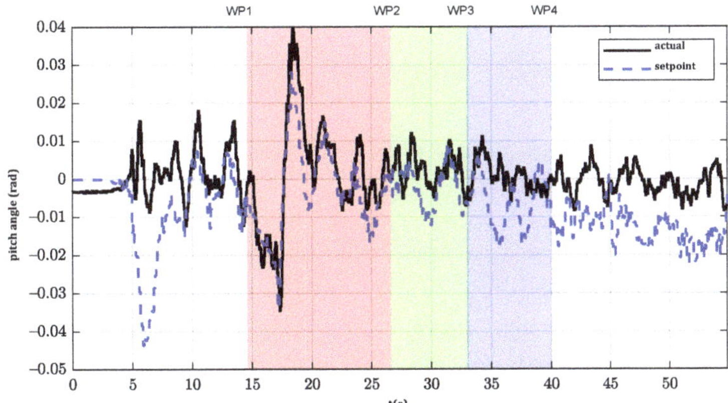

Figure 13. The pitch angle of the UAV in clip installation.

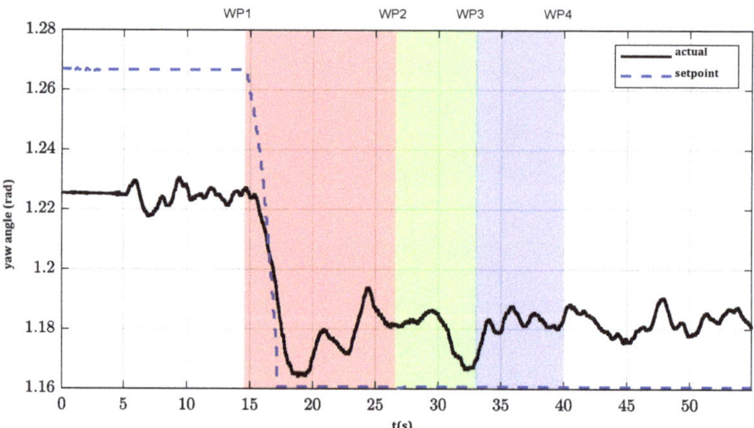

Figure 14. Yaw angle of the UAV during one of the trials.

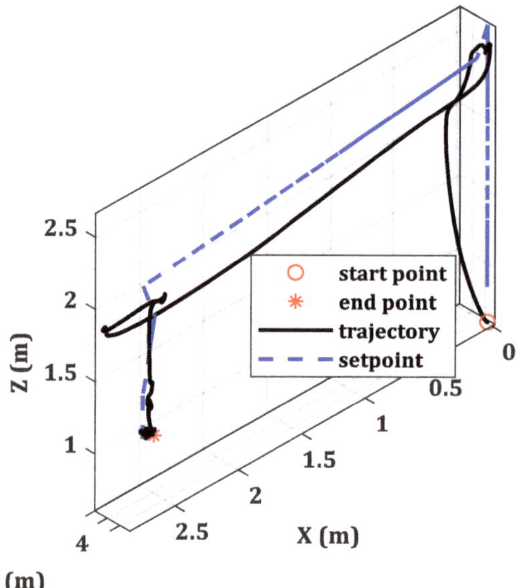

Figure 15. The 3D trajectory of the autonomous clip installation.

The operation performed during step 4 is illustrated in detail in Figure 16. First, the guide leads the cable inside the clamps (Figure 16a). The insertion operation is triggered when the cable is placed in the middle of the clamps and touching the clip (Figure 16b). The linear actuator stretches out, moving the clamps (Figure 16c) until they are fully closed around the cable (Figure 16d). Then, the actuator continues pushing the bird diverter until the cable is clipped (Figure 16e). Then, the actuator retracts, opening the clamps (Figure 16f); then, the clamp moves back (Figure 16g), and the aerial robot flies up, leaving the bird diverter deployed (Figure 16h).

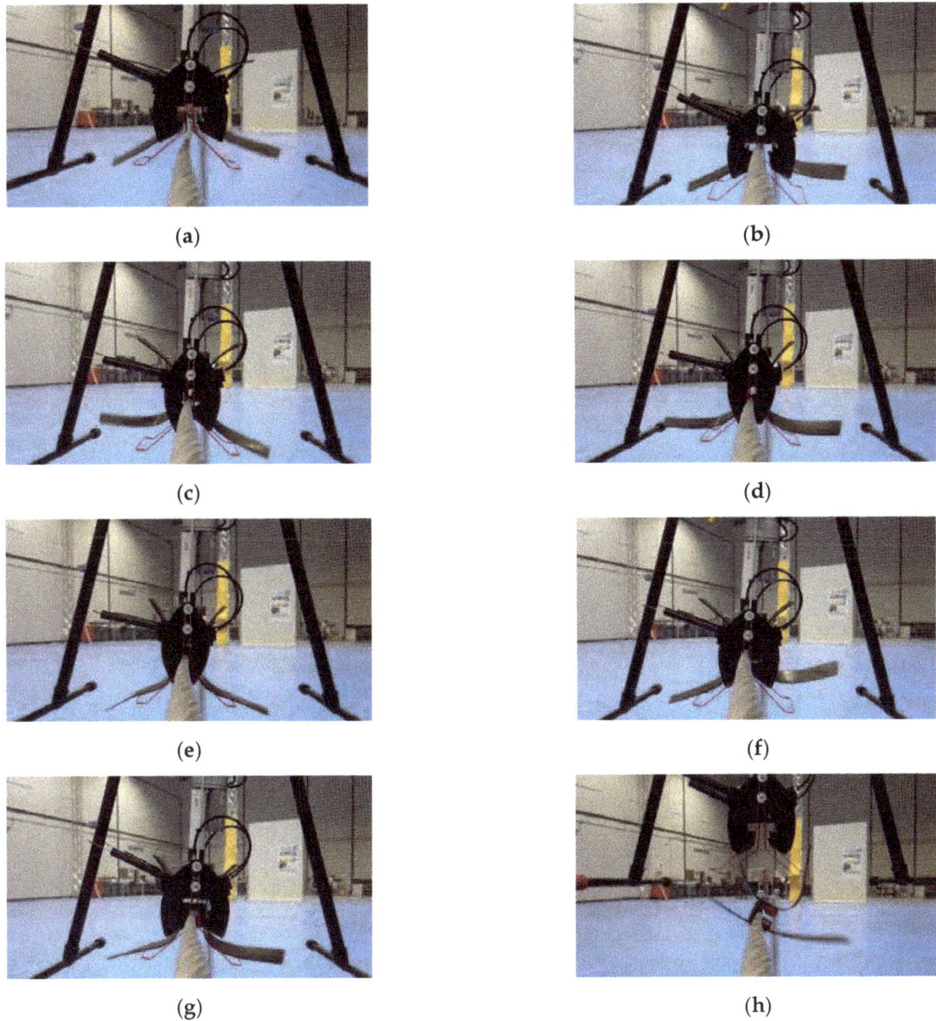

Figure 16. A series of images to show the sequence of the process; (**a**) approaching, (**b**) positioning the device, (**c**) pushing down the diverter, (**d**) fixing the diverter, (**e**) clamping, (**f**) opening, (**g**) retreating of the linear actuator, (**h**) fly back to station.

The effect of the force exerted during the installation on the hovering condition of the UAV is then analyzed. As can be seen in Figure 9, there is only a minor effect on the Z-axis error due to the vertical force during the installation, but the horizontal error (Figures 10 and 11) and orientation angles (pitch, roll, and yaw, Figures 12–14) are mainly unaffected.

Three consecutive trials have been performed to validate the repeatability and robustness and to obtain some timing data. The following three-time intervals are measured:

- Descending toward cable: the time it takes to navigate from WP2 to WP4 and until the pilot triggers the installation procedure. This time can vary because the pilot triggers the operation when he/she considers the UAV is stabilized enough in WP4.
- Clamping: the time the diverter installation tool takes to insert the bird diverter into the cable.

- Retreat: the time the linear actuator needs to retreat and open the clamps again, plus the time required to fly back to WP3. This time can also vary because the flyback command is manually triggered by the pilot.

The time details of the experiments are summarized in Table 1. The average time of the operation is 32.6 s (excluding the time required for take-off and navigation from WP1 to WP2). Descending onto the cable takes an average time of 9 s due to the constrained vertical speed and smoothness required. The clamping and retreating stages are automatic and take identical time (10 s), but, as explained before, the decision to fly back is made by the pilot, so "Retreat" values are different in Table 1. It should be noted that the "Descending Toward Cable" phase could take quite a different time in a real operation due to the long flight to cable position.

Table 1. The timing of the clip installation process.

Test Number	1	2	3	Average
Descending toward cable	11 s	7 s	9 s	9 s
Clamping	10 s	10 s	10 s	10 s
Retreat	16 s	10 s	15 s	13.6 s
Total time	37 s	27 s	34 s	32.6 s

The flight time was estimated and recorded 20 min of autonomy, measured approximately. Accounting less than 2 min for each diverter (30–40 s for installation plus fly back and new diverter uploading), a single battery would install 10 diverters. For real outdoor installation, adding a new device for holding several clips is necessary, which changes the current timing on the indoor tests. The flight to the pylon and back to the station would also be different.

4. Conclusions

This paper describes a prototype that allows for the semi-automated installation of a kind of clip-type bird diverter on power line cables. The current process is usually performed using helicopters, suspended carts, or elevators, this being costly and risky for the personnel involved. Using an autonomous aerial robot would help to reduce risks and increase efficiency. For this purpose, *a custom-made novel diverter installation tool has been designed and developed*. The novel system presented here is based on a linear actuator and two clamps attached to the aerial robot, and it allows the user to isolate the force exerted on the cable from the aerial platform. To the best of the authors' knowledge, it is the first semi-autonomous system with a customized design to conduct clip installations on power lines. A set of experiments have been performed on an indoor mock-up scenario, and the system shows a successful performance with acceptable repeatability. The diverters can be installed in less than 40 s, including some delay due to human intervention, and the aerial platform is completely stable during the operation.

Future steps: Future work will include fully autonomous installation using visual servoing to close the control loop and testing in a real de-energized power line cable. In reality, looking up when the UAV is flying near cables is difficult; it is too far away (too high) to see details with the naked eye, it is too fast to see and control with binoculars, the sky is bright and the UAV is dark, and it is very difficult/painful to work with the UAV. The GPS feedback is also not precise enough to provide sufficient accuracy for autonomous installation. As a result, the platform for the real experiment will be different in terms of feedback generation and tracking. It should be noted that the customized add-ons and the insulation, the flight controller, and the quadcopter will keep their design and mechanical structure. For these reasons, the use of onboard cameras and visual servoing is an obligation. Moreover, a new tool for multiple diverter installation is a work in progress. This would allow deploying several diverters without landing to recharge each time, thus increasing efficiency. The final objective is to perform the entire process automatically and efficiently.

The Opti-Track system is unique to indoor experiments. For the real flight and installation, the use of a ground station is necessary where it can localize the UAV for providing feedback. Onboard visual servoing is also proposed for future experimentation for recognition of the cable and giving the pilot/controller visual feedback.

Any wind gust results in bad performance. It is ideal to work in good conditions, though the study could be expanded in the future for adding active disturbance rejection systems or robust controllers.

Supplementary Materials: The following are available online at https://www.mdpi.com/article/10.3390/app11167427/s1, Video S1: The multi-copter autopilot controls the position using the feedback provided by the Opti-Track system.

Author Contributions: Conceptualization, A.R.-C., S.R.N., H.R., R.S. and J.Á.A.; methodology, A.R.-C., S.R.N., H.R., R.S. and J.Á.A.; validation, H.R. and R.S.; formal analysis, A.R.-C. and S.R.N.; investigation, A.R.-C., S.R.N., H.R., R.S. and J.Á.A.; resources, A.R.-C. and A.O.; writing—original draft preparation, S.R.N.; writing—review and editing, A.R.-C., S.R.N. and J.Á.A.; visualization, A.R.-C., S.R.N., H.R., R.S. and J.Á.A.; supervision, A.R.-C. and A.O.; project administration, A.R.-C. and A.O.; funding acquisition, A.R.-C. and A.O. All authors have read and agreed to the published version of the manuscript.

Funding: This work was supported by the European Commission H2020 Programme under the AERIAL COgnitive integrated multi-task Robotic system with Extended operation range and safety (AERIAL-CORE) project, contract number 871479.

Institutional Review Board Statement: Not applicable.

Informed Consent Statement: Not applicable.

Conflicts of Interest: There is no conflict of interest reported by the authors.

References

1. European-Commission. *Guidance on Energy Transmission Infrastructure and EU Nature Legislation*; European-Commission: Brussels, Belgium, 2018.
2. Sardaro, R.; Bozzo, F.; Fucilli, V. High-voltage overhead transmission lines and farmland value: Evidences from the real estate market in Apulia, southern Italy. *Energy Policy* **2018**, *119*, 449–457. [CrossRef]
3. Red Eléctrica de España. Available online: https://www.ree.es/es/conocenos/principales-indicadores/red-de-transporte-circuito (accessed on 15 December 2020).
4. Permanent-URL(a). Available online: https://aerial-core.eu/ (accessed on 15 December 2020).
5. Kim, J.; Ham, Y. Vision-based analysis of utility poles using drones and digital twin modeling in the context of power distribution infrastructure systems. In Proceedings of the Construction Research Congress 2020, Tempe, Arizona, 8–10 March 2020; Computer Applications; pp. 954–963.
6. Bizzarri, F.; Nitti, S.; Malgaroli, G. The use of drones in the maintenance of photovoltaic fields. *E3S Web Conf.* **2019**, *119*, 21. [CrossRef]
7. Menéndez, O.; Pérez, M.; Auat Cheein, F. Visual-based positioning of aerial maintenance platforms on overhead transmission lines. *Appl. Sci.* **2019**, *9*, 165. [CrossRef]
8. Reddy, A.; Indragandhi, V.; Ravi, L.; Subramaniyaswamy, V. Detection of Cracks and damage in wind turbine blades using artificial intelligence-based image analytics. *Measurement* **2019**, *147*, 106823. [CrossRef]
9. Myeong, W.C.; Jung, K.Y.; Jung, S.W.; Jung, Y.H.; Myung, H. Drone-type wall-climbing robot platform for structural health monitoring. In Proceedings of the 6th International Conference on Advances in Experimental Structural Engineering/11th International Workshop on Advanced Smart Materials and Smart Structures Technology, Champaign, IL, USA, 1–2 August 2015.
10. Kocer, B.B.; Tjahjowidodo, T.; Pratama, M.; Seet, G.G.L. Inspection-while-flying: An autonomous contact-based nondestructive test using UAV-tools. *Autom. Constr.* **2019**, *106*, 102895. [CrossRef]
11. Ollero, A.; Heredia, G.; Franchi, A.; Antonelli, G.; Kondak, K.; Sanfeliu, A.; Viguria, A.; Martinez-de Dios, J.R.; Pierri, F.; Cortés, J. The aeroarms project: Aerial robots with advanced manipulation capabilities for inspection and maintenance. *IEEE Robot. Autom. Mag.* **2018**, *25*, 12–23. [CrossRef]
12. Suarez, A.; Heredia, G.; Ollero, A. Lightweight compliant arm with compliant finger for aerial manipulation and inspection. In Proceedings of the 2016 IEEE/RSJ International Conference on Intelligent Robots and Systems (IROS), Daejeon, Korea, 9–14 October 2016; pp. 4449–4454.

13. Sanchez-Cuevas, P.J.; Gonzalez-Morgado, A.; Cortes, N.; Gayango, D.B.; Jimenez-Cano, A.E.; Ollero, A.; Heredia, G. Fully-actuated aerial manipulator for infrastructure contact inspection: Design, modeling, localization, and control. *Sensors* **2020**, *20*, 4708. [CrossRef]

14. Trujillo, M.Á.; Martínez-de Dios, J.R.; Martín, C.; Viguria, A.; Ollero, A. Novel aerial manipulator for accurate and robust industrial NDT contact inspection: A new tool for the oil and gas inspection industry. *Sensors* **2019**, *19*, 1305. [CrossRef] [PubMed]

15. Danko, T.W.; Chaney, K.P.; Oh, P.Y. A parallel manipulator for mobile manipulating UAVs. In Proceedings of the 2015 IEEE International Conference on Technologies for Practical Robot Applications (TePRA), Woburn, MA, USA, 11–12 May 2015; pp. 1–6.

16. Larrauri, J.I.; Sorrosal, G.; González, M. Automatic system for overhead power line inspection using an Unmanned Aerial Vehicle—RELIFO project. In Proceedings of the 2013 International Conference on Unmanned Aircraft Systems (ICUAS), Atlanta, GA, USA, 28–31 May 2013; pp. 244–252.

17. Pagnano, A.; Höpf, M.; Teti, R. A roadmap for automated power line inspection. Maintenance and repair. *Procedia Cirp* **2013**, *12*, 234–239. [CrossRef]

18. He, T.; Zeng, Y.; Hu, Z. Research of multi-rotor UAVs detailed autonomous inspection technology of transmission lines based on route planning. *IEEE Access* **2019**, *7*, 114955–114965. [CrossRef]

19. Teng, G.E.; Zhou, M.; Li, C.R.; Wu, H.H.; Li, W.; Meng, F.R.; Zhou, C.C.; Ma, L. Mini-UAV LiDAR for power line inspection. *Int. Arch. Photogramm. Remote Sens. Spat. Inf. Sci.* **2017**, *XLII-2/W7*, 297–300. [CrossRef]

20. Zhang, R.; Yang, B.; Xiao, W.; Liang, F.; Liu, Y.; Wang, Z. Automatic extraction of high-voltage power transmission objects from UAV lidar point clouds. *Remote Sens.* **2019**, *11*, 2600. [CrossRef]

21. Schofield, O.B.; Lorenzen, K.H.; Ebeid, E. Cloud to cable: A drone framework for autonomous power line inspection. In Proceedings of the 2020 23rd Euromicro Conference on Digital System Design (DSD), Kranj, Slovenia, 26–28 August 2020; pp. 503–509.

22. da Silva, M.F.; Honorio, L.M.; Marcato, A.L.M.; Vidal, V.F.; Santos, M.F. Unmanned aerial vehicle for transmission line inspection using an extended Kalman filter with colored electromagnetic interference. *ISA Trans.* **2020**, *100*, 322–333. [CrossRef] [PubMed]

23. Lu, M.; Bagheri, M.; James, A.P.; Phung, T. Wireless charging techniques for UAVs: A review, reconceptualization, and extension. *IEEE Access* **2018**, *6*, 29865–29884. [CrossRef]

24. Simic, M.; Bil, C.; Vojisavljevic, V. Investigation in wireless power transmission for UAV charging. *Procedia Comput. Sci.* **2015**, *60*, 1846–1855. [CrossRef]

25. Suarez, A.; Salmoral, R.; Zarco-Periñan, P.J.; Ollero, A. Experimental evaluation of aerial manipulation robot in contact with 15 kV power line: Shielded and long reach configurations. *IEEE Access* **2021**, *9*, 94573–94585. [CrossRef]

26. Barrientos, R.; Ponce, C.; Palacín, C.; Martín, C.A.; Martín, B.; Alonso, J.C. Wire marking results in a small but significant reduction in avian mortality at power lines: A BACI designed study. *PLoS ONE* **2012**, *7*, e32569. [CrossRef] [PubMed]

27. Ferrer, M.; Morandini, V.; Baumbusch, R.; Muriel, R.; De Lucas, M.; Calabuig, C. Efficacy of different types of "bird flight diverter" in reducing bird mortality due to collision with transmission power lines. *Glob. Ecol. Conserv.* **2020**, *23*, e01130. [CrossRef]

28. Dwyer, J.F.; Pandey, A.K.; McHale, L.A.; Harness, R.E. Near-ultraviolet light reduced Sandhill Crane collisions with a power line by 98%. *Condor* **2019**, *121*, duz008. [CrossRef]

29. Dashnyam, B.; Purevsuren, T.; Amarsaikhan, S.; Bataa, D.; Buuveibaatar, B.; Dutson, G. Malfunction rates of bird flight diverters on powerlines in the Mongolian Gobi. *Mong. J. Biol. Sci.* **2016**, *14*, 13–20. [CrossRef]

applied sciences

Article

Towards Autonomous Bridge Inspection: Sensor Mounting Using Aerial Manipulators

Antun Ivanovic [1],*, Lovro Markovic [1], Marko Car [1], Ivan Duvnjak [2] and Matko Orsag [1]

[1] Faculty of Electrical Engineering and Computing, University of Zagreb, Unska 3, 10000 Zagreb, Croatia; lovro.markovic@fer.hr (L.M.); marko.car@fer.hr (M.C.); matko.orsag@fer.hr (M.O.)

[2] Faculty of Civil Engineering, University of Zagreb, Fra Andrije Kacica-Miosica 26, 10000 Zagreb, Croatia; ivan.duvnjak@grad.unizg.hr

* Correspondence: antun.ivanovic@fer.hr

Featured Application: The main idea of this paper was to deploy a team of unmanned aerial vehicles (UAVs) to attach a sensor to a bridge using a two-component adhesive in order to perform an inspection. Constant pressure must be applied for several minutes to form a bond between two adhesives. Therefore, one UAV sprays the colored component of an adhesive while the aerial manipulator transports the sensor, detects the contact point and attaches the sensor to it. A trajectory planning algorithm was developed around the dynamic model of the UAV and the manipulator attached to it, ensuring that the end-effector is parallel to the wall normal. Finally, the aerial manipulator achieves and maintains contact with a predefined force through an adaptive impedance control approach.

Abstract: Periodic bridge inspections are required every several years to determine the state of a bridge. Most commonly, the inspection is performed using specialized trucks allowing human inspectors to review the conditions underneath the bridge, which requires a road closure. The aim of this paper was to use aerial manipulators to mount sensors on the bridge to collect the necessary data, thus eliminating the need for the road closure. To do so, a two-step approach is proposed: an unmanned aerial vehicle (UAV) equipped with a pressurized canister sprays the first glue component onto the target area; afterward, the aerial manipulator detects the precise location of the sprayed area, and mounts the required sensor coated with the second glue component. The visual detection is based on an Red Green Blue - Depth (RGB-D) sensor and provides the target position and orientation. A trajectory is then planned based on the detected contact point, and it is executed through the adaptive impedance control capable of achieving and maintaining a desired force reference. Such an approach allows for the two glue components to form a solid bond. The described pipeline is validated in a simulation environment while the visual detection is tested in an experimental environment.

Keywords: aerial robotics; inspection and maintenance; aerial manipulation; multirotor control

Citation: Ivanovic, A.; Markovic, L.; Car, M.; Duvnjak, I.; Orsag, M. Towards Autonomous Bridge Inspection: Sensor Mounting Using Aerial Manipulators. *Appl. Sci.* **2021**, *11*, 8279. https://doi.org/10.3390/app11188279

Academic Editor: Alessandro Gasparetto

Received: 23 July 2021
Accepted: 1 September 2021
Published: 7 September 2021

Publisher's Note: MDPI stays neutral with regard to jurisdictional claims in published maps and institutional affiliations.

1. Introduction

The world of unmanned aerial vehicles (UAVs) has been rapidly growing in recent years. As their design and control are perfected, these aerial vehicles have become more and more available. Nowadays, off-the-shelf ready-to-fly UAVs can be found and bought in shops, which makes them available to virtually anybody. This, in turn, has sparked a great deal of public interest in UAVs since their potential can be found in applications such as agriculture, various inspections (bridges, buildings, wind turbines), geodetic terrain mapping, the film industry, and even for hobby enthusiasts to fly and record videos from a first-person perspective. The vast majority of commercially available UAVs are equipped with a camera, while more specialized vehicles for terrain mapping or crop spraying offer a more diverse sensor suite.

All of the aforementioned systems primarily observe and gather data about the environment, while having little to no ability to interact with and change the environment. One way to augment these vehicles for physical interaction is to attach a lightweight manipulator to their body, which is the main interest of the aerial manipulation field. Although such vehicles are more complex for both modeling and control, their benefit lies in performing versatile tasks that require interaction with the environment.

In general, there are three types of bridge inspections: periodic, special and damage inspections. Periodic bridge inspections differ from country to country according to national standards, and are usually performed at least once every two to three years. Special inspections are typically used to monitor the condition of deficient elements at specific locations based on predefined requirements. Damage inspections are usually performed after events that have occurred due to environmental impacts or human actions. The aim of a bridge inspection is to evaluate and assess structural safety and reliability. Current techniques are based on traditional visual inspection with a combination of non-destructive methods (NDTs). Traditional visual inspection is performed by experienced (trained) engineers and using specialized trucks equipped with the cranes and basket, that allow inspectors to review the conditions underneath the bridge. During the inspection, the engineers are equipped with various NDT [1] tools to detect construction faults and defects such as corrosion, cracks, voids, weakening connections, and concrete delamination. Some of these NDTs require mounting small sensors to collect data, such as accelerometers, strain gauges, tilt meters and various transducers for acoustic or pressure measurements. Afterwards, the bridge is excited with vibrations, sound waves, tapping, etc., and mounted sensors record responses to these specific excitations. Furthermore, there are usually requirements for performing measurements during the bridge inspection, such as the short- and long-term monitoring of vibrations, strains, displacements, etc. Mainly, these inspections offer valuable information about the current bridge conditions, but there are a number of disadvantages. The use of trucks during inspections requires total or temporary road closures, which at the same time require safety measures to keep traffic flowing as freely as possible. In addition, inspectors often encounter challenges in reaching all portions or elements in narrow areas, such as tight spaces between girders, beams and vaults. The aforementioned significantly increases the time and overall cost of the inspection. An aerial robot, with the potential to reach these challenging locations on the bridge, could significantly reduce the time and cost of these inspections and improve worker safety. Moreover, we note that the aforementioned sensors are relatively lightweight, which makes them suitable for transportation and mounting with an aerial robot.

1.1. Concept

We envision a team of robots working together to attach sensors to bridges and similar grade separation infrastructure. In theory, such a task could be accomplished with a single aerial robot, at the cost of a complex mechanical design. The proposed team shown in Figure 1 consists of two drones. One drone applies the adhesive material, and the other attaches sensors. We envision a two-stage process using two-component adhesives which form a solid bond from two separate reactive components: the "resin" and the "hardener". The first UAV applies the resin by spraying it onto the surface, while the second one attaches the sensor with the hardener already applied before the flight.

It is important to follow the prescribed ratio of the resin and the hardener to achieve the desired physical properties of the adhesive. Only when mixed together do the two components form the adhesive. The reaction typically begins immediately after the two components are mixed and the bond strength depends both on maintaining the contact and the viscosity of the mixed adhesive during the process. Manufacturers can control the cure rate to achieve various working times (worklife) until final bond strength is achieved, ranging from minutes to weeks. Resin bases are usually more viscous than their respective hardener and are generally applied by brush, roller, applicator or spray. In this work, we propose attaching a canister of pressurized resin to the UAV, and spray it through a nozzle onto the infrastructure surface. In this scenario, the spray needs to be softer and less turbulent to reduce the amount of material lost due to bouncing and it must be colored for the detection in the second stage. Spraying with drones is not a novel concept [2,3], so without loss of generality, we will omit the details of this design and instead focus on detecting, navigating to, and sustaining contact with the sprayed surface.

In typical applications, the assemblies are usually kept in contact until the sufficient strength of the bond is achieved. When fully cured, two-component adhesives are typically tough and rigid with good temperature and chemical resistance. We rely on the robotic arm attached to the second aerial vehicle to apply a controlled contact force between the sensor and the surface. Maintaining this fixed assembly contact through the impedance control system enables us to achieve a successful curing process and create a permanent bond between the sensor and the infrastructure. After the first UAV sprays the resin onto the surface, the second aerial robot finds the sprayed part and applies the contact with the sensor's surface. Before takeoff, the surface of the sensor is brushed with a hardener. Once contact is made, it is maintained for the prescribed curing time, after which the aerial robot disembarks and leaves the sensor attached to the surface.

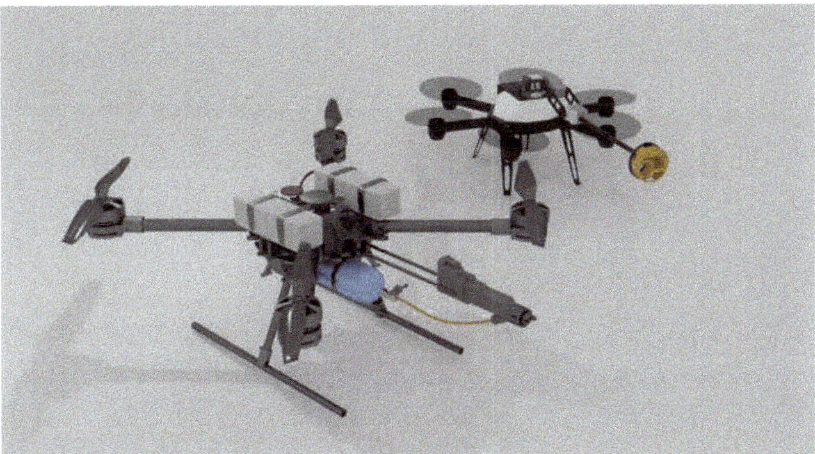

Figure 1. Two aerial robots working together to attach sensors to different parts of a bridge and similar grade separation infrastructure. The one on the left is used to spray the resin onto the surface, while the aerial robot on the right maintains contact to the surface with the sensor attached to its end-effector.

1.2. Contributions

This paper focuses on developing a method for mounting sensors on a bridge wall using an aerial manipulator. The first contribution is augmenting the model-based motion planning with the adaptive impedance controller. The motion planning method accounts for the underactuated nature of the multirotor UAV and corrects the end-effector configuration for an appropriate approach. This method also relies on the dexterity analysis which keeps the manipulator configuration within its optimal region, ensuring that the manipulator is never fully extended or contracted while mounting a sensor. The second contribution is the visual blob detection which locates and tracks the appropriate sensor mounting point. The blob detection has been experimentally verified in an indoor environment, yielding the reliable and robust tracking of the mount location, as well as the blob plane orientation. Finally, the third contribution is the simulation analysis of the system's performance, conducted on a straight and inclined wall approach. The simulation concentrates on testing the motion planning together with the impedance controller, performing a repeatability analysis and ensuring that the desired contact force is achieved.

2. Related Work

In the world of aerial inspections, a number of UAV-based solutions are being proposed by researchers. In [4], a technical survey for bridge inspections is given. Researchers in [5] present the project AERIAL COgnitive Integrated Multi-task Robotic System with Extended Operation Range and Safety (AERIAL-CORE) which focuses on power lines inspection, maintenance and installing bird diverters and line spacers. Most of these approaches are based in conjunction with new technologies to ensure faster and cheaper inspections. Nowadays, UAVs use high-resolution cameras for visual inspections and employ point cloud methods based on digital photogrammetry [6], Light Detection And Ranging (LiDAR)-based methods [7], digital image correlation [8], etc. There are also reports for visual compensation during aerial grasping [9], aerial grasping in strong winds [10], and the development of a fully actuated aerial manipulator for performing inspections underneath a bridge [11]. According to the experimental testing of contact-based bridge inspections, there is a need to develop a solution for mounting application sensors (such as accelerometers, strain gauges and tilt meters) on a bridge using a UAV. It is expected that a sophisticated system with the possibility of automatic sensor mounting will increase the frequency of measurements without interrupting traffic, ensure the safety of inspectors as well as reduce inspection time and overall costs.

As mentioned earlier, the second UAV needs to be aware of the position of the sprayed adhesive which is applied in a blob-like pattern. For this purpose, an Red Green Blue - Depth (RGB-D) camera is used due to its favorable dimensions and weight. It provides image and depth information about the environment which proves useful for object localization and UAV navigation. Such cameras were commonly found on UAVs and unmanned ground vehicles (UGVs) present at the recent MBZIRC 2020 competition. In [12,13], RGB-D information is used for color-based brick detection and localization for the wall-building challenge using UAVs and UGVs, respectively, while in [14] the authors use a Convolutional Neural Network (CNN)-based UAV detection and tracking method for the intruder UAV interception challenge. Furthermore, visual sensors proved useful in [15], where the authors performed a contact-based inspection of a flat surface with an aerial manipulator. The surface position and orientation was obtained by applying random sample consensus (RANSAC) on the RGB-D information. A thorough survey of 2D object detection methods from UAVs was given in [16]. In this paper, the authors present a modular framework for object detection in which a simple contour-based blob detector is implemented. The goal is to use RGB-D information to enable an autonomous inspection workflow. The blob position is obtained by segmenting the depth data at the points where the object is detected in the image, while RANSAC [17] is used to determine its orientation.

After the successful detection of a blob-like pattern, it is necessary to attach the inspection sensor. The first phase of the sensor attachment is achieving contact, and the second is maintaining that contact to allow for the two adhesive components to form a bond. Generally, the contact can be achieved with or without force measurements. In [18], contact with the wall is performed and maintained. Researchers in [19] performed wall contact and aerial writing experiments. The work presented in [20] modeled and exploited the effects of the ceiling effect to perform an inspection underneath a bridge. The common denominator in the former approaches is maintaining the contact without any force feedback. Although mounting a force sensor on a UAV increases both mechanical and control complexity, an immediate benefit is the ability to maintain precise contact force regardless of the environment. In [21], the researchers used a force/torque sensor to achieve compliant control while pulling a rope and a semi-flexible bar. A fully actuated UAV with a manipulator has been employed in [22] to compare force feedback control with and without the force/torque sensor. Researchers in [23] used a single degree of freedom manipulator with a force sensor mounted at the end-effector to press an emergency switch.

Relying on the blob-like pattern detection and the impedance control, a trajectory for achieving contact is required to steer the aerial manipulator towards the contact point. While mounting the sensor, it is essential that the approach and contact are perpendicular to the wall plane. This can be considered as a task constraint imposed on the planner which the aerial manipulator has to satisfy. Researchers in [24] propose a task constrained planner for a redundant robotic manipulator that enables them to do everyday tasks such as opening drawers or picking up objects. In [25], a task-constrained planner was developed for underactuated manipulators. Since multirotor UAVs are typically underactuated systems, it is necessary to address dynamics and kinematics while planning the end-effector trajectory. Aerial manipulator 6D end-effector trajectory tracking based on the differential flatness principle was presented in [26]. The underactuated nature of multirotor UAVs can cause unexpected deviations in end-effector configuration. Researchers in [27] address this particular problem by including the dynamic model of the system into the planning procedure. In our previous work [28], a trajectory planning method based on the full dynamic model of an aerial manipulator was developed. In this paper, we further augmented this method to plan for the desired force required by the impedance controller.

3. Mathematical Model

In this section, the mathematical model of the aerial manipulator is presented. The coordinate systems convention is depicted in Figure 2. Furthermore, an analysis for the manipulator dexterity and reach was performed.

3.1. Kinematics

The inertial frame is defined as L_W. The body-fixed frame L_B is attached to the center of gravity of the UAV. The position of the UAV in the world frame is given with $\mathbf{p}_W^B = \begin{bmatrix} x & y & z \end{bmatrix}^T \in$, and the attitude vector is $\Theta = \begin{bmatrix} \phi & \theta & \psi \end{bmatrix}^T$. Combining the position and attitude vectors defined the generalized coordinates of the UAV as $\mathbf{q}_B = \begin{bmatrix} (\mathbf{p}_W^B)^T & \Theta^T \end{bmatrix}^T \in \mathbb{R}^6$. Written in a matrix form T_W^B, the transformation contains both the position and orientation of the UAV obtained through an on-board sensor fusion or through an external positioning system (i.e., GPS). The notation $T_a^b \in \mathbb{R}^{4 \times 4}$ was used to denote a homogeneous transformation matrix between frames a and b.

A rigid attachment between the body of the UAV and the base of the manipulator L_0 was considered, denoted with the transformation matrix T_B^0. The manipulator used in this work was a $M = 3$ degree-of-freedom (DoF) serial chain manipulator with the end-effector attached to the last joint. The DH parameters of the arm are given in Table 1. Using this notation, one can write a transformation matrix T_0^{ee} between the manipulator base and its end-effector as a function of joint variables q_1, q_2 and q_3. For brevity, the expression for the entire matrix T_0^{ee} is left out and only the end-effector position and its approach vector equations are written using the well-known abbreviation $\cos(q_1 + q_2) := C_{12}$:

$$\mathbf{p}_0^{ee} = \begin{bmatrix} a_1(C_1 + C_{12}) + d_3 C_{123} \\ a_1(S_1 + S_{12}) + d_3 S_{123} \\ 0 \end{bmatrix}, \mathbf{z}_0^{ee} = \begin{bmatrix} C_{123} \\ S_{123} \\ 0 \end{bmatrix} \tag{1}$$

Table 1. DH parameters of the 3-DoF manipulator attached to the UAV. A virtual joint q_4^* is added to fully comply with the DH convention. Link sizes a_1 and d_3 are omitted for clarity.

Joint	θ	d	α	a
q_1	0	0	0	a_1
q_2	0	0	0	a_1
q_3	$\frac{\pi}{2}$	0	$\frac{\pi}{2}$	0
q_4^*	0	d_3	0	0

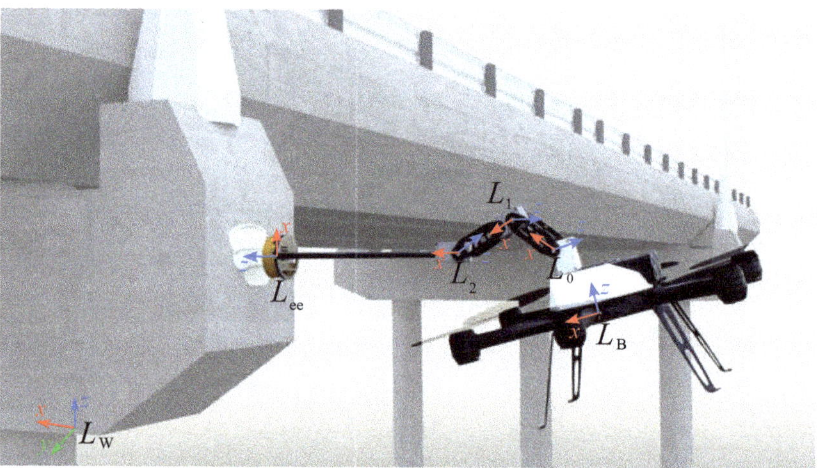

Figure 2. Coordinate systems of the world, UAV and the 3-DoF manipulator.

Putting it all together, the full kinematic chain of the aerial manipulator can be constructed as

$$T_{\mathrm{W}}^{ee} = T_{\mathrm{W}}^{\mathrm{B}} \cdot T_{\mathrm{B}}^{0} \cdot T_{0}^{ee}, \tag{2}$$

combining the fixed transformation T_{B}^{0} with $T_{\mathrm{W}}^{\mathrm{B}}$ and T_{0}^{ee} depend on UAV and manipulator motion. Since there is obvious coupling between the motion of the body and the manipulator arm, a $\beta \in [0,1]$ parameter is introduced to distribute the end-effector motion commands either to the UAV global position control or manipulator joint position control. To this end, the following distribution relationship is used:

$$\Delta \mathbf{P}_{\mathrm{UAV}} = \beta \cdot \Delta \mathbf{P}$$
$$\Delta \mathbf{P}_{arm} = (1 - \beta) \cdot \Delta \mathbf{P}, \tag{3}$$

where $\Delta \mathbf{P}$ is used to denote the desired aerial manipulator displacement expressed as the following combination body and arm motion:

$$\Delta \mathbf{P} = \Delta \mathbf{P}_{\mathrm{UAV}} + \Delta \mathbf{P}_{arm}$$
$$= \beta \cdot \Delta \mathbf{P} + (1 - \beta) \cdot \Delta \mathbf{P}. \tag{4}$$

The manipulator displacement is denoted by $\Delta \mathbf{P}_{arm}$ and the UAV displacement by $\Delta \mathbf{P}_{\mathrm{UAV}}$, where both $\Delta \mathbf{P}_{arm}$ and $\Delta \mathbf{P}_{\mathrm{UAV}}$ are expressed in the coordinate system L_0. With $\beta = 1$, the UAV motion is used to control the position of the end-effector. When $\beta = 0$, the situation is reversed and the manipulator motion is used to move the end-effector. For every other β, the end-effector motion is obtained in part by the UAV body and the manipulator arm motion.

There are obvious advantages in combining the motion of the UAV and the manipulator arm. The UAV can move in 3D space beyond the reach of the arm; however, the motion of the UAV is not as precise and dynamically decoupled. The kinematics of the arm enable the end-effector to obtain the desired approach angle $\mathbf{z}_{0}^{ee} = [\cos(\delta), \sin(\delta), 0]^T$, which under the hovering assumption, becomes equal to the global approach vector $\mathbf{z}_{W}^{ee}$ pointing towards the contact point on the infrastructure. The straightforward mathematical manipulation of Equation (1) allows for writing the constraint equation:

$$q_3 = \delta - q_1 - q_2, \tag{5}$$

which ensures that the manipulator points in the right direction, where δ is the desired manipulator inclination in the body x–z plane.

To find the optimal manipulator pose during contact, the dexterity $\mathfrak{D}$ and the reach $\mathfrak{R}$ of the pose were taken into account, while considering that the joints are as far as possible from their physical limits $\mathfrak{L}$. Since the motion of the arm is constrained with its approach axis condition, a reduced form of a Jacobian matrix was used $\mathbf{J} = \left[\frac{\delta \mathbf{p}_0^{ee}}{\delta q_1}, \frac{\delta \mathbf{p}_0^{ee}}{\delta q_2}\right]$ to derive the pose dexterity index $\mathfrak{D} = \left|\mathbf{J}^T \cdot \mathbf{J}\right|$ and determine how far the current pose is from the null space of the manipulator [29]. The reach of the pose $\mathfrak{R} = (\mathbf{p}_0^{ee})^T \cdot \mathbf{p}_0^{ee}$ was also taken into account, since the goal was to keep the end-effector and the contact point as far away from the UAV body. Finally, the following equation is defined:

$$\mathfrak{L} = \frac{(q_1^2 - Q_{1max}^2)(q_2^2 - Q_{2max}^2)}{Q_{1max}^2 Q_{2max}^2}, \tag{6}$$

to measure how far away the given configuration is (i.e., q_1, q_2) from the joint limits Q_{1max}, Q_{2max}. Normalizing $\mathfrak{D}, \mathfrak{R}$ and $\mathfrak{L}$ enables combining the three conditions into a single manifold $\mathfrak{M} = \mathfrak{D} \cdot \mathfrak{R} \cdot \mathfrak{L}$ and find the optimal configuration $\mathbf{q}_M^* = \begin{bmatrix} q_1^* & q_2^* & q_3^* \end{bmatrix}^T$ for the desired approach angle δ. The described method is depicted in Figure 3 for the specific case of the approach angle $\delta = 0°$, but can be extended to any value of the approach angle through Equation (5).

As a side note, the manipulator attachment on the top of the UAV body was chosen to be able to reach surfaces underneath the bridge. Although this shifts the center of gravity upwards, the stability of the system is not compromised since the manipulator is constructed of lightweight materials.

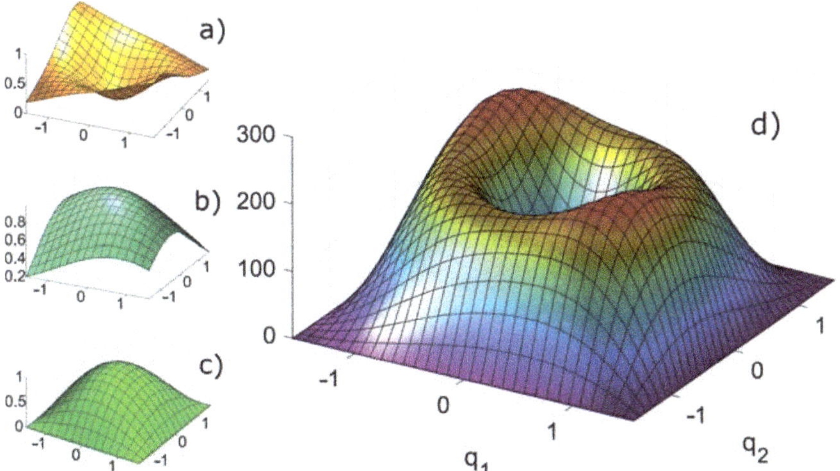

Figure 3. The visual decomposition of dexterity $\mathfrak{D}$, reach $\mathfrak{R}$, limit $\mathfrak{L}$ and the overall combined surface. This analysis is performed for $\delta = 0°$: (**a**) the dexterity $\mathfrak{D}$ surface shows the measure of how far the manipulator is from the null space. Values around zero are closer to the null space; (**b**) the reach $\mathfrak{R}$ surface shows how far the end-effector can move in a certain configuration. This value tends towards zero as the arm approaches a folded configuration; (**c**) the limit $\mathfrak{L}$ depicts how far a certain configuration is from the physical limits of the manipulator joints; and (**d**) the combined manifold $\mathfrak{M}$ of the formerly described surfaces. Higher values offer better trade-off between dexterity, reach and limit, defining the optimal manipulator configuration $\mathbf{q}_M^*$.

3.2. Dynamics

The most complicated task of the aerial manipulator is attaching the sensor to a wall and maintaining the required force reference while the two-component adhesive hardens.

To successfully perform such a task, the coupled UAV-manipulator system dynamics have to be addressed for the precise end-effector configuration planning.

Considering the UAV dynamics only, the derivative of the generalized coordinates can be defined as $\dot{\mathbf{q}}_{\mathrm{B}} = \left[(\dot{\mathbf{p}}_{\mathrm{W}}^{\mathrm{B}})^T \quad (\omega_{\mathrm{W}}^{\mathrm{B}})^T \right]^T \in \mathbb{R}^6$. Here, the $(\dot{\mathbf{p}}_{\mathrm{W}}^{\mathrm{B}})^T$ is the linear velocity of the body of the UAV in the world frame and the $(\omega_{\mathrm{W}}^{\mathrm{B}})^T$ represents the angular velocity of the UAV in the world frame. The UAV's propulsion system consists of n_p propellers rigidly attached to the body. Each propeller produces force and torque along the z_{B} axis. The vector of the propeller rotational velocities is simply defined as

$$\boldsymbol{\Omega}_{\mathrm{UAV}} = \left[\Omega_1 \quad \ldots \quad \Omega_{n_p} \right]^T \in \mathbb{R}^{n_p}. \tag{7}$$

Force and torque produced by each propeller are non-linear functions depending on the rotational velocity $\boldsymbol{\Omega}_{\mathrm{UAV}}$. Rather than using the rotational velocities as control inputs, they can be mapped to a more convenient space. Namely, the mapped control input space can be written as

$$\mathbf{u}_{\mathrm{UAV}} = \mathbf{K} \cdot diag(\boldsymbol{\Omega}_{\mathrm{UAV}}) \cdot \boldsymbol{\Omega}_{\mathrm{UAV}}, \tag{8}$$

where $\mathbf{K} \in \mathbb{R}^{4 \times n_p}$ is the mapping matrix and $\mathbf{u}_{\mathrm{UAV}} = \left[u_1 \quad u_2 \quad u_3 \quad u_4 \right]^T$, where u_4 represents the net thrust and u_1, u_2 and u_3 are moments around the body frame axes.

As stated earlier, the manipulator consists of three rotational DoFs. Therefore, the joint positions of the manipulator are defined as $\mathbf{q}_{\mathrm{M}} = \left[q_1 \quad q_2 \quad q_3 \right]^T$. The rotational velocity of each joint is a time derivative of joint positions $\dot{\mathbf{q}}_{\mathrm{M}} = d\mathbf{q}_{\mathrm{M}}/dt$. The torque of each joint is considered the control input of the manipulator $\mathbf{u}_{\mathrm{M}} = \left[\tau_1 \quad \tau_2 \quad \tau_3 \right]^T$.

The resulting generalized coordinates of the aerial manipulator can be written as $\mathbf{q} = \left[\mathbf{q}_{\mathrm{UAV}} \quad \mathbf{q}_{\mathrm{M}} \right]^T \in \mathbb{R}^9$, and the velocities can be obtained in the same manner as $\dot{\mathbf{q}} = \left[\dot{\mathbf{q}}_{\mathrm{UAV}} \quad \dot{\mathbf{q}}_{\mathrm{M}} \right]^T \in \mathbb{R}^9$. The resulting control inputs of the system can be expressed as $\mathbf{u} = \left[\mathbf{u}_{\mathrm{UAV}} \quad \mathbf{u}_{\mathrm{M}} \right]^T \in \mathbb{R}^7$. Finally, the full system dynamics can be written as

$$\mathbf{M}(\mathbf{q})\ddot{\mathbf{q}} + \mathbf{c}(\mathbf{q}, \dot{\mathbf{q}}) + \mathbf{g}(\mathbf{q}) = \mathbf{u}, \tag{9}$$

where $\mathbf{M}(\mathbf{q}) \in \mathbb{R}^{9 \times 7}$ is the inertia matrix, $\mathbf{c}(\mathbf{q}, \dot{\mathbf{q}}) \in \mathbb{R}^7$ is the vector of centrifugal and Coriolis forces, $\mathbf{g}(\mathbf{q}) \in \mathbb{R}^7$ is the gravitational term.

4. Control System

The overall control of the aerial manipulator consists of several nested control loops. The complete controller overview, with motion planning and blob detection blocks, is depicted in Figure 4.

4.1. Aerial Manipulator Control

At the inner most level, the UAV is controlled through cascade attitude and rate controllers. The input to these controllers is the desired orientation and based on the state, the output is the vector of the rotors' angular velocities. The second level of control, which uses the inner attitude control loop, consists of two additional cascades, the position and the velocity control. These controllers receive a referent position and velocity feed-forward value to generate the desired vehicle orientation and thrust. The manipulator joints are controlled through standard Proportional, Integral, Derivative (PID) controllers; however, in a real-world setting, servo motors with integrated control are typically used.

As mentioned earlier, it is important to track the desired force after contact with a wall is achieved. To accomplish this, an adaptive impedance controller is employed to generate an appropriate setpoint for the position controller. This controller receives a trajectory supplied by the mission planner, which steers the aerial manipulator towards the sensor mounting target on the bridge.

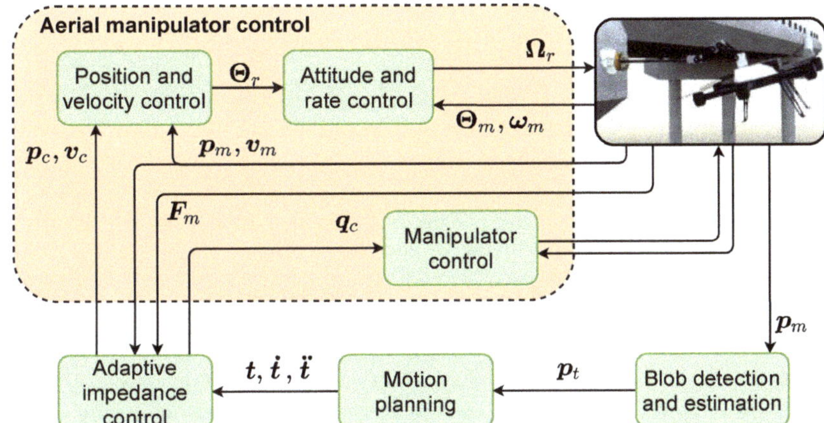

Figure 4. The overall functional schematic of the system. The aerial manipulator control subsystem is necessary for the controlling position and attitude of the UAV, and manipulator joints. On top of this controller, the adaptive impedance control is employed in order to track the desired force. Motion planning generates an appropriate trajectory based on the target point supplied by the blob detection algorithm.

4.2. Adaptive Impedance Control

The objective of the adaptive impedance controller is to ensure a stable physical interaction between the aerial manipulator and the environment [30]. As mentioned earlier, the standard UAV control scheme is based on position and attitude controllers. When interacting with the environment, the desired contact force must be considered. The position controlled system can be extended to follow the desired force by introducing an impedance filter. The design of such a filter is explained here for a single DoF.

The behavior of the system is defined by the target impedance as

$$e(t) = m(\ddot{x}_c(t) - \ddot{x}_r(t)) + b(\dot{x}_c(t) - \dot{x}_r(t)) + k(x_c(t) - x_r(t)), \tag{10}$$

where m, b and k are constants, $x_r(t)$ is the referent position, provided to the impedance filter as an input, and $x_c(t)$ is the output of the impedance filter representing the position command. The filter is designed as a linear second-order system with a dynamic relationship between the position and the contact force tracking error $e(t)$ so that it mimics a mass-spring–damper system. The contact force tracking error is defined as follows:

$$e(t) = f_r(t) - f(t), \tag{11}$$

where $f_r(t)$ is the other filter input defining the referent force, and $f(t)$ is the measured (exerted) contact force. If the environment is modeled as a first-order elastic system (equivalent spring system) with unknown stiffness k_e, the measured force can be approximated as

$$f(t) = k_e(x(t) - x_e(t)), \tag{12}$$

where $x(t)$ is the position of the manipulator and $x_e(t)$ is the position of the environment in an unexcited state. By substituting Equation (12) in Equation (11), the position of the aerial manipulator can be expressed as follows:

$$x(t) = \frac{f_r(t) - e(t)}{k_e} + x_e(t). \tag{13}$$

Assuming that the commanded position value can be achieved by the aerial manipulator, i.e., $x = x_c$, the substitution of Equation (13) in Equation (10), the system in the steady state can be described as follows:

$$e(t) = \frac{k \cdot k_e}{k + k_e}\left(\frac{f_r(t)}{k_e} + x_e(t) - x_r(t)\right). \tag{14}$$

For a contact force error of zero in the steady state, the following must hold:

$$x_r(t) = \frac{f_r(t)}{k_e} + x_e(t). \tag{15}$$

In other words, the position setpoint has to be designed in such a way that it compensates for the displacement of the environment due to the exerted contact force. To ensure this, a value of the unknown environment stiffness k_e is needed. Furthermore, k_e plays a fundamental role in the stability of the impedance filter Equation (10), which ultimately affects the stability of the aerial manipulator while in contact with the environment. A stable contact between the aerial manipulator and the environment can be ensured using the Hurwitz stability criterion, by designing the system with $b/m > 0$ and $(k + k_e)/m > 0$. However, since k_e is unknown, an adaptation law for the position setpoint that guarantees the contact stability while compensating for this hidden, unknown parameter is proposed.

The adaptation law is derived starting from Equation (15). An adaptive parameter $\kappa(t)$ is introduced so that:

$$x_r(t) = \kappa(t)f_r(t) + x_e(t). \tag{16}$$

It can be shown using the Lyapunov stability analysis that the following adaptation dynamics equation for $\kappa(t)$ will yield a stable system response:

$$k\dot{\kappa}(t) + b\ddot{\kappa}(t) + m\dddot{\kappa}(t) = -\gamma\sigma(t) + \gamma_d\dot{\sigma}(t). \tag{17}$$

We refer the interested reader to the proof which can be found in the Appendix A.

5. Motion Planning

As discussed in Section 1.1, the main concept of this paper was to use a team of two UAVs, each applying one component of the adhesive. To apply the "resin" component, the UAV has to plan a collision-free trajectory and position itself in front of the target area to start spraying. This is fundamentally different from mounting a sensor coated with "hardener". In the latter case, apart from planning a collision-free trajectory, the manipulator-endowed UAV has to apply pressure for a certain amount of time for the two components to mix.

From the perspective of motion planning, the planner needs to be augmented to include a manipulator with three degrees of freedom, contact force and the weighing parameter β. To successfully maintain the pressure, the planner relies on the impedance controller described in Section 4.2. Furthermore, one of the requirements when mounting the sensor on the wall is for the sensor to be perpendicular to the wall. Therefore, it is necessary to take the underactuated nature of the multirotor UAVs into account during the motion planning. Namely, the errors in the planned end-effector configuration were mainly induced due to the roll and pitch angles while executing the planned motion. In our previous work [28], we developed a model-based motion planner for aerial manipulators that is capable of correcting the aforementioned end-effector deviations. In this paper, the idea from [28] was extended to consider the impedance control when obtaining the full state of the aerial manipulator.

5.1. Waypoint Configuration

When dealing with an aerial manipulator, exerting some contact force inevitably yields a high dimensional waypoint configuration. We define a single waypoint as a set of UAV and joint poses, together with the force reference and motion distribution factor β:

$$\mathbf{w} = \begin{bmatrix} \mathbf{q}_B^T & \mathbf{q}_M^T & \mathbf{f}_r^T & \beta \end{bmatrix}^T \in \mathbb{R}^{13}, \tag{18}$$

where $\mathbf{q}_B \in \mathbb{R}^6$ and $\mathbf{q}_M \in \mathbb{R}^3$ are generalized coordinates of the UAV and the manipulator defined in Section 3.1. The force reference vector $\mathbf{f}_r = \begin{bmatrix} f_x & f_y & f_z \end{bmatrix}^T \in \mathbb{R}^3$ and weighing scalar parameter β are required by the impedance controller. Furthermore, the impedance controller assumes a step change of these values. Ideally, the change should occur at the moment of contact since no force can be exerted without contact. Therefore, these values are only changed at the final waypoint.

Apart from the desired force and the parameter β, the final waypoint must contain the UAV position and orientation, as well as the manipulator joint configuration. Specifying these values relies on the blob detection algorithm presented in Section 6. Namely, the algorithm outputs the position and orientation of the detected blob in the world frame. Following the manipulator dexterity and reach analysis described in Section 3.1, the optimal manipulator configuration $\mathbf{q}_M^*$ is obtained based on the provided plane normal. The optimal manipulator configuration is then used as the desired configuration for the final waypoint. This way, during operation, the manipulator never reaches a fully extended or contracted pose, which allows the impedance controller to command both the arm and the UAV to achieve and maintain the desired force.

5.2. Trajectory Planning

There are three phases in the trajectory planning procedure. First, an initial trajectory is planned based on the provided waypoints. Second, the initial trajectory is sent to a simulated model in order to obtain the full state of the aerial manipulator during the trajectory execution. Third, the end-effector configuration is corrected based on the full state of the vehicle, and the final trajectory is sent to the target aerial manipulator.

Initial Trajectory

To execute a smooth motion towards the desired waypoint, we use a suitable time-optimal path planning by reachability analysis (TOPP-RA) trajectory planner [31]. The TOPP-RA algorithm searches for the time-optimal trajectory and is based on a "bang-bang" principle on the generalized torque of each DoF. The planner is capable of receiving the input waypoints of an arbitrary dimension and output a smooth trajectory. Each DoF has to be provided with dynamical constraints in terms of velocity and acceleration, which are respected during the trajectory generation process.

As mentioned, the input to the TOPP-RA trajectory is the path of a set of $n \geq 2$ waypoints:

$$\mathscr{P} = \left\{ \mathbf{w}_i \mid \mathbf{w}_i \in \mathbb{R}^{13}, i \in (0, 1, \dots, n) \right\}. \tag{19}$$

Based on the dynamical constraints, the output of the TOPP-RA planner is a sampled trajectory:

$$\mathscr{T}_s = \left\{ \mathbf{t}(kT_s) \mid \mathbf{t}(kT_s) \in \mathbb{R}^{3 \times 13}, k \in (0, \dots, n_t) \right\}, \tag{20}$$

where $\mathbf{t} = \begin{bmatrix} (\mathbf{w})^T & (\dot{\mathbf{w}})^T & (\ddot{\mathbf{w}})^T \end{bmatrix}^T \in \mathbb{R}^{3 \times 13}$ is a single sampled trajectory point consisting of position, velocity and acceleration; T_s is the sampling time; and n_t is the number of points in the sampled trajectory. Note that each trajectory point contains both roll and pitch angles. Although these angles can be planned through the TOPP-RA algorithm, they are omitted at this point because of the underactuated nature of the multirotor UAV. Nevertheless, they are used later in the paper when the model corrections are applied.

The impedance controller expects a step change in the force and weighing parameter β referent values. To satisfy this requirement, large constraints for velocity and acceleration are imposed for these DoFs. However, because other DoFs have constraints below their physical limit, the overall force and β trajectory has a slower, dynamically smooth profile. These profiles also have overshoots and undershoots which are not acceptable because they are not within the hard constraints required for β. To tackle this problem, a simple piecewise constant velocity interpolation was applied to the force and β. This way, a large velocity constraint produces a step change which is a suitable input to the impedance controller. A visual example of the difference between the TOPP-RA and piecewise constant velocity interpolation is depicted in Figure 5.

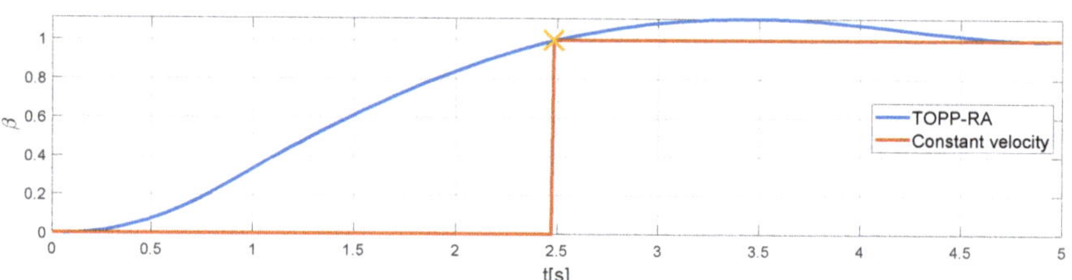

Figure 5. Visual comparison between TOPP-RA and the piecewise constant velocity interpolation. Waypoints for both trajectories are kept the same, and around $t = 2.5$ s, the middle waypoint(yellow cross) is reached. Although the dynamical constraints are the same, TOPP-RA takes other degrees of freedom into account and produces a trajectory with overshoot which is not suitable for parameter β.

5.3. Model-Based Corrections

The initial trajectory from Equation (20) is planned without any consideration about the underactuated nature of the multirotor UAV. To obtain the unknowns, namely roll and pitch angles, the initial trajectory can be executed in a simulation environment. The chosen simulation environment is, in our case, *Gazebo*, because it is realistic and supports the robotics operating system (ROS), which is the backbone of our implementation. The simulated aerial manipulator is based on the mathematical model described in Section 3. The standard cascade PID controllers are employed for low-level attitude and high-level position control. The impedance controller is built on top of the position controller and provides a position reference based on the input trajectory. More details about the simulation environment are provided in Section 7.

The first step is executing the initial trajectory in the aforementioned simulation environment. While executing, the roll and pitch angles are recorded as they are needed for obtaining the full state of the UAV. Rearranging Equation (2) and plugging the unknown roll and pitch angles in the full state of the UAV, the transform of the end-effector in the manipulator base frame can be obtained:

$$T_{L_0}^{ee} = (T_{\text{B}}^{L_0})^{-1} \cdot (T_{\text{W}}^{\text{B}})^{-1} \cdot T_{\text{W}}^{ee}. \tag{21}$$

Usingthe inverse kinematics of the manipulator, joint values $\mathbf{q}_{\text{M}}$ for the desired end-effector configuration are obtained. This way, the null space of the aerial manipulator is used for the end-effector correction. Note that due to the configuration of the manipulator, an exact solution of the inverse kinematics will not always exist. In such a case, an approximate closest solution is used instead.

The final trajectory is constructed by replacing the initial $\mathbf{q}_{\text{M}}$ with the corrected values. This trajectory is afterwards sent to the target aerial manipulator.

The careful reader should note that the developed three DoF manipulator operates on the x and z position in the body frame, as well as the pitch angle. This allows the impedance controller to maintain the orientation perpendicular to the wall, while compensating for the

UAV body motion in the x and z axes. However, the system will experience disturbances and control errors which will act on the roll and pitch angle, and the lateral movement along the body y axis. We can address these issues either with mechanical dampers or by adding additional degrees of freedom to the manipulator, which will be explored in future work.

6. Blob Detection

This section presents the methods we propose to detect the hardener blob position and orientation. A modular object detection framework, as shown in Figure 6, is designed to ensure a reliable blob pose detection. Since the detection is to be done on board the UAVs, RGB-D cameras are selected. Therefore, the inputs to the framework are images and organized point clouds obtained from the visual sensor. The remainder of this section introduces the individual components of the framework and adds implementation details where necessary.

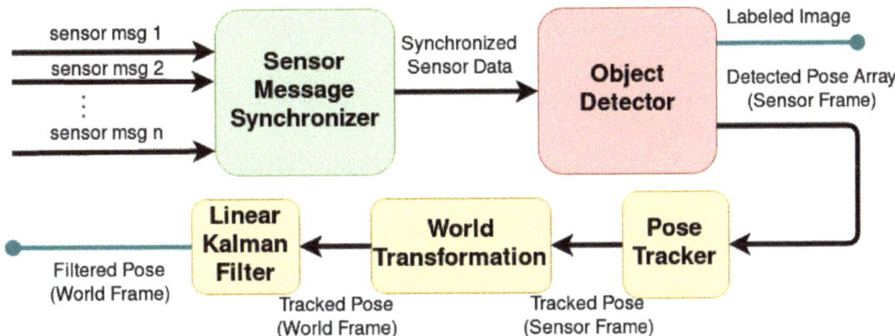

Figure 6. The pipeline for the modular object detection framework. Inputs to the system are an arbitrary number of sensor messages (image, depth, point cloud, etc.) along with sensor–world frame transformations. Output is the detected blob pose in the world frame. The synchronizer and detector are modular components, while the linear Kalman filter, world transformation and pose tracker stay invariant.

The sensor message synchronizer is responsible for the time-based synchronization of the given sensor message streams. In the case of blob detection, a module that synchronizes images and organized pointclouds from an RGB-D camera is derived. This is necessary since the algorithm detects the blob in both 2D image space and 3D point clouds, which are not necessarily sampled simultaneously. The underlying implementation uses ROS libraries to synchronize messages with an approximate time policy.

An object detector attempts to find a set of object poses using synchronized sensor data. The module used in this paper detects blob poses and is implemented in the following way. First, all the blob positions and radii are found in the image frame using the standard blob detection functionality found in the OpenCV libraries. Second, the depth information corresponding to the detected blobs is isolated from the organized point cloud. Finally, blob positions are calculated as centroids of the corresponding depth positions, while the orientation is obtained through the random sample consensus (RANSAC) algorithm from the Point Cloud Library (PCL).

The remaining framework components are independent from synchronizer and detector modules. The pose tracker is used to track the obtained object's poses through multiple frames based on the closest Euclidean distance criterion. This component solves the issue of multiple objects being visible, as it always outputs the pose of the currently tracked object. Moreover, it increases the robustness of the system since it remembers the object poses for a certain number of frames, which allows some leniency with the detector.

The goal of the world transformation component is to transform the tracked pose from the sensor to the world frame using the estimated odometry from an external source that any UAV should have access to. Additionally, since the blob poses are to be sent as references to the trajectory planner, it is important to correctly compute the blob orientation. Since the blob is a flat surface, there are two equally correct possible orientations that can be detected. Therefore, the blob orientation is chosen as follows:

$$
R_{\text{blob}} = \begin{cases} R_{\text{blob}} & \text{if } \mathbf{r}_{1B} \cdot \mathbf{r}_{1\text{blob}} \geq 0 \\ R_{\text{blob}} \cdot R_{180} & \text{otherwise} \end{cases},
\tag{22}
$$

where $\mathbf{r}_{1B}$ is the heading component of the UAV rotation matrix expressed in world coordinates $R_B = \begin{bmatrix} \mathbf{r}_{1B} & \mathbf{r}_{2B} & \mathbf{r}_{3B} \end{bmatrix}$ and $\mathbf{r}_{1\text{blob}}$ is the heading component of the blob rotation matrix expressed in world coordinates $R_{\text{blob}} = \begin{bmatrix} \mathbf{r}_{1\text{blob}} & \mathbf{r}_{2\text{blob}} & \mathbf{r}_{3\text{blob}} \end{bmatrix}$ and $R_{180} = diag(-1, -1, 1)$.

Finally, a linear Kalman filter with a constant velocity model is used to further increase the robustness of the system and provide smoother blob position estimates. The constant velocity model for each axis is given as follows:

$$
\mathbf{x}_{k+1} = F_k \mathbf{x}_k + \mathbf{w}_k, \qquad F_k = \begin{bmatrix} 1 & T_s \\ 0 & 1 \end{bmatrix},
\tag{23}
$$

where T_s is the discretization step, $\mathbf{x}_k \in \mathbb{R}^2$ is the state vector containing the position and velocity along the corresponding axis and $\mathbf{w}_k \in \mathbb{R}^2$ is the process noise. The observation model along a single axis is given as follows:

$$
z_k = H_x \mathbf{x}_k + v_k, \qquad H_k = \begin{bmatrix} 1 & 0 \end{bmatrix},
\tag{24}
$$

where $z_k \in \mathbb{R}$ is the position observation along the corresponding axis and $v_k \in \mathbb{R}$ is the measurement noise.

If the detector is unable to provide measurements and the pose tracker removes the pose from the tracking set, the linear Kalman filter is still able to provide blob position estimates.

Experimental validation of the described methods is performed in an indoor Optitrack environment with an Intel Realsense D435 RGB-D camera. To ensure ground truth is available for detection validation, reflective markers are attached to both the camera and the blob. In order to determine the transformation between the camera optical frame and the reflective markers attached to the camera, an optimization-based calibration approach is used as described in [32].

Results are shown in Figures 7 and 8. The experiments are performed with the UAV in constant motion while looking at the general direction of the painted blob. Figure 7 shows a relative difference between the ground truth UAV motion in the world frame and the UAV motion as observed from the detected blob frame. Figure 8 presents the comparison of ground truth and detected blob positions expressed in the world frame. It is important to note that camera calibration errors can manifest themselves as static offsets between the detected and ground truth blob positions in Figure 8. However, in this case, the visual detection provided a reliable blob tracking results which is a direct consequence of careful camera calibration.

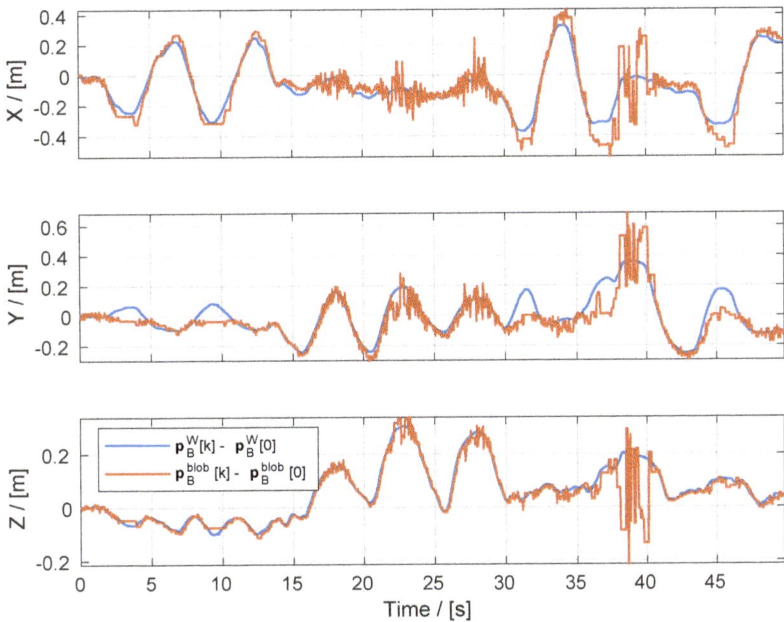

Figure 7. This figure shows the comparison of normalized motion of the UAV body frame as observed from the Optitrack world frame and as observed from the detected blob frame labeled $\mathbf{p}_B^W[k] - \mathbf{p}_B^W[0]$ and $\mathbf{p}_B^{blob}[k] - \mathbf{p}_B^{blob}[0]$, respectively.

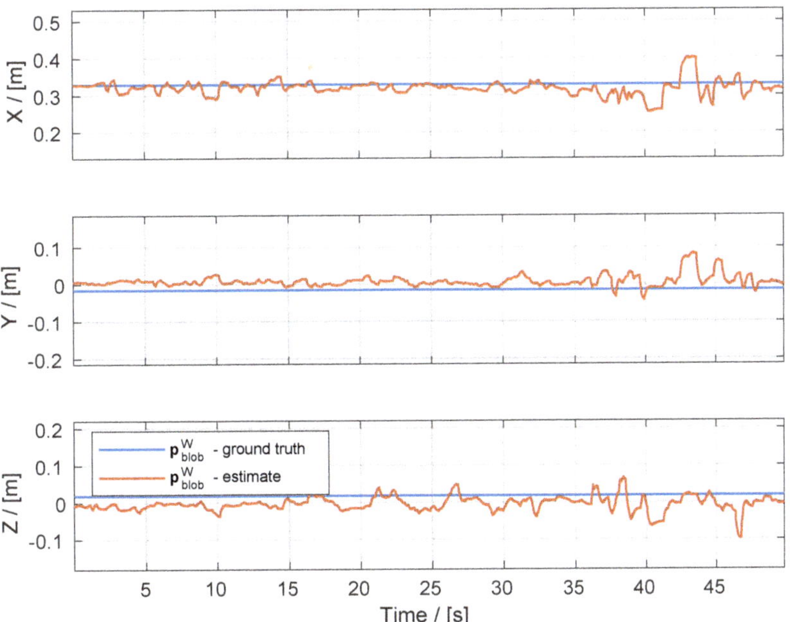

Figure 8. This figure shows the comparison of the ground truth blob position and the blob position estimate expressed in the Optitrack world frame. The root mean squared error across all three axes is 4.6 cm.

7. Simulation

The environment used for simulating the UAV and manipulator dynamics, as well as the contact with the environment, is the widely accepted Gazebo simulator. It is realistic and highly modular, with a large community and a support for the robot operating system (ROS), which is also the primary implementation environment for impedance control, motion planning and blob detection. Through ROS, Gazebo has a large variety of developed plugins realistically simulating various sensors and actuators. All simulations were conducted with Linux Ubuntu 18.04 operating system and ROS Melodic middleware installed.

The UAV is modeled as a single rigid body with n_p propellers mounted at the end of each arm. As propulsion units, these propellers generate thrust along the z axis of the UAV body. To simulate the propeller dynamics, the `rotors_simulator` package is used. It contains a plugin that models thrust based on the user-provided propeller parameters [33]. Furthermore, to obtain the UAV attitude and position, IMU and odometry plugins are mounted on the vehicle. The manipulator was mounted on the body of the UAV and consists of three joints connected with links. A rod type tool is mounted as the end-effector, with a force-torque sensor required by the impedance controller. Furthermore, a monocular camera with an infrared projector is also mounted for the blob detection.

7.1. End-Effector Motion Distribution Analysis

Given some end-effector configuration, the inverse kinematics is responsible for finding the UAV position and yaw angle, as well as the manipulator joint values that satisfy the desired configuration. The parameter β from Equation (3) defines a ratio of how much the manipulator joints and UAV position and orientation contribute to achieving the desired end-effector configuration, as described in Section 3.1. Recalling the values, $\beta = 1$ only moves the UAV in the direction of the desired end-effector configuration; and $\beta = 0$ uses the inverse kinematics of the manipulator to achieve the desired configuration.

To determine the influence of β on the overall system, an analysis was conducted with different β values. The desired end-effector configuration was chosen to be in contact with a plane perpendicular to the bridge wall which required the force reference along the x axis. The waypoints for the trajectory planner were kept the same across all trials, and only β was changed. The results of this analysis are depicted in Figure 9. As can be observed, all trials produced very similar results with oscillating force upon contact and eventually reaching the desired reference, providing us with no obvious conclusion regarding how to select the optimal β. However, following the dexterity analysis from Section 3.1, and only relying on the manipulator motion might drive the system close to its limits due to the UAV body movement. On the other hand, the motion of the UAV induces disturbances in the end-effector pose control. The manipulator is therefore responsible for compensating errors introduced by the motion of the UAV body. Taking all of the aforementioned into account, the value is chosen as $\beta = 0.5$ so that both the manipulator and the UAV are simultaneously used to maintain a steady contact force.

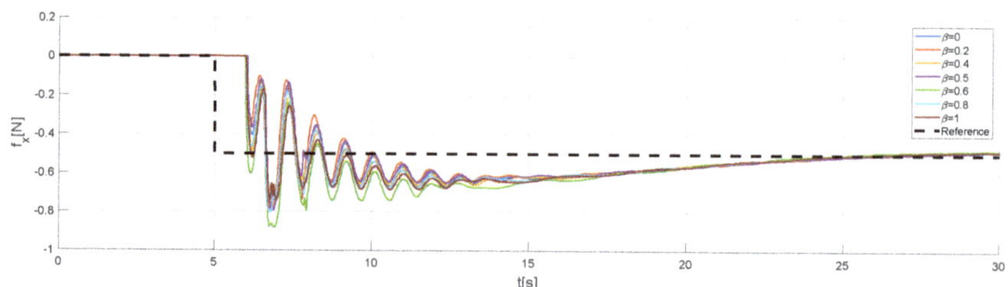

Figure 9. Force response comparison for different values of the parameter β. The analysis was conducted on a plane perpendicular to the ground where the force reference along the x axis is required.

7.2. Bridge Sensor Mounting

Since the concept of this paper was to mount inspection sensors on a bridge, the simulation trials were tailored in the same direction. After spraying the first component, it is necessary to achieve and maintain a stable contact while the second adhesive component on the sensor dries. Since the manipulator is attached above the propellers, the workspace of the manipulator is limited to contact above the UAV or on the plane perpendicular to the ground.

Naturally, the first set of simulation trials were conducted by holding the desired force on a plane perpendicular to the ground. In this case, the contact force only acts along the x axis and the response is depicted in Figure 10. The time delay between the planned and executed contact is present due to the impedance filter which slows down the dynamics of the referent trajectory. After the initial contact, there are some oscillations and an overshoot which diminish over time and the desired force reference is achieved.

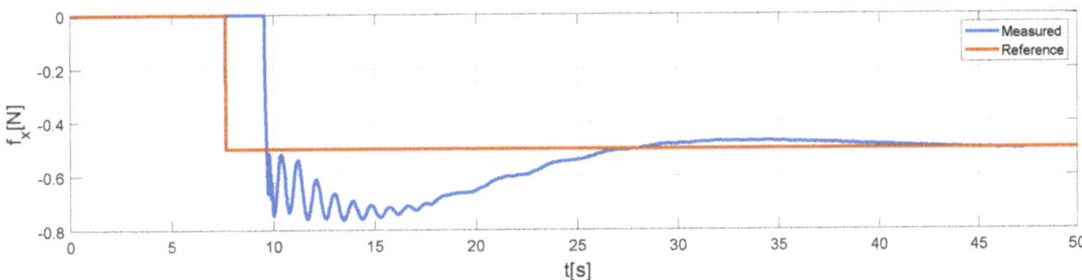

Figure 10. Force response in case of a contact plane perpendicular to the ground, $\delta = 0°$.

The second set of simulation trials included an inclined contact plane. This requires the UAV approach from below the plane and achieving contact perpendicular to the plane. Since the plane is inclined for $\delta = 68°$, the planned force referent values have components in both the x and z axes, as shown in Figure 11. Similarly to the previous example, the force response has some oscillations around the instance of contact, but it eventually settles and reaches the desired force reference.

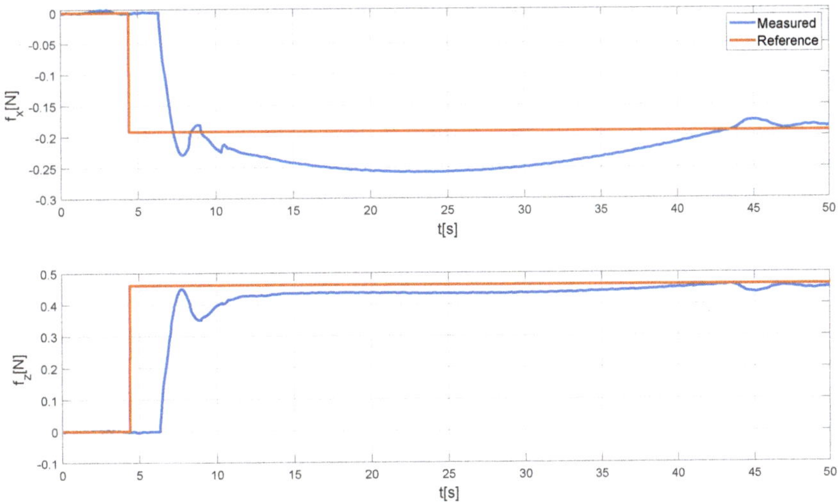

Figure 11. Force response in the case of the contact plane inclined for $\delta = 68°$.

The simulation tests for $\delta = 0°$ and $\delta = 68°$ were performed $n = 10$ times for each case, as depicted in Figure 12. The left portion of the figure is a dot product between the normal of the blob $\mathbf{r}_t$ and the end-effector orientation vector $\mathbf{r}_{ee}$. If the value of the dot product $\mathbf{r}_t \cdot \mathbf{r}_{ee} = 1$, the two vectors are parallel which results in a successful approach. For both angles, the dot product is very close to 1 and the orientation error is negligible. On the right, the distance between the center of the target and contact point is shown. The error distance is in both cases less than 0.1 m, which ensures the relatively high precision of sensor mounting, well within margins for the bridge inspection. The accompanying video of simulation tests can be found on our YouTube channel [34].

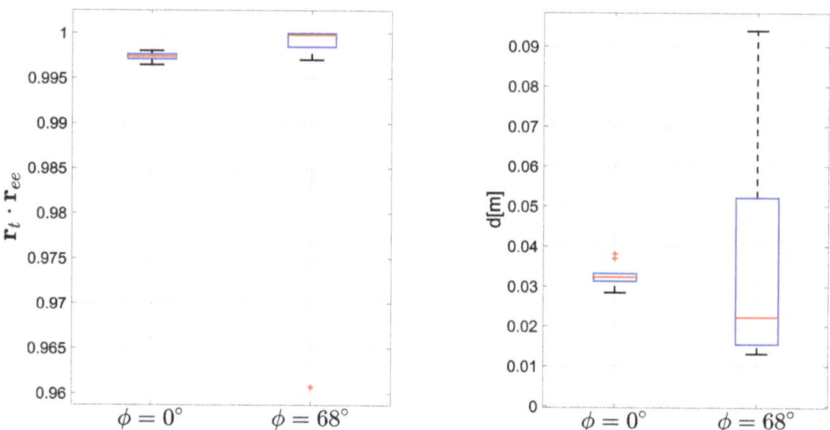

Figure 12. (**Left**): box and whiskers plot of the dot product between the blob plane normal vector $\mathbf{r}_t$ and the end-effector orientation vector $\mathbf{r}_{ee}$; and (**Right**): box and whiskers plot of the distance from the target point after contact.

8. Conclusions

This paper presents a step towards autonomous bridge inspection by investigating the possibility of mounting various inspection sensors using an aerial manipulator. Currently, inspectors use specialized trucks with cranes and baskets in order to access the area underneath the bridge. This inevitably leads to road closure which poses an inconvenience for both inspectors and traffic. To alleviate this problem, the aforementioned aerial manipulators can be used to access difficult-to-reach areas of the bridge. As mounting sensors require forming a bond between the wall and sensor, we envision using a two-component adhesive with a short cure time. Since the aerial manipulator has to achieve and maintain contact with the sensor mount point, short cure times are desirable because of te limited flight time of these platforms. Nevertheless, current flight times of outdoor multirotors reach up to 30 min, which ensures enough time for the two adhesive components to form the bond.

Although preliminary, the results of this paper seem promising. The visual detection was extensively tested and reliably tracks the blob position. The adaptive impedance controller is capable of maintaining the required force. Even though there are some oscillations and settling times in the force response, in practical use, it does not make much difference since the curing time of the adhesive is at least several minutes. The trajectory planner was augmented to plan in the force space which allows for setting the force reference step change before the contact. The simulation results show the high repeatability of the overall system which gives us the confidence to perform experiments in a real-world environment.

Our first step in future work was to perform experiments in a controlled laboratory environment. The outdoor environment poses a different set of challenges including the lower accuracy positioning system and unpredictable disturbances, i.e., wind gusts. Since these factors will inevitably reflect on the overall end-effector accuracy, we are looking into augmenting the manipulator to be able to compensate for lateral movements, as well as roll and yaw angles. To further increase the system's accuracy, the developed visual tracker will be used to improve feedback around the tracked blob on the bridge wall in real-world experiments.

Author Contributions: All authors contributed equally to this work. Conceptualization, M.O. and I.D.; methodology, A.I., L.M. and M.C.; software, A.I., L.M. and M.C.; investigation, A.I.; writing—original draft preparation, A.I.; writing—review and editing, L.M., I.D. and M.O.; supervision, I.D. and M.O.; project administration, I.D. and M.O.; funding acquisition I.D. and M.O. All authors have read and agreed to the published version of the manuscript.

Funding: This work was supported by the European Commission Horizon 2020 Programme through the project under G.A. number 820434, named Energy-Aware BIM Cloud Platform in a Cost-Effective Building Renovation Context—ENCORE. Furthermore, this research was part of the scientific project Autonomous System for Assessment and Prediction of Infrastructure Integrity (ASAP), financed by the European Union through the European Regional Development Fund—The Competitiveness and Cohesion Operational Programme (KK.01.1.1.04.0041).

Institutional Review Board Statement: Not applicable.

Informed Consent Statement: Not applicable.

Data Availability Statement: Not applicable.

Conflicts of Interest: The authors declare no conflict of interest.

Appendix A

Here, we show detailed proof that the adaptation law (17) is stable. Assuming that THE referent force $f_r(t)$ is constant, $f_r(t) = F_r$, the time derivatives of Equation (16) are:

$$\dot{x}_r(t) = \dot{\kappa}(t)F_r + \dot{x}_e(t),$$

$$\ddot{x}_r(t) = \ddot{\kappa}(t)F_r + \ddot{x}_e(t),$$

$$\text{(A1)}$$

while the derivatives of Equation (13) yield:

$$\dot{x}(t) = \frac{-\dot{e}(t)}{k_e} + \dot{x}_e(t),$$

$$\ddot{x}(t) = \frac{-\ddot{e}(t)}{k_e} + \ddot{x}_e(t).$$

$$\text{(A2)}$$

By susbstituting Equations (A1) and (A2) into Equation (10), the dynamics of the contact force error can be obtained as

$$m\ddot{e}(t) + b\dot{e}(t) + (k + k_e)e(t) = g(t), \tag{A3}$$

where:

$$g(t) = F_r[k(1 - k_e\kappa(t)) - k_e(b\dot{\kappa}(t) + m\ddot{\kappa}(t))], \tag{A4}$$

and $x = x_c$. The adaptation law thus determines the dynamics of the adaptation parameter $\kappa(t)$ and defines the dynamics of the contact force error. Formally, the adaptation law should enforce $g(t) \rightarrow g^*$ such that $e(t) \rightarrow 0$ and $\kappa(t) \rightarrow 1/k_e$.

For the Lyapunov candidate:

$$V(t) = \frac{1}{2}\left[p_1 e(t)^2 + p_2 \dot{e}^2(t)\right] + \frac{1}{2\gamma}[g(t) - g^*]^2, \tag{A5}$$

with p_1, p_2 and γ as positive parameters, the condition $\dot{V}(t) \leq 0$ yields:

$$\frac{2}{\gamma}g(t)\dot{g}(t) + 2g(t)[p_1 e(t) + p_2 \dot{e}(t)] - \frac{2}{\gamma}\dot{g}(t)g^* \leq 0. \tag{A6}$$

After reordering, we obtain:

$$\dot{g}(t)[g(t) - g^*] \leq -\gamma\sigma(t)g(t), \tag{A7}$$

where $\sigma(t) = [p_1 e(t) + p_2 \dot{e}(t)]$. By choosing:

$$\dot{g}(t) = -\gamma\sigma(t) + \gamma_d\dot{\sigma}(t), \tag{A8}$$

where γ_d is a positive constant, Lyapunov condition Equation (A7) becomes:

$$g(t) \leq g^*, \tag{A9}$$

i.e., for the adaptation law to be stable, $g(t)$ should be bounded. Since $x_r, \dot{x}_r$ and $\ddot{x}_r$ are bounded, so are $e, \dot{e}$ and $\ddot{e}$. Therefore, $g(t)$ is also bounded, i.e., the condition in Equation (A9) is satisfied. The adaptation law is finally obtained by taking the derivative of Equation (A3), and substituting $\dot{g}(t)$ with Equation (A8), yields the (17). Parameters γ and γ_d dictate the adaptation dynamics. Based on the measured contact force, the error adaptation law Equation (17) estimates the adaptation parameter κ (reciprocal value of the environment stiffness), which is then used in Equation (16) for calculating the referent position x_r.

References

1. Kashif Ur Rehman, S.; Ibrahim, Z.; Memon, S.A.; Jameel, M. Nondestructive test methods for concrete bridges: A review. *Constr. Build. Mater.* **2016**, *107*, 58–86. [CrossRef]
2. Hunt, G.; Mitzalis, F.; Alhinai, T.; Hooper, P.A.; Kovac, M. 3D printing with flying robots. In Proceedings of the 2014 IEEE International Conference on Robotics and Automation (ICRA), Hong Kong, China, 31 May–7 June 2014; pp. 4493–4499. [CrossRef]
3. Vempati, A.S.; Kamel, M.; Stilinovic, N.; Zhang, Q.; Reusser, D.; Sa, I.; Nieto, J.; Siegwart, R.; Beardsley, P. PaintCopter: An Autonomous UAV for Spray Painting on Three-Dimensional Surfaces. *IEEE Robot. Autom. Lett.* **2018**, *3*, 2862–2869. [CrossRef]
4. Jeong, E.; Seo, J.; Wacker, J. Literature Review and Technical Survey on Bridge Inspection Using Unmanned Aerial Vehicles. *J. Perform. Constr. Facil.* **2020**, *34*, 04020113. [CrossRef]
5. Cacace, J.; Orozco-Soto, S.M.; Suarez, A.; Caballero, A.; Orsag, M.; Bogdan, S.; Vasiljevic, G.; Ebeid, E.; Rodriguez, J.A.A.; Ollero, A. Safe Local Aerial Manipulation for the Installation of Devices on Power Lines: AERIAL-CORE First Year Results and Designs. *Appl. Sci.* **2021**, *11*, 6220. [CrossRef]
6. Zollini, S.; Alicandro, M.; Dominici, D.; Quaresima, R.; Giallonardo, M. UAV Photogrammetry for Concrete Bridge Inspection Using Object-Based Image Analysis (OBIA). *Remote Sens.* **2020**, *12*, 3180. [CrossRef]
7. Bolourian, N.; Hammad, A. LiDAR-equipped UAV path planning considering potential locations of defects for bridge inspection. *Autom. Constr.* **2020**, *117*, 103250. [CrossRef]
8. Reagan, D.; Sabato, A.; Niezrecki, C. Unmanned aerial vehicle acquisition of three-dimensional digital image correlation measurements for structural health monitoring of bridges. In *Proceedings of the Nondestructive Characterization and Monitoring of Advanced Materials, Aerospace, and Civil Infrastructure, Portland, OR, USA, 25–29 March 2017*; Wu, H.F., Gyekenyesi, A.L., Shull, P.J., Yu, T.Y., Eds.; International Society for Optics and Photonics, SPIE: Bellingham, DC, USA, 2017; Volume 10169, pp. 68–77. [CrossRef]
9. Chen, H.; Quan, F.; Fang, L.; Zhang, S. Aerial Grasping with a Lightweight Manipulator Based on Multi-Objective Optimization and Visual Compensation. *Sensors* **2019**, *19*, 4253. [CrossRef] [PubMed]
10. Zhang, G.; He, Y.; Dai, B.; Gu, F.; Yang, L.; Han, J.; Liu, G. Aerial Grasping of an Object in the Strong Wind: Robust Control of an Aerial Manipulator. *Appl. Sci.* **2019**, *9*, 2230. [CrossRef]
11. Sanchez-Cuevas, P.J.; Gonzalez-Morgado, A.; Cortes, N.; Gayango, D.B.; Jimenez-Cano, A.E.; Ollero, A.; Heredia, G. Fully-Actuated Aerial Manipulator for Infrastructure Contact Inspection: Design, Modeling, Localization, and Control. *Sensors* **2020**, *20*, 4708. [CrossRef] [PubMed]
12. Baca, T.; Penicka, R.; Stepan, P.; Petrlik, M.; Spurny, V.; Hert, D.; Saska, M. Autonomous Cooperative Wall Building by a Team of Unmanned Aerial Vehicles in the MBZIRC 2020 Competition. *arXiv* **2020**, arXiv:2012.05946.
13. Vatavuk, I.; Polic, M.; Hrabar, I.; Petric, F.; Orsag, M.; Bogdan, S. Autonomous, Mobile Manipulation in a Wall-building Scenario: Team LARICS at MBZIRC 2020. *J. Field Robot.* **2021**, submitted for publication.
14. Barišić, A.; Petric, F.; Bogdan, S. Brain over Brawn—Using a Stereo Camera to Detect, Track and Intercept a Faster UAV by Reconstructing Its Trajectory. *arXiv* **2021**, arXiv:2107.00962.
15. Markovic, L.; Car, M.; Orsag, M.; Bogdan, S. Adaptive stiffness estimation impedance control for achieving sustained contact in aerial manipulation. In Proceedings of the 2021 IEEE International Conference on Robotics and Automation, Xi'an, China, 30 May–5 June 2021.

16. Cazzato, D.; Cimarelli, C.; Sanchez-Lopez, J.L.; Voos, H.; Leo, M. A Survey of Computer Vision Methods for 2D Object Detection from Unmanned Aerial Vehicles. *J. Imaging* **2020**, *6*, 78. [CrossRef]

17. Fischler, M.A.; Bolles, R.C. Random Sample Consensus: A Paradigm for Model Fitting with Applications to Image Analysis and Automated Cartography. *Commun. ACM* **1981**, *24*, 381–395. [CrossRef]

18. Fumagalli, M.; Naldi, R.; Macchelli, A.; Carloni, R.; Stramigioli, S.; Marconi, L. Modeling and control of a flying robot for contact inspection. In Proceedings of the 2012 IEEE/RSJ International Conference on Intelligent Robots and Systems, Algarve, Portugal, 7–12 October 2012; pp. 3532–3537. [CrossRef]

19. Alexis, K.; Darivianakis, G.; Burri, M.; Siegwart, R. Aerial Robotic Contact-based Inspection: Planning and Control. *Auton. Robot.* **2016**, *40*, 631–655. [CrossRef]

20. Jimenez-Cano, A.E.; Sanchez-Cuevas, P.J.; Grau, P.; Ollero, A.; Heredia, G. Contact-Based Bridge Inspection Multirotors: Design, Modeling, and Control Considering the Ceiling Effect. *IEEE Robot. Autom. Lett.* **2019**, *4*, 3561–3568. [CrossRef]

21. Cataldi, E.; Muscio, G.; Trujillo, M.A.; Rodriguez, Y.; Pierri, F.; Antonelli, G.; Caccavale, F.; Viguria, A.; Chiaverini, S.; Ollero, A. Impedance Control of an aerial-manipulator: Preliminary results. In Proceedings of the 2016 IEEE/RSJ International Conference on Intelligent Robots and Systems (IROS), Daejeon, Korea, 9–14 October 2016; pp. 3848–3853. [CrossRef]

22. Nava, G.; Sablé, Q.; Tognon, M.; Pucci, D.; Franchi, A. Direct Force Feedback Control and Online Multi-Task Optimization for Aerial Manipulators. *IEEE Robot. Autom. Lett.* **2020**, *5*, 331–338. [CrossRef]

23. Meng, X.; He, Y.; Li, Q.; Gu, F.; Yang, L.; Yan, T.; Han, J. Contact Force Control of an Aerial Manipulator in Pressing an Emergency Switch Process. In Proceedings of the 2018 IEEE/RSJ International Conference on Intelligent Robots and Systems (IROS), Madrid, Spain, 1–5 October 2018; pp. 2107–2113. [CrossRef]

24. Stilman, M. Global Manipulation Planning in Robot Joint Space With Task Constraints. *IEEE Trans. Robot.* **2010**, *26*, 576–584. [CrossRef]

25. Cefalo, M.; Oriolo, G. Task-constrained motion planning for underactuated robots. In Proceedings of the 2015 IEEE International Conference on Robotics and Automation (ICRA), Seattle, WA, USA, 26–30 May 2015; pp. 2965–2970. [CrossRef]

26. Yu, Y.; Lippiello, V. 6D Pose Task Trajectory Tracking for a Class of 3D Aerial Manipulator From Differential Flatness. *IEEE Access* **2019**, *7*, 52257–52265. [CrossRef]

27. Tognon, M.; Cataldi, E.; Chavez, H.A.T.; Antonelli, G.; Cortés, J.; Franchi, A. Control-Aware Motion Planning for Task-Constrained Aerial Manipulation. *IEEE Robot. Autom. Lett.* **2018**, *3*, 2478–2484. [CrossRef]

28. Ivanovic, A.; Car, M.; Orsag, M.; Bogdan, S. Exploiting Null Space in Aerial Manipulation through Model-In-The-Loop Motion Planning. In Proceedings of the 2020 International Conference on Unmanned Aircraft Systems (ICUAS), Athens, Greece, 1–4 September 2020; pp. 686–693. [CrossRef]

29. Zargarbashi, S.; Khan, W.; Angeles, J. The Jacobian condition number as a dexterity index in 6R machining robots. *Robot. Comput.-Integr. Manuf.* **2012**, *28*, 694–699. [CrossRef]

30. Car, M.; Ivanovic, A.; Orsag, M.; Bogdan, S. Position-based adaptive impedance control for a UAV. In Proceedings of the 2018 International Conference on Unmanned Aircraft Systems (ICUAS), Dallas, TX, USA 12–15 June 2018; pp. 957–963. [CrossRef]

31. Pham, H.; Pham, Q. A New Approach to Time-Optimal Path Parameterization Based on Reachability Analysis. *IEEE Trans. Robot.* **2018**, *34*, 645–659. [CrossRef]

32. Maric, B.; Polic, M.; Tabak, T.; Orsag, M. Unsupervised optimization approach to in situ calibration of collaborative human–robot interaction tools. In Proceedings of the 2020 IEEE International Conference on Multisensor Fusion and Integration for Intelligent Systems (MFI), Karlsruhe, Germany, 14–16 September 2020; pp. 255–262. [CrossRef]

33. Furrer, F.; Burri, M.; Achtelik, M.; Siegwart, R. Robot Operating System (ROS): The Complete Reference (Volume 1). In *Studies in Computational Intelligence*; Chapter RotorS—A Modular Gazebo MAV Simulator Framework; Springer International Publishing: Cham, Switzerland, 2016; pp. 595–625. [CrossRef]

34. LARICS. Aerial Manipulation Approach for Autonomous Bridge Inspection. Available online: https://youtu.be/bBaxPMZEV04 (accessed on 23 July 2021).

 *applied sciences*

Article

Aerial Robotic Solution for Detailed Inspection of Viaducts

Rafael Caballero [1,*], Jesús Parra [1], Miguel Ángel Trujillo [1,*], Francisco J. Pérez-Grau [1], Antidio Viguria [1] and Aníbal Ollero [2]

1   Advanced Center for Aerospace Technologies (FADA-CATEC), 41309 Sevilla, Spain; jparra@catec.aero (J.P.); fjperez@catec.aero (F.J.P.-G.); aviguria@catec.aero (A.V.)
2   GRVC Robotics Labs, University of Seville, 41092 Sevilla, Spain; aollero@us.es
*   Correspondence: rcaballero@catec.aero (R.C.); matrujillo@catec.aero (M.Á.T.)

Abstract: The inspection of public infrastructure, such as viaducts and bridges, is crucial for their proper maintenance given the heavy use of many of them. Current inspection techniques are very costly and manual, requiring highly qualified personnel and involving many risks. This article presents a novel solution for the detailed inspection of viaducts using aerial robotic platforms. The system provides a highly automated visual inspection platform that does not rely on GPS and could even fly underneath the infrastructure. Unlike commercially available solutions, our system automatically references the inspection to a global coordinate system usable throughout the lifespan of the infrastructure. In addition, the system includes another aerial platform with a robotic arm to make contact inspections of detected defects, thus providing information that cannot be obtained only with images. Both aerial robotic platforms feature flexibility in the choice of camera or contact measurement sensors as the situation requires. The system was validated by performing inspection flights on real viaducts.

Keywords: inspection; maintenance; UAV; aerial robotics; aerial robotic manipulation; viaduct; LIDAR; photogrammetry; contact

Citation: Caballero, R.; Parra, J.; Trujillo, M.Á.; Pérez-Grau, F.J.; Viguria, A.; Ollero, A. Aerial Robotic Solution for Detailed Inspection of Viaducts. *Appl. Sci.* **2021**, *11*, 8404. https://doi.org/10.3390/app 11188404

Academic Editor: Yosoon Choi

Received: 6 August 2021
Accepted: 7 September 2021
Published: 10 September 2021

Publisher's Note: MDPI stays neutral with regard to jurisdictional claims in published maps and institutional affiliations.

1. Introduction

The inspection of viaducts and bridges is a very time-consuming and resource-intensive activity. It requires heavy involvement from highly qualified and specifically trained personnel. Additionally, these inspections pose health and safety risks that are mainly derived from working at height and the difficulty of the operation. Current inspection methodologies involve the use of climbing operators, who, by means of ropes, hang from the structure and perform the measurements required by the inspectors to evaluate its current state (see Figure 1a). These works present many potential accident risks due to the difficulty and technical level required to access certain complicated areas at heights, the possible physical fatigue of the workers, human errors in the safety of the operation, or even problems with the use of specific measuring tools.

An alternative method is the use of heavy machinery, like cherry pickers, truck-mounted lifts, and cranes (see Figure 1b). This machinery requires new specialized personnel to operate it and perform inspections and does not eliminate the problem of having to expose people to work at heights. The surfaces to be inspected are usually located at a high altitude, and under it, several types of obstacles could be found, such as traffic of vehicles or trains, water flows, or rough terrain. This means that the use of machinery has to be done from the top surface of the structure, interrupting its service and increasing the operational costs of the inspection.

Currently, the highly qualified staff required for inspections is normally composed of civil engineers working for engineering firms specialized in structures. Every little deformation, crack, or defect can be the cause of a potentially bigger critical problem, so they must be identified as soon as possible. An example that requires high accuracy is the measurement of cracks width with an error smaller than 0.1 mm. Moreover, crack depth can only be measured using contact sensors. However, visual inspection is the most extended way of assessing the preliminary status of the viaduct before deciding if specialized equipment is needed. This means that, in order to find the smallest defects, the inspector must be very close to them. Furthermore, human subjectivity or the lack of experience could lead to an undervaluation of the severity of a defect.

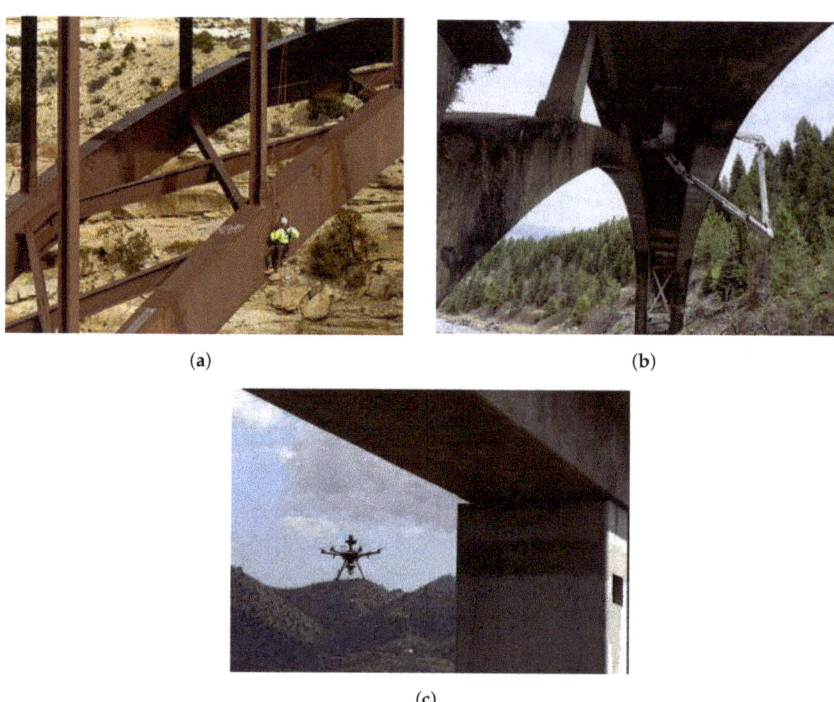

(a)

(b)

(c)

Figure 1. Examples of current inspection methodologies. (**a**) Current method for the viaduct inspection with rope access, from Ayres Associates, Inc. [1]. (**b**) Current method for the viaduct inspection using specialized machinery, from Forsgren Associates, Inc. [2]. (**c**) Proposed new methodology using the flight platform AERO-CAM in Álora, Málaga (Spain) performing an inspection under a viaduct deck.

This whole process presents many inefficiencies that can be minimized using aerial robotic technologies combined with computer vision algorithms (e.g., artificial intelligence) and other computerized technologies supporting the post-processing of the acquired data. This article proposes an aerial robotic solution for gathering all the needed data to analyze the status of a viaduct. Our proposed solution drastically improves the safety of the inspections, as it does not involve the work at height of any human being or the use of heavy machinery. It reduces inspection times and costs by reducing the number of specialized people required to perform the inspection and avoiding the need to interrupt the use of the structure. On the other hand, it improves the quality of the data obtained, since it is the inspectors themselves who indicate the points to be inspected by the aerial robotic platforms, which can always be quickly sent back to obtain more information if necessary. In addition, the system is flexible to use the different sensors needed: either cameras or sensors that require physical contact with the structure.

1.1. Aerial Robots for Inspection

The use of autonomous unmanned aerial vehicles (UAVs) to capture images and their subsequent use in infrastructure analysis is currently on the rise. After an exhaustive search of the projects developed in the research world, relevant applications have been found that make use of autonomous UAVs for image capture in the inspection of railway tracks [3], where autonomous flights are carried out to follow the tracks, capturing images for subsequent analysis. Other applications, like mining inspection obtained 3D maps [4], make it possible to evaluate the earthworks carried out; wind turbines inspection for evaluation of deformations or damage [5]; inspection of civil infrastructures capturing images and their subsequent analysis through the use of neural networks to identify possible cracks or landslides [6]. Finally, in [7], bridge inspection is studied in a similar way as proposed in this article, where images are captured autonomously using a UAV platform and then analyzed using photogrammetry software. The main limitation raised is that they use Global Navigation Satellite System (GNSS) positioning to automate flights, while in the system proposed in this article, the aircraft that captures images does not navigate with GNSS but with onboard sensors such as LIDAR. The physical characteristics of the viaducts cause the GNSS signal to be partially or totally degraded when flying near or under them. These degradation problems with global positioning signals are discussed in [8], where the most critical ones affecting this article are signal masking and multipath. Therefore, the degradation of this signal leads to localization problems in the UAV in which it can drift its flight or even make sudden changes in its positioning, seriously compromising the safety and integrity of the operation. In addition, GNSS poses problems of repeatability of inspection operations, since the number and position of available satellites varies over time.

There are currently some commercial systems for infrastructure inspection such as the one offered by Skydio [9], which makes use of several onboard cameras to navigate and perform the inspection autonomously. Unlike Skydio's system, our system is able to reference the inspection to a global coordinate system that can be used from the construction of the viaduct until the end of its life. This feature makes our methodology better suited to the current workflow of inspectors, who already use total stations to check the displacements of structures against global references defined during construction. Additionally, thanks to the global coordinate system, the solution offers the possibility for more than one aircraft to navigate and perform an inspection while maintaining the same references. By using a generic gimbal, our system is much more flexible than Skydio's system in the choice and configuration of the required camera sensor, as it is not limited to the built-in camera. In addition, our system uses an aerial robot that performs inspection by making physical contact with the structure. This provides information that cannot be acquired by pictures exclusively, such as precise measurements of crack depth and width, material hardness, concrete humidity, etc.

In relation to this inspection by contact with an aerial robot, there are different lines of research with a number of projects that focus on maintaining stability during physical contact [10,11]. In [12] the predecessor aerial manipulator of the one used in this work is presented, which was patented [13] and awarded with the EU Radar Innovation Award 2017 [14]. In [15], an aerial vehicle that operates overhead using a rigid arm, and that is even capable of keeping in contact with the surface [16], is presented. In [17] an aerial vehicle that operates at the bottom as well as at the front, is presented. Aerial manipulators capable of operating in either direction are presented in [18]. Stable contact operations have been achieved using a pusher trirotor in [19] or, a quadrotor in [20,21]. In [22] a long rigid tool exerting force against a surface is applied.

1.2. Aerial Robots Localization

Different sensors can be used to achieve effective positioning of aerial robots. Total stations can be used to localize an UAV in motion with respect to a reference system, as in [23]. This presents several problems, such as the high cost of this tool, as well as the dependence on flying continuously in line of sight with it or having an uninterrupted wireless communication system with the aerial robot. In addition, total stations only provide position information, not orientation.

On the other hand, the localization problem can also be solved through monocular or stereo visual cameras, as in [24,25]. The problem with visual cameras is that they are totally dependent on external light conditions. This is especially problematic when navigating near infrastructure because of the shadows and light changes it can cause. Other localization systems use a LIDAR as the main sensor [26]. These sensors do not depend on external light conditions, as they use lasers to measure distances. While their application in autonomous driving is on the rise, their use in aerial robots is still limited because they require more available payload and a more powerful onboard computer than cameras. In [27], a comparison of several algorithms applied to aerial robots is made.

1.3. Article Introduction

This article proposes a novel solution for the detailed inspection of viaducts using aerial robotics. This solution is an alternative to current inspection methodologies, improving safety, costs, time, and data quality. Given the amount of details involved in this solution, this article first provides a general overview and then focuses on the technical and experimental aspects of the visual inspection.

The rest of the article is organized as follows. Section 2 describes the proposed viaduct inspection system and the two aircraft involved. Section 3 presents a localization solution that provides one of the aircraft with autonomous capabilities to perform a visual inspection of a viaduct. Section 4 outlines the localization and inspection experiments and presents the results used to assess the performance of the proposed solution. Finally, the conclusions and future work are summarized in Section 5.

2. System Description

The proposed viaduct inspection system offers a comprehensive solution to check and to evaluate the condition of these infrastructures through its integrated tools. A workflow has been created that meets the needs to speed up, reduce the cost, and increase the safety of these inspections. All the tasks are carried out with aerial robotic platforms whose characteristics are chosen according to the task to be performed. These tasks can be general and/or detailed photographs or make physical contact with the structure to take measurements with sensors as required.

The workflow is shown in Figure 2 and is as follows. Given a viaduct of interest on which an inspection should be performed, the proposed workflow begins with the creation of a mission. During this phase, it is required to acquire a 3D map of the structure in which the inspector can select the areas and points of interest. This 3D map is not only useful for the creation and subsequent visualization of the mission, but it is also required for the global localization of the aircraft. It is obtained with the help of a robotic total station that establishes an arbitrary coordinate system and performs scans to obtain the 3D points around it. It is important to capture these data from different points of view to obtain a complete point cloud of the viaduct. To facilitate the subsequent use of this map, the reference system should be aligned with the *ENU* axes (**x** = East, **y** = North, **z** = Top). If possible, it is desirable to obtain an approximate GNSS coordinate of the origin of the point cloud to locate it globally. Otherwise, this can be done manually. The use of the total station is then limited to one time only. Once the map has been created, it can be reused in all subsequent inspections, provided that the viaduct has not suffered significant changes.

With the points and areas of interest selected on the 3D map created, this information is sent to the aerial platform, which translates it to its local coordinate system and creates the route of waypoints and actions necessary to carry out the inspection autonomously capturing overlapping pictures. This mission is then a first general visual inspection of the structure to locate any possible defect. This check is performed by taking general pictures of the structure with high resolution in an automated way using the visual inspection platform described in Section 2.1.

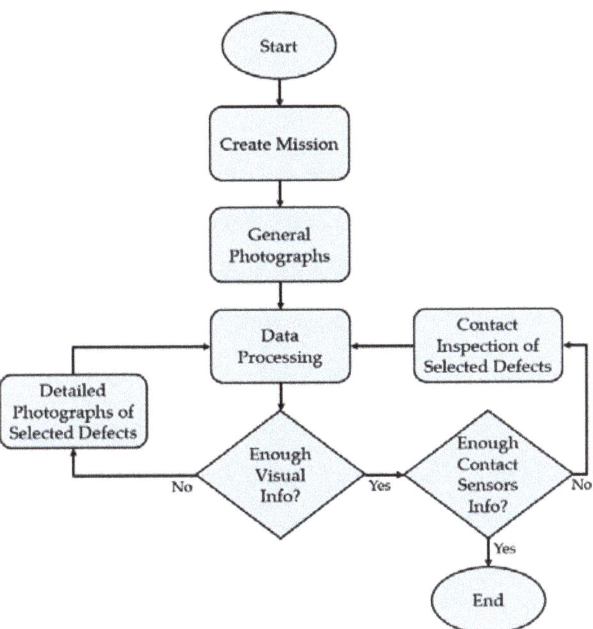

Figure 2. General viaduct inspection workflow.

After the first general pictures are taken, they are analyzed to check the condition of the viaduct, paying special attention to those areas where a defect is suspected. This analysis can be performed manually by an inspector or automatically by applying an automatic image defect detection algorithm, like [28,29]. After the analysis, a decision should be made to determine if more detailed information on the defects found or suspect areas are required. If so, more detailed visual or contact information may be obtained using the visual or contact inspection platforms, respectively. In case more visual information is required, the mission previously created by the inspector can be reused but using another camera configuration that better collects the required information. For example, a different lens can be used for the camera with a longer focal length to obtain better details of the specific area. However, if the previous mission does not meet the requirements of the new visual inspection, the inspector can create a new one with the 3D map and select the previously found defects.

When the missions are finished and the visual information obtained is sufficient, our proposed solution also considers the use of a specific platform for contact inspections. This platform is described in Section 2.2 and has a robotic arm with an end effector on which a sensor can be installed. In case of finding defects in the visual inspection that require a deep analysis with specific sensors that require physical contact with the structure, this platform is sent to those defects and captures data. When the contact inspection data is analyzed, a decision is made as to whether further visual or contact information is required or whether the inspection is terminated.

In short, the proposed workflow is an iterative process in which one can always return to a suspect area to obtain more detailed information. All the inspections are carried out by aerial robots specifically designed for each purpose. The viaduct inspection system comprises two platforms that work sequentially as described previously. The following sections describe these UAVs, showing their configuration and capabilities.

2.1. Visual Inspection

The visual inspection UAV (see Figure 3) is known as AERO-CAM. This aircraft is specialized in taking very high quality images of a structure. This UAV is equipped with a stabilized camera which takes the images of desired areas.

Figure 3. AERO-CAM robot for visual inspection UAV.

The AERO-CAM platform is built from a DJI Matrice 600 Pro on which the necessary components to operate have been installed. Since the DJI is a commercial platform, the system is easily replicable. The standard configuration of the UAV has been preserved, with both the autopilot and the rotors and blades being those recommended by DJI. The autopilot includes a GPS/GNSS receiver, a 9-axis IMU, a magnetometer, and a barometric altimeter. In addition, it carries a Lightware Laser Altimeter for precision landing. Regarding the camera system, the UAV is equipped with a Gremsy T3V2 gimbal [30] mounted in the slot available above the platform and carrying a Sony Alpha 7 camera [31]. This mounting location allows the gimbal-camera set to have a better available field of view and can even take pictures pointing completely upwards, as opposed to mounting it on the bottom of the UAV like most commercial camera drone systems. This is especially useful when performing inspections under a viaduct as the UAV will be able to take pictures of the bottom part of the deck. Depending on the space available for the flight and the amount of detail to be obtained in each image, the camera can be equipped with different lenses. This camera is managed by a Raspberry Pi Model 3B+ that implements a software developed using using Sony's Camera Remote SDK [32].

To provide the platform with autonomous capabilities, the UAV mounts an Ouster OS0-128 LIDAR sensor [33] under the avionics with a custom anti-vibration structure. All previously mentioned sensors are connected together with the autopilot to the onboard computer, which is an Intel NUCi7. Finally, an Ubiquiti Rocket M5 is used for ground communications and to connect the Rapsberry Pi and Intel NUC via Wi-Fi.

Both onboard computers run Ubuntu 18.04 and ROS Melodic and have their clocks synchronized for greater accuracy in capturing images with metadata. The software of the platform is programmed as nodes that communicate with each other. Figure 4 shows the scheme of processes that operate in the system.

With all this equipment, the AERO-CAM is able to perform completely autonomous visual inspections even in GNSS-denied environments. It is capable of taking off and landing on its own, as well as carrying out the mission created from the 3D map of the viaduct. These missions are composed of many waypoints that have an image associated with them. Each time the UAV reaches a waypoint, it moves the gimbal and captures the corresponding image autonomously.

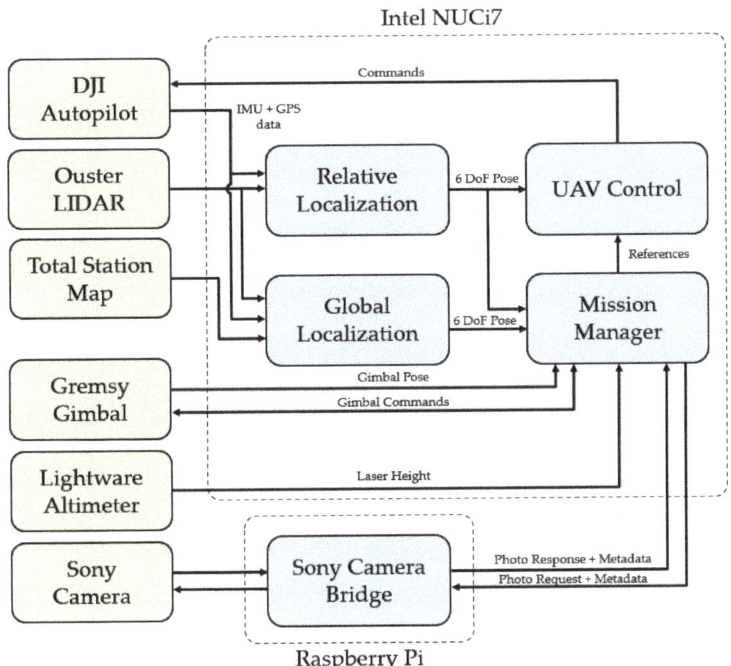

Figure 4. AERO-CAM general software architecture.

2.2. Contact Inspection

The aerial contact inspection robot is named AeroX [12] (see Figure 5). It is a specialized aircraft capable of contacting static surfaces. This UAV is composed of two different platforms: the aerial platform and the Robotic Mobile Contact Platform (RMCP), which will be in charge of the Ultrasonic Testing (UT) inspection for measuring the cracks' depth. The RMCP is attached at the end of the contact device of the aerial platform.

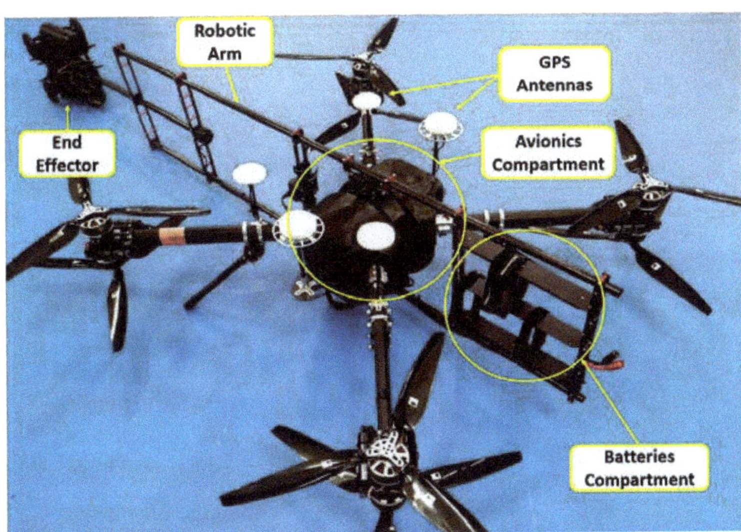

Figure 5. AeroX robot for contact inspection.

AeroX is a novel aerial robotic manipulator that performs physical contact inspection with unprecedented capabilities. It is composed of a robotic vehicle, a six degree-of-freedom (DoF) robotic arm, and a robotic end-effector equipped with wheels and inspection sensors. AeroX has a semi-autonomous operation, which provides interesting advantages in contact inspection. In the free-flight mode, the pilot guides the robot until performing contact with its end-effector on the surface to be inspected. During contact, AeroX is in its fully-autonomous GNSS-free contact-flight mode, in which the robot keeps its relative position with respect to the surface contact point using only its internal sensors. During autonomous flight, the inspector—with uninterrupted contact—can move the end-effector on the surface to accurately select the points to be inspected with sensors that require to be in contact with or very close to the surface.

The AeroX controller is able to efficiently compensate perturbations thanks to its design, which transmits the surface contact forces and perturbations to the robot center of mass and allows small movements of the aerial part of the robot in every DoF to absorb other perturbations such as wind. AeroX adopts a 4 coaxial rotor configuration and a simple and efficient design which provides high stability, maneuverability, and robustness to rotor failure. It can perform contact inspection on surfaces at any orientation, including vertical, inclined, horizontal top or horizontal bottom, and its operation can be easily integrated into current maintenance operations in many industries.

Although AeroX is part of the proposed solution for viaduct inspection, the technical and experimental development of this article focuses on the AERO-CAM localization algorithms. For more information about AeroX, please refer to [12].

3. Localization Solution

The proposed solution for the visual inspection of the viaduct requires the creation of a previous 3D map using a total station. This map will be a point cloud that identifies the reference coordinate origin for the entire inspection system. To create this map, operators should ensure that the *ENU* coordinate system is followed. This map can be reused in future inspections of the viaduct.

The UAV system has its own localization and navigation algorithm that provides the transform $\{T_{LD}\}$ whose origin is the take-off point $\{L\}$. Since this location may vary, the complete system requires a second localization system that establishes the 3D transformation, $\{T_{GL}\}$, between the initial UAV pose and the global reference system, $\{G\}$, expressed in *ENU* coordinates at the origin of the map created by the total station. These transforms can be visualized in Figure 6.

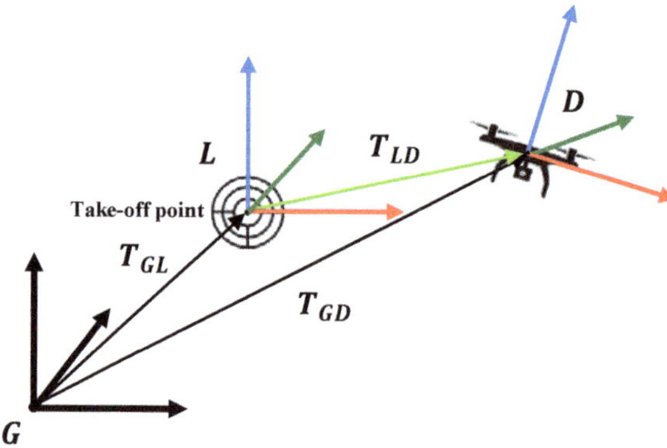

Figure 6. Full transformation system.

Therefore, the AERO-CAM platform has two parallel localization processes to perform the automatic inspection of the viaducts. The following subsections explain the details of both processes.

3.1. Global Localization System

The function of the global localization system is to find the transform, $\{T_{GL}\}$, which establishes the connection between the global reference system of the viaduct 3D map and the UAV localization system as expressed in Equation (1). Finding this transform is crucial, as it will allow the aerial robot to safely navigate to those areas of interest selected by the inspector without maintaining the same take-off position between flights. This process eliminates total station dependency after the initial 3D map has been created/acquired. In addition, since the viaduct can be found in an inaccessible area, the take-off position may not be replicated between flights. This can occur even on inspections on different days where changing terrain or weather conditions make it impossible to replicate the take-off position accurately.

$$T_{GL} = T_{GD} * T_{LD}^{-1} \tag{1}$$

$$T_{GD} = T_{GL} * T_{LD} \tag{2}$$

This global localization system is designed to calculate the transform at the start of each mission, just before the aerial robot takes off. Therefore, the transform, $\{T_{GL}\}$, is fixed and will only vary during the flight if another transform with better accuracy has been obtained. During flight, this system continues to calculate the transform between the UAV's current position and the base map, $\{T_{GD}\}$, as expressed in Equation (2), so that if the accuracy of the transform improves, it gets updated. This last case can be also visualized in Figure 6. This in-flight update is only applied with the confirmation of the inspector on the ground, who personally checks whether the mean square error calculated by the global positioning algorithm is better.

The global localization can be executed on a ground computer asynchronously, since this calculation need not be instantaneous. In this case, the onboard computer sends the data to the ground computer, which performs the calculations and sends the results back to the aerial robot. This update has no direct impact on the relative localization of the UAV, as this is not affected by the change of $\{T_{GL}\}$. This update has no direct impact on the ongoing flight of the UAV, as its relative localization and control are not affected. However, the mission waypoints, which are referenced to $\{G\}$, are updated in the onboard computer. Therefore, the UAV changes its target points to more accurate ones.

To find the correspondence between the 3D map generated by the total station and the data from the onboard sensors, we apply an algorithm that makes use of the geometric characteristics of the point clouds. Firstly, the point clouds are preprocessed to filter out the sparse data noise by applying a filter that removes outliers if the number of neighbors in a given radius (e.g., 0.1 m) is smaller than a given number, typically 15. Secondly, the algorithm performs a distributed downsampling by applying a voxel grid filter and tries to remove the ground points. The process of eliminating the ground points is carried out by creating a parametrizable grid of squares that is filled with the z-value of the lowest point within each square. For each square, all points with z-values between the minimum and a given threshold (1.5 m) are removed. Then, the algorithm calculates the FPFH (Fast Point Feature Histogram) descriptors [34] of the remaining distributed points. These features encode the geometric properties of the k-nearest neighbors of a given point using the average curvature of the multidimensional histogram around that point. Among its advantages, these features are invariant in position and a certain level of noise. After this feature extraction process, the Random Sample Consensus (RANSAC) algorithm is applied to find a first approximation between both inputs. The result is then corrected according to the problem-specific assumptions outlined below and refined via the Iterative Closest Point (ICP) algorithm. These correction and refinement steps are applied twice to further adjust the result. They are only refined twice as doing it more has not shown a substantial improvement of the result but an increase of computational load and computational time. Depending on whether the initial guess is reliable and if, at the instant of processing, the UAV is close to the structure, the RANSAC stage can be exchanged for the ICP algorithm directly to obtain better results. To identify these refinement stages, they are named ICP1, ICP2, and ICP3, with ICP1 being the one that can be exchanged for RANSAC, as explained before. Figure 7 shows a block diagram of the main steps of the algorithm.

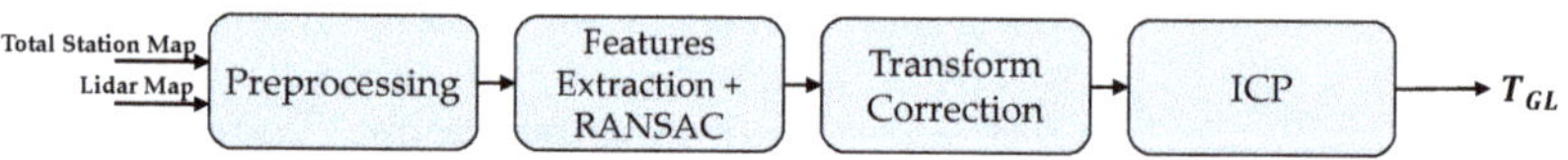

Figure 7. Global localization algorithm.

To improve this process, some assumptions are made that simplify the problem and work in all possible scenarios:

- The total station 3D map is filtered manually so that it only contains information about the viaduct. This prevents unwanted objects from appearing.
- The total station map's origin and the UAV's take-off point are aligned with the horizon, so they both have 0 degrees in pitch and roll.
- The absolute yaw orientation of the UAV is known thanks to an onboard magnetometer. Therefore, the approximate rotation matrix between the LIDAR and the total station map is known beforehand.
- The GNSS position of the total station map's origin is known approximately. If the drone has GNSS coverage at the take-off point, an initial guess for its position can also be obtained.

The introduction of the above assumptions mainly corrects the orientation before starting the alignment process, thus reducing the problem to almost pure translation. In addition, even minimal GNSS coverage at the take-off point provides an initial guess that makes the problem converge more accurately and faster. In case no GNSS coverage is available, the approximate coordinate of the take-off point with respect to the total station 3D map can be entered manually.

In addition to the above assumptions, in case of significant changes between reality and the reference map obtained with the total station due to catastrophes or severe structural failures, the discordant areas of the reference map should be removed. Alternatively, a map of the new state of the structure can be created with the same reference origin as the previous one.

3.2. Relative Localization System

The purpose of the relative localization system is to find the transform $\{T_{LD}\}$, which describes the motion of the aerial robot from its take-off point. This take-off point will be located near the viaduct, on a flat surface parallel to the horizon so that the UAV can take off safely. This localization is performed using only current readings from the onboard sensors and does not require any prior data. It is desirable that this localization is as accurate as possible and minimizes drift over time, as much as possible, to avoid a significant divergence between the UAV's perceived and actual poses.

The relative localization system makes use of the LIDAR and a 9-axis IMU to calculate the UAV's pose at each instant. The algorithm operates at high frequency in real time, updating the pose at the same frequency as the IMU, which in the case of AERO-CAM is 400 Hz. The LIDAR is set to an operating frequency of 10 Hz. This algorithm is executed entirely onboard the aerial robot in the equipped Intel NUC. Despite running in real time, this algorithm has the highest processing load among the programs executed. It is of vital importance to the system, as it provides localization feedback to the UAV control algorithm, so that it can ensure a stable flight while navigating autonomously to the desired target points. The localization algorithm is based on LIO-SAM [35] and its general architecture is adapted to AERO-CAM as shown in Figure 8.

This architecture establishes a tightly coupled fusion between the LIDAR and the IMU, building a factor graph in which the measurements made by the sensors are integrated to build and optimize the map, as shown in Figure 9. The factor graph is optimized using smoothing and mapping a Bayes tree with iSAM2 [36]. The IMU pre-integration is based on [37]. Since the double integration of IMU measurements leads to large drift, the architecture proposes its short-term integration instead, correcting its bias thanks to the localization at lower frequency in the built map using the information of the LIDAR point cloud. In order to process everything in real time, the algorithm discards LIDAR readings if they are not sufficiently displaced (typically 1 m and 0.2 radians) with respect to the previous reading (known as LIDAR keyframes). In this way, a lot of redundant information that would otherwise increase the computational load is discarded. Between LIDAR keyframes, the IMU readings are integrated, converging in a node of the graph that would be the state of the location at that given instant. Unlike the original algorithm, the adaptation for AERO-CAM does not introduce GPS/GNSS factors since the signal quality is totally impaired during the inspection flight due to the structure itself. Another difference with the original algorithm is that the loop closure option is disabled to avoid possible jumps in the odometry. The main reason is that this odometry is used to close the control loop so as to avoid as many peaks and spikes as possible, as it is safety critical to smooth the flight near the viaduct during the inspection. However, this particularization of the algorithm can lead to larger drifts in the calculated odometry. To overcome this problem, inspection flights are assumed to have a controlled duration with a planned route close to the viaduct, thus providing a rich point cloud which will help to minimize drift.

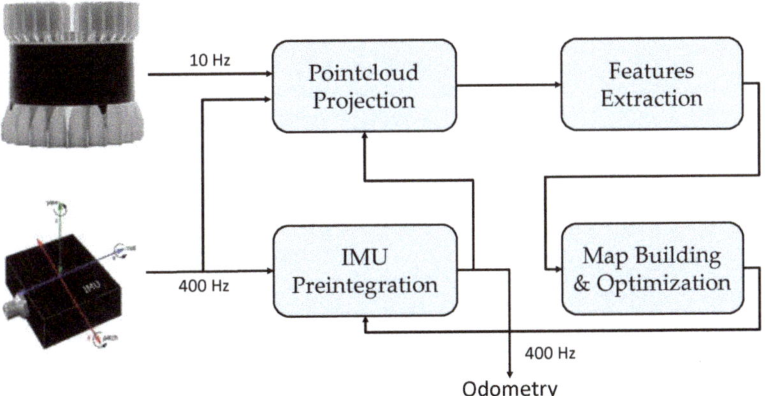

Figure 8. LIO-SAM adapted architecture.

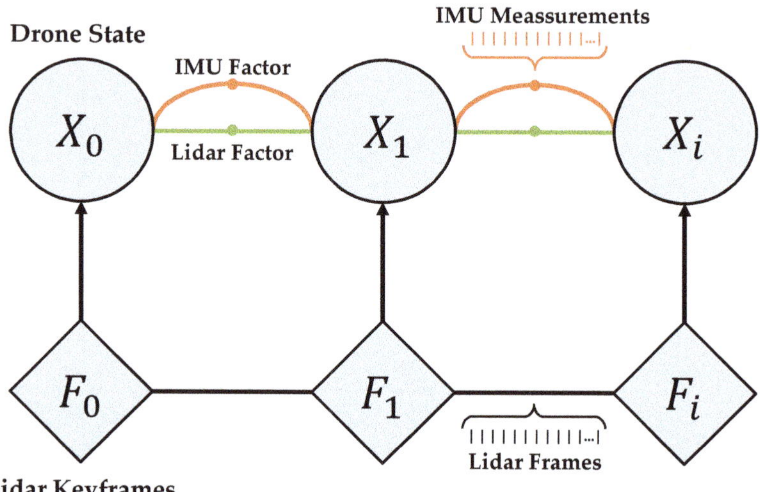

Figure 9. LIO-SAM adapted factor graph.

As already mentioned, the result of all this processing is the relative localization of the aerial robot with a high frequency (400 Hz) that serves the control algorithm to proceed with the AERO-CAM. This publication does not intend to go into the details of the original LIO-SAM implementation. For more details, please refer to [35].

4. Experimental Results

The experimentation phase of this article was carried out with the AERO-CAM platform, performing the various experiments described next. Real flights around and under two viaducts were performed to evaluate the localization solution. On the one hand, there is the railway viaduct *Arroyo del Espinazo* in Álora, Málaga (Spain). This viaduct was inaugurated in 2006 and is currently in use. It has a length of 1.2 km with a maximum pillar height of 93 m and a width of 14 m. The pillars are equidistantly distributed and have a hollow square cross-section. On the other hand, there is the road viaduct *Puente de las Navas* in Algodonales, Cádiz (Spain). This viaduct was built in the 1980s and is still active with the A-384 road passing over it; therefore, it withstands daily traffic. It is approximately 350 m long and consists of cylindrical pillars supporting in pairs three longitudinal beams on which the deck rests. Both viaducts are in a good state of conservation, presenting small aesthetic defects in the concrete during the inspections without danger. The utility of the experimental inspections is focused on predictive maintenance, being able to return in the future to carry out the same inspection and compare the evolution. The flights in Álora were pilot assisted in order to perform realistic routes while those in Algodonales were fully autonomous. A preliminary map of the viaducts was created using a Leica Nova MS50 total station.

The trajectory followed in these experiments consists of a take-off close to the viaduct and a flight inspection of different areas that may include changes in altitude. Figure 10 shows some of these trajectories. In order to obtain the ground truth of the trajectory followed by the platform, a prism was installed on it and, using the Leica total station, the position with respect to the origin of the viaduct map was tracked. Note that the total station only provides position data, as it cannot estimate orientation. This position data is provided at 20 Hz.

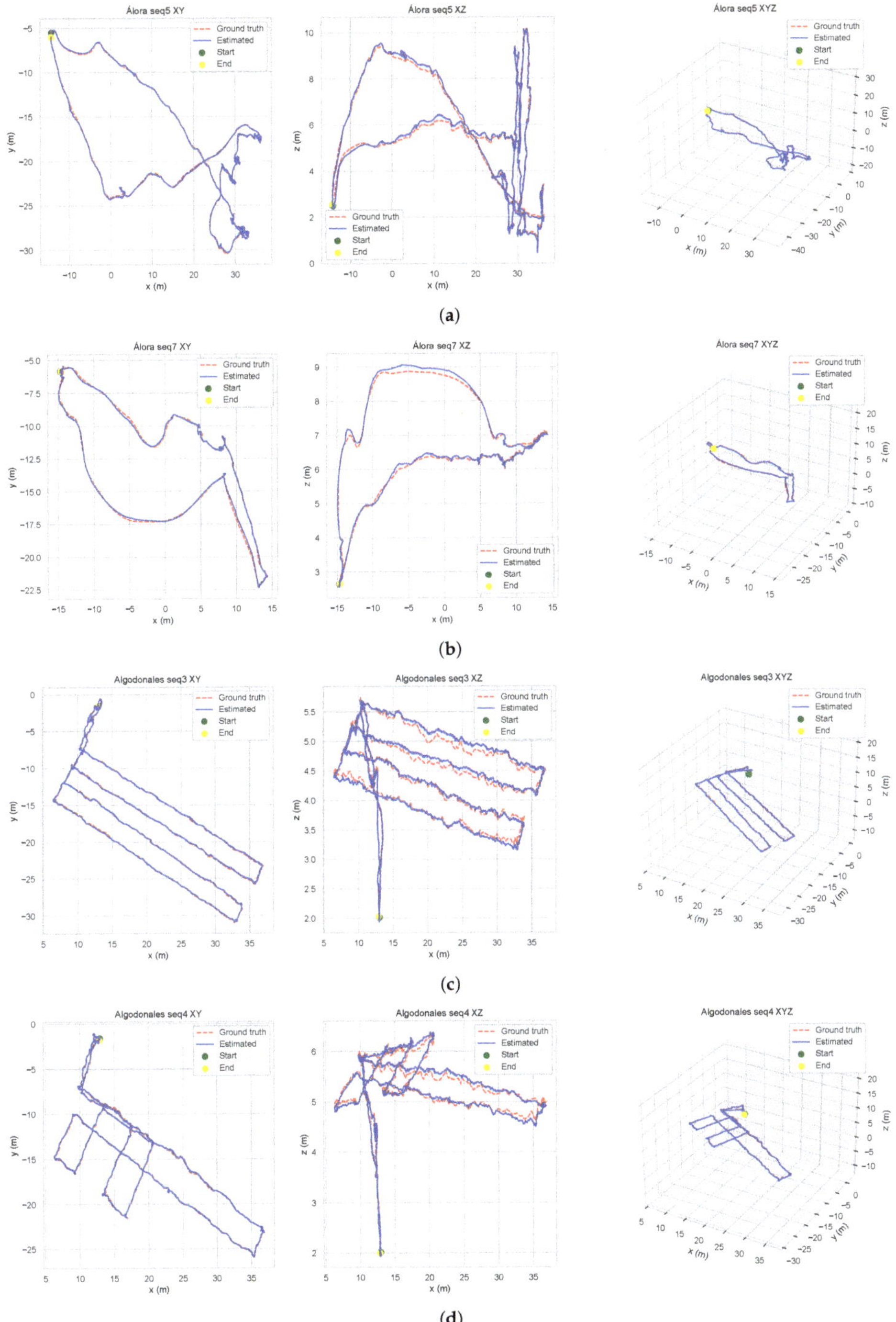

Figure 10. Examples of trajectories followed from take-off to landing. The maneuvers were performed under a deck of the Algodonales and Álora viaducts. The graphs on the left illustrate X-Y, center X-Z and right X-Y-Z. (**a**) Alora sequence 5, (**b**) Alora sequence 7, (**c**) Algodonales sequence 3, (**d**) Algodonales sequence 4.

4.1. Global Localization

The experiments to test the global localization algorithm consisted of extracting LIDAR readings from the aerial platform for different time instants and inputting them for computation. These instants include moments before take-off and during flight and landing. To illustrate this process, Figure 11 shows two alignment examples at take-off.

On the one hand, the initial alignment of the total station 3D map and the LIDAR data readouts are shown, taking into account the assumptions introduced in Section 3.1. On the other hand, both point clouds aligned with the algorithm results are shown.

During the experimentation, the execution of the global localization algorithms was carried out on a laptop with a 4-core Intel Core i7-8564U CPU and 8 GB of RAM. The execution and convergence time of the algorithm varied between 12 and 25 s for each resulting transform. This duration is not a problem since the first iteration is performed before take-off. The rest of them can be performed during the flight and update the transform at convenience, as explained in Section 3.1.

The performance evaluation of the global localization algorithm is carried out by studying the mean error between matches after each ICP step. All tested cases converge to a valid solution. Tables 1 and 2 show the metrics obtained for Algodonales and Álora, respectively. As explained in Section 3, the ICP1 step is not always executed, so it is denoted in the tables as "*" when there is no data. The so called "Proportional Correspondence" metrics (Prop. Corr.) shows the number of correspondences in that ICP stage divided by the size of the LIDAR point cloud at that instant in time. The MSE metrics represents the mean square error of the correspondences after applying the transform obtained at that stage.

Figure 11. Global localization examples at take-off. Red points are the LIDAR reading. (a) Algodonales, (b) Álora.

Table 1. Algodonales global localization metrics.

Seq	ICP1		ICP2		ICP3	
	Prop. Corr.	MSE	Prop. Corr.	MSE	Prop. Corr.	MSE
take-off_0	*	*	0.38	0.1551	0.34	0.0252
middle_0	*	*	0.45	0.3004	0.33	0.0354
ending_0	*	*	0.58	0.2141	0.56	0.0230

Table 2. Álora global localization metrics.

Seq	ICP1		ICP2		ICP3	
	Prop. Corr.	MSE	Prop. Corr.	MSE	Prop. Corr.	MSE
take-off_0	0.98	0.0615	0.98	0.0328	0.96	0.0051
take-off_1	0.97	0.0060	0.97	0.0060	0.97	0.0044
beginning_0	*	*	0.83	0.0370	0.86	0.0082
middle_0	*	*	0.73	0.0312	0.73	0.0087
ending_0	*	*	0.65	0.0283	0.65	0.0107

Additionally, during these instants, the position given by the total station—which has the same reference system as the 3D map—was obtained. This position serves as a ground truth to check the output of the global localization algorithm, since the output of this algorithm should correspond with the reading of the total station. Again, since the total station does not provide orientation, only the transitional part is considered. Tables 3 and 4 show the obtained results.

Table 3. Comparative table between the ground truth obtained by the total station and the output of the global location algorithm for the Algodonales viaduct.

Seq	Ground Truth [m]			Algorithm Estimation [m]			Distance [m]			
	x	y	z	x	y	z	Δx	Δy	Δz	Total
take-off_0	−9.30	−27.64	0.51	−9.20	−27.59	0.34	−0.10	−0.05	0.17	0.20
middle_0	−0.29	−27.75	2.29	−0.38	−27.63	2.12	0.09	−0.12	0.17	0.23
ending_0	14.73	−31.02	2.39	14.70	−31.18	2.27	0.03	0.16	0.12	0.20

Table 4. Comparative table between the ground truth obtained by the total station and the output of the global location algorithm for the Álora viaduct.

Seq	Ground Truth [m]			Algorithm Estimation [m]			Distance [m]			
	x	y	z	x	y	z	Δx	Δy	Δz	Total
take-off_0	−13.99	−5.20	2.92	−14.10	−5.35	2.37	0.11	0.15	0.55	0.58
take-off_1	−10.24	−7.39	4.97	−10.59	−7.37	4.43	0.35	−0.02	0.54	0.64
beginning_0	−5.19	−7.75	5.13	−5.46	−7.87	4.72	0.27	0.12	0.41	0.51
middle_0	4.62	−9.74	5.59	4.60	−9.78	5.10	0.02	0.04	0.49	0.49
ending_0	23.45	−20.41	5.24	23.37	−20.35	4.76	0.08	−0.06	0.48	0.49

The results are considered valid since the algorithm is able to converge correctly in the proposed realistic cases. The advantage of the global localization system is that, if run before take-off, the operator can visually validate the obtained result and proceed with the inspection if there is no problem. Tables 3 and 4 show how the final 3D error is between 0.2 and 0.64 m for the tested cases, the z-axis (vertical) being the most affected. The results are considered good since the uncertainty of the point clouds and of the algorithm itself must be taken into account. While the total station error is in the order of millimeters (always proportional to the distance), the LIDAR points have an error of ±1.5–5 cm (both errors according to the manufacturers), which may influence the result.

4.2. Relative Localization

In the experiments to test the relative localization, the position data estimated by the algorithm were compared with the ground truth from the total station. It should be taken into account that the latter can only provide positions without orientation, so only the translation part is compared. The comparison between the two sets was made with a time association of the positions and a scale-free alignment was performed with the Umeyama algorithm [38]. The EVO framework [39] was used to facilitate this task. The metrics used are the APE (*Absolute Position Error*), to evaluate the global consistency, and the RPE (*Relative Position Error*), to evaluate the local one. For the RPE, an increment of 0.5 m was selected for the calculations. For both metrics, data such as maximum peak, mean, standard deviation, and RMSE (*Root Mean Square Error*) were obtained. Tables 5 and 6 show the results.

Table 5. Algodonales dataset description and localization errors.

Seq	Description		APE [m]				RPE [m] ($\Delta = 0.5$ m)			
	Dist [m]	Time [s]	Max	Mean	Std	RMSE	Max	Mean	Std	RMSE
seq0	175.65	376.7	0.49	0.11	0.07	0.13	1.1	0.25	0.2	0.32
seq1	227.92	397.85	0.45	0.13	0.06	0.14	0.94	0.4	0.2	0.45
seq2	182.02	419.7	0.36	0.12	0.05	0.13	1.0	0.54	0.27	0.61
seq3	276.3	526.14	1.04	0.36	0.16	0.4	1.62	0.56	0.25	0.61
seq4	251.91	515.85	0.52	0.12	0.06	0.14	0.99	0.54	0.23	0.59

Table 6. Álora dataset description and localization errors.

Seq	Description		APE [m]				RPE [m] ($\Delta = 0.5$ m)			
	Dist [m]	Time [s]	Max	Mean	Std	RMSE	Max	Mean	Std	RMSE
seq0	93.18	260.52	0.44	0.12	0.08	0.14	0.49	0.16	0.1	0.19
seq1	204.47	486.57	0.85	0.14	0.09	0.17	1.35	0.16	0.08	0.18
seq2	119.83	334.97	0.38	0.12	0.06	0.13	0.35	0.13	0.06	0.14
seq4	280.93	547.72	0.66	0.17	0.08	0.18	10.36	0.18	0.31	0.36
seq5	324.76	644.37	0.73	0.16	0.09	0.18	0.48	0.17	0.11	0.2
seq6	108.96	211.71	0.79	0.13	0.12	0.18	0.57	0.23	0.12	0.26
seq7	113.4	160.51	0.75	0.18	0.13	0.22	0.69	0.25	0.11	0.27

4.3. Inspection Result

The results obtained after performing a mission show the tracking of the planned trajectory by taking images of the viaduct. Each image obtained saves metadata containing the exact pose and instant in which they were taken with respect to the 3D map of the viaduct. In this way, it is always possible to review the inspection performed and to know the exact location to which each image belongs, with respect to the 3D map. Figure 12 illustrates one of the experiments performed. Specifically, it refers to Algodonales sequence 2, where a flight was performed under the viaduct deck, along the external part of the viaduct. Each vertical arrow indicates the pose of the camera for each acquired image. The green line represents the trajectory of the AERO-CAM.

The pictures shown in Figure 12b,c show an example of the visual information obtained with the AERO-CAM. Both pictures show different parts of the lower part of the viaduct deck, which is a difficult area to access. Figure 12b focuses on one of the cross beams, while picture Figure 12c shows the outer side of the deck. As explained above, the quality and level of detail of these pictures depend on the choice and configuration of the camera as well as the distance to the structure configured in the mission. In this case, an 85 mm lens has been used with a distance to the structure of about 2 m. This results in pictures with a resolution of 9504 × 6336 pixels in which the detail density per pixel is very high.

The distribution of the pictures along the structure is also crucial for a possible later analysis by performing a reconstruction of the structure through photogrammetry. All pictures have enough overlap between them to make this possible. The overlap is not only beneficial for possible photogrammetry but also allows the same point to be analyzed from different pictures, adding more redundancy and information to the system.

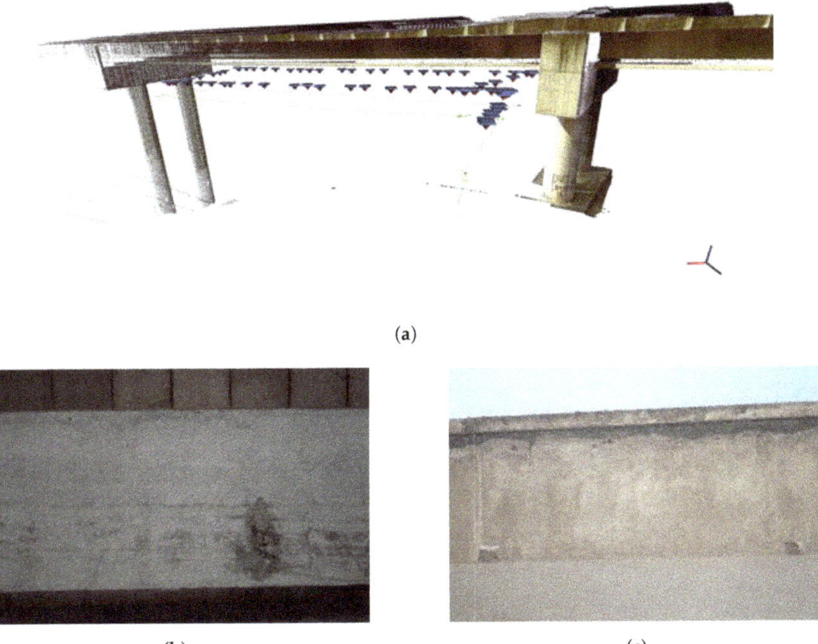

(a)

(b)   (c)

Figure 12. Algodonales sequence 2 inspection result. (**a**) shows the 3D model with the followed route (green) and the location of the acquired pictures (blue planes). (**b,c**) show two examples of the acquired pictures.

5. Conclusions

This work demonstrates that it is possible to perform an inspection of a viaduct with aerial robots as an alternative to current methodologies, saving time and cost and improving the safety and quality of data obtained. The design and development of the AERO-CAM and AeroX inspection platforms was quite successful, since together they cover the sensor needs of a viaduct: either looking for possible defects or analyzing existing ones.

In addition, the provision of autonomous capabilities to the platforms, especially to the AERO-CAM, greatly facilitates the work and provides more flexibility than conventional methods. These capabilities also reduce the number of specialized people needed to operate these platforms, thus improving their safety and speed of use.

Thanks to the choice of a LIDAR-type sensor for the autonomous capabilities of the AERO-CAM, the platform is fully versatile to operate in variable lighting situations, either due to weather conditions or to possible shadows and lighting changes that may be caused by the viaducts themselves. In addition, the system does not rely on a total station for the flight of the robotic platforms. Likewise, the camera installed in its gimbal can be configured to adapt to the different level of detail required or even be replaced by another one without having to redesign the platform.

As future developments, although the localization presented in this article provides good and sufficient results in terms of accuracy and calculation speed, the AERO-CAM does not consider taking off autonomously from areas where its LIDAR cannot see the structure, as the global localization would fail. This means that, until the system is able to locate itself, the pilot has to fly the UAV to the viaduct. Therefore, other strategies can be developed to complement the localization system to overcome these edge cases.

On the other hand, another line of work is the inclusion of a detect and avoid system to provide the AERO-CAM with more advanced intelligent capabilities when executing missions. Currently, the system relies on the fact that the mission designed by the inspector is free of obstacles and the waypoints keep a safe distance from the viaduct, but it could make use of the LIDAR readings to detect potential hazards on the route and re-plan it in real time.

Author Contributions: Conceptualization, R.C., M.Á.T., F.J.P.-G., A.V. and A.O.; Investigation, R.C., J.P., M.Á.T. and F.J.P.-G.; Methodology, R.C.; Software, R.C. and J.P.; Validation, R.C. and J.P.; Project administration, M.Á.T., F.J.P.-G., A.V. and A.O.; writing—original draft preparation, R.C.; writing—review and editing, R.C., J.P., M.Á.T., F.J.P.-G., A.V. and A.O. All authors have read and agreed to the published version of the manuscript.

Funding: This work was funded by the RESIST (H2020-2019-769066) and PILOTING (H2020-2020-871542) projects funded by the European Commission.

Institutional Review Board Statement: Not applicable.

Informed Consent Statement: Not applicable.

Data Availability Statement: The data presented in this study are available on request from the corresponding author. The data are not publicly available yet due to ongoing private work.

Acknowledgments: Authors would like to thank David Tejero for his support in the development of the global localization algorithm, Jorge Mariscal for his review of this article and the interesting discussions, and the GRVC from the University of Seville for collaborating with their total station. In addition, the authors would like to thank Ferrovial for allowing and providing access to infrastructure for the experimentation phase.

Conflicts of Interest: The authors declare no conflict of interest.

Abbreviations

The following abbreviations are used in this manuscript:

APE	Absolute Pose Error
DoF	Degree of Freedom
ENU	East North Up
FPFH	Fast Point Feature Histogram
GNSS	Global Navigation Satellite System
GPS	Global Positioning System
ICP	Iterative Closest Point
IMU	Inertial Measurement Unit
LIDAR	Light Detection and Ranging
MDPI	Multidisciplinary Digital Publishing Institute
MSE	Mean Squared Error
RANSAC	Random Sample Consensus
RMCP	Robotic Mobile Contact Platform
RMSE	Root Mean Square Error
RPE	Relative Pose Error
UAV	Unmanned Aerial Vehicle
UT	Ultrasonic Testing

References

1. Ayres Associates, Inc. Bridge Inspection with Rope Access. Available online: https://www.ayresassociates.com/project/eagle-canyon-bridge-inspection (accessed on 26 August 2021).
2. Forsgren Associates, Inc. Viaduct Inspection with a Crane. Available online: https://www.forsgren.com/bridge-management-services-inspections (accessed on 14 July 2021).
3. Saini, A.; Singh, D. DroneRTEF: Development of a novel adaptive framework for railroad track extraction in drone images. *Pattern Anal. Appl.* **2021**, 1–20. [CrossRef]
4. Erenoglu, O. UAV for 3D Morphological Mapping Applications: A Case Study of Koru Mining Site, Canakkale, NW Turkey. *Teh. Vjesn.* **2021**, *28*, 1044–1050. [CrossRef]
5. Khadka, A.; Afshar, A.; Zadeh, M. Strain monitoring of wind turbines using a semi-autonomous drone. *Wind. Eng.* **2021**, 1–12. [CrossRef]

6. Chen, K.; Reichard, G.; Xu, X.; Akanmu, A. Automated crack segmentation in close-range building façade inspection images using deep learning techniques. *J. Build. Eng.* **2021**, *43*, 102913. [CrossRef]
7. Popescu, C.; Mirzazade, A.; Ohlsson, U.; Sas, G. *Bridge Inspections Using Unmanned Aerial Vehicles—A Case Study in Sweden*; Luleå Tekniska Universitet: Luleå, Sweden, 2021.
8. MacGougan, G.; Lachapelle, G.; Nayak, R.; Wang, A. Overview of GNSS signal degradation phenomena. In Proceedings of the International Symposium Kinematic Systems in Geodesy, Geomatics and Navigation, Banff, AB, Canada, 5–8 June 2001.
9. Skydio Inc. Available online: https://www.skydio.com/inspection (accessed on 29 July 2021).
10. Jimenez-Cano, A.E.; Braga, J.; Heredia, G.; Ollero, A. Aerial manipulator for structure inspection by contact from the underside. *IEEE Int. Conf. Intell. Robot. Syst.* **2015**, *2015*, 1879–1884. [CrossRef]
11. Suarez, A.; Heredia, G.; Ollero, A. Lightweight compliant arm with compliant finger for aerial manipulation and inspection. *IEEE Int. Conf. Intell. Robot. Syst.* **2016**, *2016*, 4449–4454. [CrossRef]
12. Trujillo, M.; Martínez-De Dios, J.; Martín, C.; Viguria, A.; Ollero, A. Novel aerial manipulator for accurate and robust industrial NDT contact inspection: A new tool for the oil and gas inspection industry. *Sensors* **2019**, *19*, 1305. [CrossRef] [PubMed]
13. Trujillo, M.A.; Viguria, A.; Marquez, J.C.; Petrus, A.L.; Jimenez, A.; Garcia, J.J.; Ollero, A. Aeronave con Dispositivo de Contacto. EU Patent ES2614994A1, 2 March 2018.
14. EU Innovation Radar Prize. Available online: https://www.innoradar.eu/innoradarprize (accessed on 28 July 2021).
15. Kocer, B.B.; Tjahjowidodo, T.; Pratama, M.; Seet, G.G.L. Inspection-while-flying: An autonomous contact-based nondestructive test using UAV-tools. *Autom. Constr.* **2019**, *106*, 102895. [CrossRef]
16. Sanchez-Cuevas, P.J.; Ramon-Soria, P.; Arrue, B.; Ollero, A.; Heredia, G. Robotic system for inspection by contact of bridge beams using UAVs. *Sensors* **2019**, *19*, 305. [CrossRef]
17. Kamel, M.; Alexis, K.; Siegwart, R. Design and modeling of dexterous aerial manipulator. *IEEE Int. Conf. Intell. Robot. Syst.* **2016**, *2016*, 4870–4876. [CrossRef]
18. Park, S.; Her, J.; Kim, J.; Lee, D. Design, modeling and control of omni-directional aerial robot. *IEEE Int. Conf. Intell. Robot. Syst.* **2016**, *2016*, 1570–1575. [CrossRef]
19. Papachristos, C.; Alexis, K.; Tzes, A. Efficient force exertion for aerial robotic manipulation: Exploiting the thrust-vectoring authority of a tri-tiltrotor UAV. In Proceedings of the IEEE International Conference on Robotics and Automation, Hong Kong, China, 31 May–7 June 2014; pp. 4500–4505. [CrossRef]
20. Darivianakis, G.; Alexis, K.; Burri, M.; Siegwart, R. Hybrid predictive control for aerial robotic physical interaction towards inspection operations. In Proceedings of the IEEE International Conference on Robotics and Automation, Hong Kong, China, 31 May–7 June 2014; pp. 53–58. [CrossRef]
21. González-deSantos, L.M.; Martínez-Sánchez, J.; González-Jorge, H.; Navarro-Medina, F.; Arias, P. UAV payload with collision mitigation for contact inspection. *Autom. Constr.* **2020**, *115*, 103200. [CrossRef]
22. Gioioso, G.; Ryll, M.; Prattichizzo, D.; Bulthoff, H.H.; Franchi, A. Turning a near-hovering controlled quadrotor into a 3D force effector. In Proceedings of the IEEE International Conference on Robotics and Automation, Hong Kong, China, 31 May–7 June 2014; pp. 6278–6284. [CrossRef]
23. Delgado, J.P.; Soria, P.R.; Arrue, B.C.; Ollero, A. Bridge Mapping for Inspection Using an UAV Assisted by a Total Station. *Adv. Intell. Syst. Comput.* **2017**, *694*, 309–319. [CrossRef]
24. Qin, T.; Li, P.; Shen, S. VINS-Mono: A Robust and Versatile Monocular Visual-Inertial State Estimator. *IEEE Trans. Robot.* **2018**, *34*, 1004–1020. [CrossRef]
25. Campos, C.; Elvira, R.; Rodríguez, J.J.; Montiel, J.M.; Tardós, J.D. ORB-SLAM3: An Accurate Open-Source Library for Visual, Visual—Inertial, and Multimap SLAM. *IEEE Trans. Robot.* **2021**. [CrossRef]
26. Zhang, J.; Singh, S. LOAM: Lidar Odometry and Mapping in Real-time. In Proceedings of the Robotics: Science and Systems, Berkeley, CA, USA, 12–16 July 2014.
27. Milijas, R.; Markovic, L.; Ivanovic, A.; Petric, F.; Bogdan, S. A Comparison of LiDAR-based SLAM Systems for Control of Unmanned Aerial Vehicles. In Proceedings of the 2021 International Conference on Unmanned Aircraft Systems (ICUAS), Athens, Greece, 15–18 June 2021; pp. 1148–1154. [CrossRef]
28. Mundt, M.; Majumder, S.; Murali, S.; Panetsos, P.; Ramesh, V. Meta-learning Convolutional Neural Architectures for Multi-target Concrete Defect Classification with the COncrete DEfect BRidge IMage Dataset. *Proc. IEEE Comput. Soc. Conf. Comput. Vis. Pattern Recognit.* **2019**, *2019*, 11188–11197.
29. Chow, J.K.; Su, Z.; Wu, J.; Tan, P.S.; Mao, X.; Wang, Y.H. Anomaly detection of defects on concrete structures with the convolutional autoencoder. *Adv. Eng. Inform.* **2020**, *45*, 101105. [CrossRef]
30. Gremsy T3-Gremsy. Available online: https://gremsy.com/products/gremsy-t3v3 (accessed on 15 June 2021).
31. Sony Alpha 7. Available online: https://www.sony.es/electronics/camaras-lentes-intercambiables/ilce-7-body-kit (accessed on 16 July 2021).
32. Sony-Camera Remote SDK. Available online: https://support.d-imaging.sony.co.jp/app/sdk/en/index.html (accessed on 15 June 2021).
33. Ouster OS0-128. Available online: https://ouster.com/products/os0-lidar-sensor/ (accessed on 15 June 2021).
34. Rusu, R.B.; Blodow, N.; Beetz, M. *Fast Point Feature Histograms (FPFH) for 3D Registration*; Institute of Electrical and Electronics Engineers (IEEE): Piscataway, NJ, USA, 2009; pp. 3212–3217. [CrossRef]

35. Shan, T.; Englot, B.; Meyers, D.; Wang, W.; Ratti, C.; Rus, D. LIO-SAM: Tightly-coupled lidar inertial odometry via smoothing and mapping. In Proceedings of the IEEE International Conference on Intelligent Robots and Systems, Las Vegas, NA, USA, 25–29 October 2020; pp. 5135–5142. [CrossRef]
36. Kaess, M.; Johannsson, H.; Roberts, R.; Ila, V.; Leonard, J.J.; Dellaert, F. ISAM2: Incremental smoothing and mapping using the Bayes tree. *Int. J. Robot. Res.* **2012**, *31*, 216–235. [CrossRef]
37. Forster, C.; Carlone, L.; Dellaert, F.; Scaramuzza, D. On-Manifold Preintegration for Real-Time Visual-Inertial Odometry. *IEEE Trans. Robot.* **2015**, *33*, 1–21. [CrossRef]
38. Umeyama, S. Least-Squares Estimation of Transformation Parameters between Two Point Patterns. *IEEE Trans. Pattern Anal. Mach. Intell.* **1991**, *13*, 376–380. [CrossRef]
39. Grupp, M. Evo: Python Package for the Evaluation of Odometry and SLAM. 2017. Available online: https://github.com/MichaelGrupp/evo (accessed on 14 July 2021).

Article

Drone-Based Vibration Monitoring and Assessment of Structures

Sabrina Carroll [1], Joud Satme [2], Shadhan Alkharusi [3], Nikolaos Vitzilaios [1,*], Austin Downey [1,3] and Dimitris Rizos [3]

1. Department of Mechanical Engineering, University of South Carolina, Columbia, SC 29208, USA; src2@email.sc.edu (S.C.); austindowney@sc.edu (A.D.)
2. Department of Electrical Engineering, University of South Carolina, Columbia, SC 29208, USA; jsatme@email.sc.edu
3. Department of Civil and Environmental Engineering, University of South Carolina, Columbia, SC 29208, USA; shadhan@email.sc.edu (S.A.); rizos@engr.sc.edu (D.R.)
* Correspondence: vitzilaios@sc.edu

Abstract: This paper presents a novel method of procuring and processing data for the assessment of civil structures via vibration monitoring. This includes the development of a custom sensor package designed to minimize the size/weight while being fully self-sufficient (i.e., not relying on external power). The developed package is delivered to the structure utilizing a customized Unmanned Aircraft System (UAS), otherwise known as a drone. The sensor package features an electropermanent magnet for securing it to the civil structure while a second magnet is used to secure the package to the drone during flight. The novel B-Spline Impulse Response Function (BIRF) technique was utilized to extract the Dynamic Signature Response (DSR) from the data collected by the sensor package. Experimental results are presented to validate this method and show the feasibility of deploying the sensor package on structures and collecting data valuable for Structural Health Monitoring (SHM) data processing. The advantages and limitations of the proposed techniques are discussed, and recommendations for further developments are made.

Keywords: Structural Health Monitoring; Unmanned Aircraft System; drone; damage detection; electropermanent magnet; B-spline impulse response function; Dynamic Signature Response

Citation: Carroll, S.; Satme, J.; Alkharusi, S.; Vitzilaios, N.; Downey, A.; Rizos, D. Drone-Based Vibration Monitoring and Assessment of Structures. *Appl. Sci.* **2021**, *11*, 8560. https://doi.org/10.3390/app11188560

Academic Editors: Alejandro Suarez, Jonathan Cacace and Matko Orsag

Received: 11 August 2021
Accepted: 10 September 2021
Published: 15 September 2021

Publisher's Note: MDPI stays neutral with regard to jurisdictional claims in published maps and institutional affiliations.

1. Introduction

This work presents an innovative process by which the health status of difficult-to-access structures (e.g., bridges) can be determined, utilizing a wireless sensor package for data acquisition, deployed by an Unmanned Aircraft System (UAS) (commonly known as a drone), which provides data for Dynamic Signature Response (DSR) techniques for damage detection. This sensor technology is a compact and low-cost sensor package with an onboard power supply that allows wireless acceleration data collection, eliminating the need for wired connection to the power grid. The sensor package is equipped with an electropermanent magnet that facilitates the rapid installation to various points on the underside of a metal structure or any structure outfitted with metal plates. The sensor package is deployed and retrieved using a drone designed to operate under structures and deliver the packages safely to regions that may not otherwise be readily accessible (such as the underside of bridges that cross rivers or busy highways). The data collected by the sensor package is then analyzed with damage detection algorithms that process the vibration data and extract the DSR, which depends only on the structural condition. Based on the correlation between data collected at a prior, known, and healthy, state and the current state, Level 1 damage detection (the presence of damage in the structure) can be determined [1].

This paper is organized as follows: Section 2 provides the relevant background on structural health assessment using drones and vibration-based sensing algorithms. In Section 3, we present the development of the novel sensor package, the UAS for sensor deployment, and the data processing algorithm for the structural vibration data collected. The experimental validation of the proposed system along with experimental results are presented and discussed in Section 4. Finally, in Section 5, we conclude the paper and discuss future work.

2. Background

2.1. Accelerometer-Based Sensing

Accelerometers have countless uses, from flipping the screen orientation on smartphones to keeping planes on their trajectory in flight. Accelerometers also play a crucial role in Structural Health Monitoring (SHM). Researchers have created many methods of determining the condition of a structure using the vibration signature [2]. However, hardware implementations of such methods are limited due to multiple factors including size, cost, and power requirements. The high-sensitivity, low-noise electronics required to make such measurements are relatively large, expensive, and require experienced personnel to operate [2]. Attempts to make more compact systems and increase the resolution and sensitivity are included in the references of this work [3,4]. However, these compact systems can be labor-intensive and dangerous to deploy in remote areas. Such areas also make it impossible to connect these systems to the power grid, creating a need for wireless and power-independent sensor packages.

Accelerometers are electromechanical devices that detect the change in acceleration in a given direction. Piezoelectric (PE) and Integrated Electronics Piezoelectric (IEPE) accelerometers use internal crystals and the piezoelectric effect to generate a charge related to the acceleration introduced to the system. There are also piezoresistive accelerometers that vary their resistance in relation to an input acceleration. The third common type of accelerometers are the Capacitive Microelectromechanical Systems (MEMSs). These accelerometers have a suspended microstructure, embedded in the Integrated Circuit (IC), with a small mass that is free to vibrate and measure the change of capacitance of the microstructure due to acceleration [5]. DC coupling of capacitive MEMS accelerometers allows for the detection of low-frequency signals, which is critical when measuring the vibration signature of large structures, which typically resonate at frequencies below 1 Hz. Furthermore, MEMSs are capable of detecting steady-state acceleration [6,7] in addition to allowing the integration and double integration of their outputs to determine velocity and displacement, which is important for quantifying the stresses and loads on structures. Figure 1 compares performance versus qualitative cost estimates for the three main types of accelerometers. For this work, a MEMS accelerometer was chosen for its reliability, affordability, and ease-of-use.

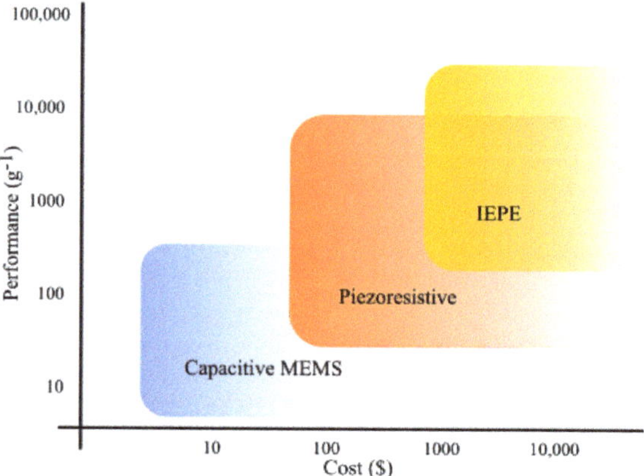

Figure 1. Comparison of performance versus cost for the three main types of accelerometers.

2.2. Unmanned Aircraft Systems in SHM

One of the major challenges in SHM is accessing remote areas or difficult-to-access segments of structures. UAS (drones) are increasingly used as a solution to this problem. Thus far, much of the research considering drones in SHM applications has focused on the use of onboard cameras for image acquisition [8,9]. The assessment is then performed by a qualified engineer through visual inspection of the images with or without the use of image processing software. Another approach is to use a drone to hold a reflector prism against the bridge for total station measurements instead of holding the prism manually [10]. In this case, the prism is affixed directly to the UAS in a configuration described as a *helping hand* [11]. Moreover, Unmanned Aerial Manipulators (UAMs) are a new type of drone that achieve higher precision by utilizing a gripper or multifingered end effector to provide additional degrees of freedom over *helping hands* [11].

Jimenez-Cano et al. [10] presented an application of drones for SHM in which a dynamic model and controller were developed to achieve contact between the manipulator arm and the underside of a structure. They proposed to use an ultrasound sensor, attached to the end effector, to assess the depth of cracks in the structure. During the experiments, the proposed aerial manipulator was able to maintain continuous contact with the structure, which is necessary for the assessment; however, the experiments were performed without the actual ultrasound sensor. In another study by Bodie et al. [12], a contact-based UAS measured the electrical potential difference between materials in a sample of concrete as a method of nondestructive testing.

In addition to these contact-based methods, there is ongoing research into the deployment and retrieval of sensors by drones. For SHM applications, these sensors can be any of the variety of options considered in the previous subsection. For example in [13], a drone was developed with a hooked gripper able to attach a dummy package with a preprepared adhesive to a target by using the arm to apply pressure to the package against the desired surface. The package was then removed using a loop attached to the package, with which the hook can engage, and then, reverse thrust was used to remove it from the surface. Closer to this work, drones were attached directly to civil structures using electropermanent magnets for the collection of vibration data from civil structures [14]. In this case, an electropermanent magnet was utilized to maintain contact between the drone and the underside of the structure until the onboard systems have completed the data collection. However, the sensor package was integrated with the UAS and not treated as a separate system for deployment.

2.3. Vibration-Based SHM

SHM refers to activities used to diagnose structures over time. Structural damage can be detected by investigating the dynamic properties of the structure. The change in the modal parameters (e.g., frequencies, mode shapes, and modal damping) indicates the change of the physical properties of the structure (e.g., mass, damping, stiffness) [15]. Methods for SHM can be divided into local and global methods; local methods are used in locations where damage is expected, while global methods are used to detect damage through monitoring changes in the structure's dynamic fingerprint [16]. Vibration-based damage detection methods are classified into four levels based on what the method is able to do [17]:

- Level 1: indicates that damage is present;
- Level 2: identifies the location of the damage;
- Level 3: quantifies the severity of the damage;
- Level 4: finds the remaining lifetime of the structure.

Research developments in early damage detection focus on model-based and data-based methods. Model-based methods assume that a detailed numerical model of the structure is available, and the model parameters are updated based on measurements obtained from the physical structure [15,18]. Data-driven methods use in situ monitoring to identify changes in the structure's response characteristics and correlate those changes to the location (Level 2) and severity (Level 3) of damage [18–20]. Contactless techniques are also being developed to capture the pure dynamic response of the structure without added masses or stiffening due to the attached sensors [21].

In this work, a data-driven method is investigated to detect Level 1 damage using time-domain responses. The B-Spline Response Function (BIRF) technique [22–25] was implemented to extract the DSR of the structure from experimental vibration measurements. The DSR is a unique characteristic of the structure and will remain the same if the geometry, materials, and boundary conditions of the structure are not altered. The DSR is independent of the actual loads that caused the vibration. Therefore, the DSR represents the current state and condition of a structure. Comparing the DSR before and after damage can be used for damage detection.

3. System Development

The goal of this work was to develop a system that can identify structural damage using vibration data collected by wireless sensor packages deployed with drones. This section presents the three main elements of this system: sensor package, drone, and method for damage identification.

3.1. Sensor Package

The sensor package consisted of three components: (i) electronics for data gathering, storage, and wireless control, (ii) a docking mechanism based on an electropermanent magnet, and (iii) the power supply (battery). These components are presented in Figures 2 and 3, while Table 1 summarizes the main features of the sensor package. A compact 3D-printed frame was built to house these components, forming the standalone sensor package shown in Figure 4 (black box).

Table 1. Summary of the sensor package features.

Sample Size	Rate of Data Collection	Resolution	Sensitivity	Steady-State Power	Package Weight
74,000 Samples	28 kHz	16-bit	5400 LSB/g	0.296 W	380 g

The sensor package features the ARM Cortex-M7 processor onboard the Teensy 4.0 high-performance microcontroller. The system utilizes the 600 MHz clock speed of the Teensy 4.0 and the Serial Communication Interface (SPI) to communicate between the modules and achieve a data collection rate up to 28 kHz. The sensor package also includes three control options (Figure 2): a PWM port used for wireless control, a digital 5 V as a hard-wired trigger, and manual controls for testing and calibration purposes.

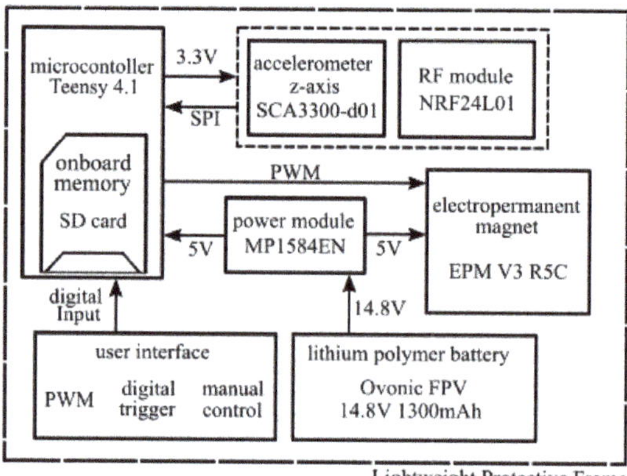

Figure 2. Block diagram of the sensor package circuit schematic.

Figure 3. Constituent components of the sensor package.

Power is supplied using a DC-DC buck converter (MP1584EN manufactured by Monolithic Power Systems) and a lithium polymer battery (Lumenier 7.4V 1000 mAh), as shown in Figure 3. This battery currently runs the sensor package for about 11 h on a single charge. The deployment period can be prolonged by utilizing standby mode where unused modules are turned off to conserve energy. In cold weather though, testing has shown that batteries tend to discharge relatively faster, in turn requiring a more restrictive power conservation schedule, limiting the number of tests conducted per deployment.

The docking mechanism consists of an electropermanent magnet (NicaDrone EPM V3 R5C) able to mount the sensor package onto metal surfaces (Electropermanent Magnet #1 in Figure 4). The package collects acceleration data using a MEMS accelerometer (Murata SCA3300-d01) mounted to the metal frame of the electropermanent magnet to ensure contact with the structure being examined. The system can collect up to 74,000 samples as one continuous dataset, before having to transfer the data onto the onboard nonvolatile memory (SD card), a process that takes approximately 3.288 s.

Figure 4. The docking and retrieval system of the sensor package featuring two electropermanent magnets: #1 on the top of the package, used to mount it to the underside of structures; and #2 on the drone docking platform, used to secure, transfer, deliver, and retrieve the package.

3.2. UAS Platform

The standalone sensor package does not require connection to the power grid, so it can be deployed to any structure. Traditional (human operator) deployment and retrieval though may be challenging in remote locations. For example, in remote bridges, a lift from the ground is required to reach the underside (which may be impractical if the bridge crosses a body of water or a busy highway) or a crane on the topside of the bridge, which can position the operator underneath the structure. Both methods are risky and require costly equipment and roads to be shut down to make the underside of the structure accessible. A drone platform has been developed to overcome these limitations and securely deploy and retrieve sensor packages in difficult-to-access areas.

The drone platform was based on the DJI F550, a hexacopter with six standard 2312E motors, shown in Figure 5 [26]. Pixhawk 4 was used as the flight controller, and the system was operated via the ArduPilot Mission Planner. Onboard sensors included an IMU, accelerometers, gyroscope, magnetometer, and barometer to provide data for a variety of parameters including drone orientation and altitude. There was also an external GPS used for global positioning when the signal was available. The FrSky X8R radio was used as the onboard receiver to communicate with the FrSky 2.4 GHz Taranis X9D radio controller for manual flight commands. The sensor package, as previously described, was attached to the UAS via a second electropermanent magnet (#2 in Figure 4), which was controlled using one of the RC servo ports available on the power management board of the Pixhawk 4.

Figure 5. Overview of the custom Unmanned Aircraft System (UAS or drone) developed for sensor deployment and retrieval.

The system was designed to work as follows: On the ArduPilot Mission Planner, two channels from the FrSky Taranis were set to send high, low, and neutral commands to the two magnets. During the high command, the magnet begins the charging sequence, which can only be completed once it is in contact with a ferromagnetic surface. Once the magnet is mounted on the metallic surface, this command can be shifted to neutral as the electropermanent magnet will remain magnetized until the low command is sent. When the low command is sent, it will disengage the magnet.

During operations, the sensor package is delivered onboard the drone (Magnet #2 engaged) to the desired point on the structure where Magnet #1 arms and connects the sensor package and drone to the structure. Once the package is secured in place, Magnet #2 disengages, and the drone detaches and flies away. The sensor package then periodically collects and stores acceleration data for DSR processing. When this operation is complete, the drone conducts a retrieval sequence by reversing the docking procedure. A summary of the status of both magnets in each phase of operations is given in Table 2.

Table 2. Operation status for Electropermanent Magnets #1 and #2 for each sensor package and UAS position.

Electropermanent Magnet #1	Electropermanent Magnet #2	Sensor Package	UAS
Unmagnetized	Magnetized	Attached to UAS	At delivery
Magnetized	Unmagnetized	Attached to structure	Package deployed

To assist with the retrieval of the sensor package, the UAS was outfitted with a 3D printed alignment tool (docking platform in Figure 4). The alignment tool features angled walls, which help to center the sensor package over the electropermanent magnet and prevent the sensor package from coming into contact with the propellers.

3.3. Method of Damage Identification

The Dynamic Signature Response (DSR) of a system is a characteristic response of the system and was first introduced by Rizos et al. [23] as the B-Spline Impulse Response Function (BIRF) for soil–structure interaction analysis. The BIRF is a time-varying function that is defined between any two points in a mechanical system. It represents the time history of the response of one of the two points due to a unit amplitude B-spline excitation applied to the other point [23,24,27,28]. The B-spline functions are piecewise smooth polynomials of order k and belong to a family of base functions used in data interpolation and approximation. They are defined with respect to a time knot sequence $t_n, n = 1, 2, \ldots,$ by the recursive formula,

$$B_n^k(t) = \begin{cases} 1, & \text{if } t_n < t < t_{n+1} \\ 0, & \text{elsewhere} \end{cases} \qquad k = 1, \tag{1}$$

$$B_n^k(t) = \left(\frac{t - t_n}{t_{n+k-1} - t_n} \right) B_n^{k-1}(t) + \left(\frac{t_{n+k} - t}{t_{n+k} - t_{n+1}} \right) B_{n+1}^{k-1}(t), \quad k > 1. \tag{2}$$

The B_n^k-spline polynomials are of degree $k - 1$ and have $k - 2$ continuous derivatives. While the BIRFs are computed efficiently in numerical models of physical systems [23,28,29], it is almost impossible to measure the BIRFs directly in physical systems due to the difficulty of accurately reproducing the B-spline loading on the physical system. The BIRF between two points of a mechanical system, however, can be extracted from vibration tests that acquire the time history record of a response, R, and excitation, f, at the two points. For example, the fourth-order-based BIRF is computed in a discrete form at a time step N as:

$$BIRF_N = \left(R_N - \sum_{n=1}^{N-2} BIRF_n f_{N-n+2} + BIRF_{N-1}(f_3 - f_1) \right) / (f_2 + 2f_1), \tag{3}$$

where subscripts indicate the time step at which the corresponding functions are evaluated. Details of the method along with verification and validation studies were presented in Alkharousi et al. [30,31].

Structural assessment procedures through dynamic testing focus on identifying changes in the dynamic properties of the system through modal analysis and/or model updating techniques. In Equation (3), the BIRFs are independent of any external excitations and are considered the DSRs that capture the current state of the dynamic system. Any changes in the current state of the structure are captured as changes in its DSRs. Therefore, if the response time history, R, and the corresponding excitation, f, are known through vibration test measurements conducted before and after a damage-inducing event, the difference in the corresponding DSR is an indication of damage [30,31].

The computation of the BIRF shown in Equation (3) represents in essence a discrete deconvolution of the response signal, and the BIRF can be perceived as an equivalent time domain transfer function of the system that captures all frequencies in the range defined by the sampling time interval. It is noted and had been demonstrated that the analysis of only a short duration segment of the recorded response is adequate for change detection through the DSR algorithm, compared to mainstream methods that require, in most cases, the entire response record from multiple sensors. However, the noise inherent to the time history response records may have to be removed through filtering before the DSR is extracted. Nevertheless, the computing demand of the proposed method is very low, minimizing the postprocessing times, leading potentially to damage detection in real time, and facilitating the integration to the damage detection algorithms in the overall proposed system.

4. Experimental Validation and Results

In order to validate the developed system (drone, sensor package, damage detection algorithm), we created a laboratory experimental testbed and designed experiments to collect and analyze data. The goal was to show the feasibility of the drone-based deployment and retrieval of our sensor package, as well as the extraction of damage information from the collected vibration data.

4.1. Test Apparatus

A test apparatus was constructed in the lab to replicate a bridge with a metallic surface, as shown in Figure 6. The structure was 6.5 ft tall and 6.5 ft wide (2 m × 2 m) with a 2 ft by 4 ft (60 cm × 120 cm) steel plate mounted in the center to serve as a docking platform. Figure 7 shows a schematic of this plate. An electromagnetic shaker (Ling Dynamic Systems V408) was permanently mounted on top of the plate and used to excite the structure during our experiments. The test apparatus was placed inside a drone cage where the OptiTrack Motion Capture system was used for rigid body tracking. This system can provide accurate pose data (position and orientation) for our drone flights.

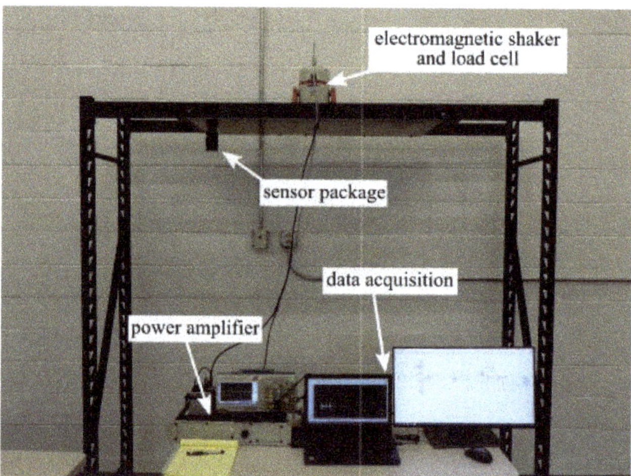

Figure 6. Overview of the test apparatus with key components annotated.

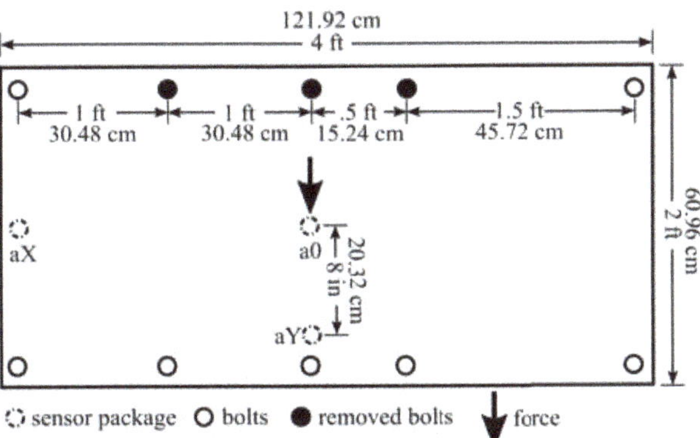

Figure 7. Schematic of the steel plate structure used in the experiments featuring the sensor package placement and the relative location of the supporting bolts used to simulate damage.

4.2. Drone-Based Deployment Validation

In this experiment, we aimed to validate the use of the drone for the sensor package deployment. The experiment was designed as follows: The drone approaches the structure and conducts the docking sequence described in Section 3.2, then mounts the sensor package to the structure. After sufficient time for data collection has passed, the UAS approaches the structure again and reverses the docking sequence to retrieve the package. When the package has been retrieved, the entire system lands, and the test is complete. During this experiment, the electromagnetic shaker was set to excite the structure at 11.3 Hz while we collected acceleration data from the sensor package and displacement data from the OptiTrack Motion Capture system. Due to the underside of the structure being outside of the OptiTrack field of view, only the UAS position and orientation were tracked using the OptiTrack system.

The results are presented in Figure 8. The first 25 s of the test were used to coordinate the data collection and ensure that the electropermanent magnets were properly set before takeoff. Adjustments were performed to properly align the UAS with the structure before making contact at 38 s, at which point, the UAS maintained contact until the 48 s mark (area II) to ensure the engagement of the electropermanent magnet on the sensor package and the disengaging of the electropermanent magnet on the UAS. Once the magnets were set, the UAS left the sensor package on the structure from the 38 s mark until the 75 s mark (area III). During this time, the UAS hovered near the structure to allow the sensor package to perform data collection without interference from the hanging mass of the UAS.

After allowing sufficient time for data collection, the UAS began the retrieval process by making contact with the package at 75 s. Contact was again maintained for 10 s to confirm the change in the electropermanent magnet state (area IV). For retrieval, the magnet on the UAS was engaged to maintain hold of the sensor package during landing prior to the magnet on the sensor package disengaging, allowing the system to begin its descent. This descent was performed immediately after securing the sensor package, and the drone landed at 94 s, completing the flight test.

The acceleration of the sensor package presented in Figure 8 substantiated these movements showing intense vibration when the package was onboard the UAS and significantly smaller acceleration while in contact with a structure, indicating successful deployment. Experimental results showed the feasibility of the method, which can now be further developed [26].

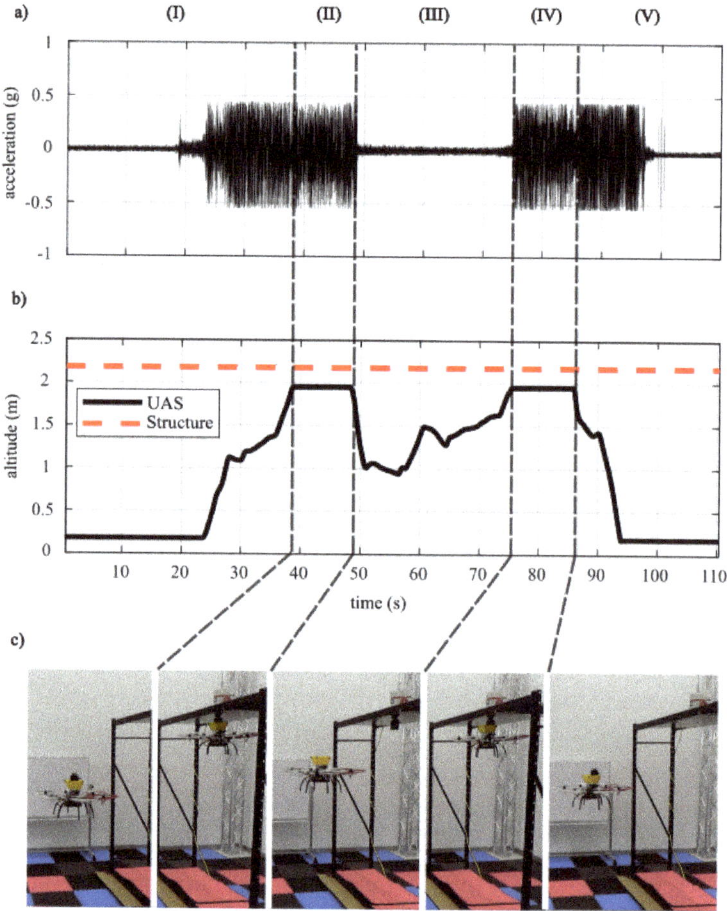

Figure 8. UAS deployment of the sensor package showing (**a**) acceleration of the sensor package, (**b**) displacement of the UAS, and (**c**) photos of the sensor deployment and retrieval experiment.

4.3. Sensor Package Validation

This experiment aimed at validating the accuracy of acceleration data collected by the sensor package. For this purpose, the sensor package was manually mounted along with a PCB accelerometer onto the steel plate, which was excited by the electromagnetic shaker (as shown in Figure 6). Using a data acquisition module, the excitation signal was swept from 1–20 Hz over a test period of 50 s. Due to the limited sample size of the sensor package (74,000 samples), a sampling frequency of 1480 Hz was used as a result of the limited buffer size onboard the sensor package. The validation was conducted utilizing a frequency response function of the sensor package data over the reference PCB accelerometer, shown in Figure 9. The results indicated an exponentially increasing error as the frequency approached zero. At lower frequencies (<5 Hz), the acceleration signal was below the sensitivity of the sensor package (5400 LSB/g), resulting in the large error shown. This could be mitigated in future designs by utilizing a higher accuracy accelerometer and decreasing the damping caused by the sensor package frame.

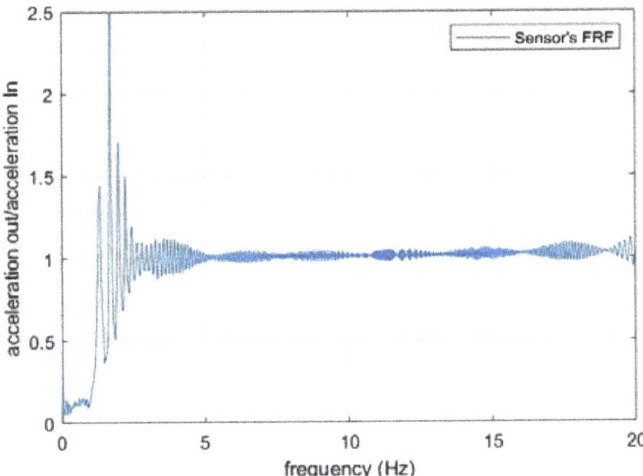

Figure 9. Frequency response function of the sensor package for 1–20 Hz excitation.

4.4. Damage Detection Validation

This experiment was designed to validate the ability of the sensor package to collect meaningful data for SHM applications. Since the system was developed to detect Level 1 damage, the sensor package needed to collect data before and after the structure had sustained damage. To simulate damage, three bolts supporting the steel plate were removed, as shown in Figure 7. This experiment also utilized the electromagnetic shaker to generate harmonic excitation forces. Additionally, a load cell (Ling Dynamic Systems V408) was placed between the shaker and the plate to record the force transmitted to the structure to provide ground truth measurements. The sensor package was then manually mounted directly underneath the electromagnetic shaker and load cell, in the position a0 shown in Figure 7. The shaker was first actuated at 7 Hz for 4 s and then stopped to allow the structure to vibrate freely. This was repeated at 14 Hz for both the undamaged and damaged structure.

The collected data underwent preprocessing in which the data from each experiment were smoothed through a 50-point rolling average and resampled to a total of 200 points to be considered as the response, R_N, in Equation (3). The associated forcing function, f_N, in Equation (3) can be either the load cell record or the voltage signal that drives the linear actuator [31]. In this paper, the load cell record was utilized. Equation (3) can now be implemented on each record to extract the DSR of the structure. It has been shown that damage can be detected by processing even a short segment of the record [31], provided that the time step of the acquisition is adequate to capture the maximum desired frequency affecting the response.

Figure 10 shows a 400-step segment of the DSR of the undamaged structure. Each DSR was extracted from the acceleration time history sampled in each of the three tests with the 14 Hz excitation force. As expected, the three DSR signals were almost identical and showed a high correlation ($R^2 = 0.989$). This indicated that the state of the structure had not changed between acquisitions, and therefore, no damage was present. Figure 11 compares the DSR of the damaged and undamaged structure to the 14 Hz excitation force. These two BIRFs showed no correlation, indicating a change in the structural condition; thus, the presence of damage is detected.

In Figures 10 and 11, the damage detection algorithm operated on a 0.8 s window of the free vibration structural response with an acquisition step of 2×10^{-3} s. This allowed for approximately 10 sampling points within a response cycle, and thus, this was the minimum acquisition step required to capture frequencies up to approximately 300 Hz. It is noted that the structural response record was resampled to 1/4th of the original sampling frequency. It is also noted that an even shorter segment of the structural response could potentially be used in the damage detection algorithm.

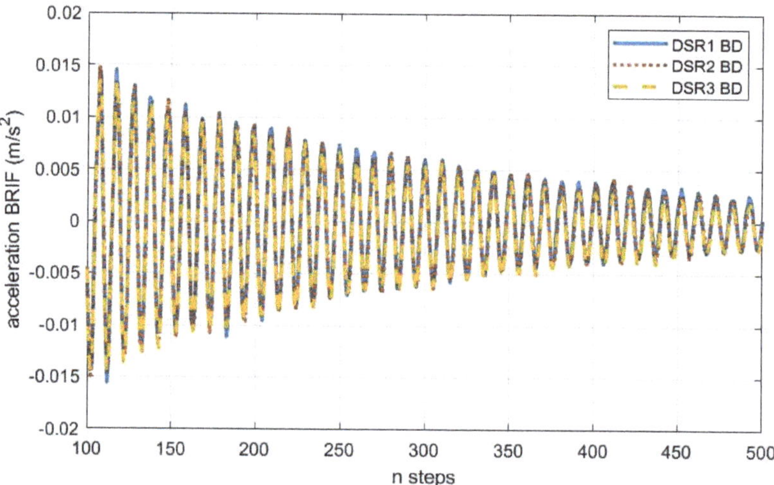

Figure 10. The Dynamic Signature Response (DSR) extracted from 400 steps of the free vibration phase of the response of the undamaged structure excited by a 14 Hz harmonic force.

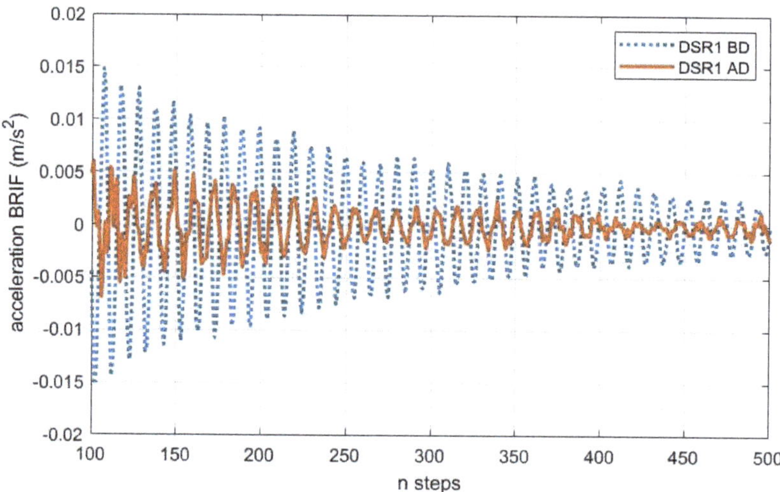

Figure 11. The DSR extracted from 400 steps of the free vibration phase of the response of the undamaged and damaged structure excited by a 14 Hz harmonic force.

Table 3 summarizes the R^2 correlation coefficients of the BIRFs extracted from all tests for the two excitation frequencies on the undamaged and damaged structure. It was observed that all BIRFs correlated well ($R^2 > 0.8$) when extracted from responses of either the undamaged or damaged structure, for both frequencies. However, the correlation was completely lost ($R^2 \approx 0$) when the BIRFs of the undamaged and damaged structures were compared, regardless of the excitation.

Appl. Sci. **2021**, 11, 8560

Table 3. Correlation coefficients between BIRFs of the undamaged and damaged structure for the 7 Hz and 14 Hz excitation frequencies.

| | | | Before Damage | | | | | | After Damage | | | | | |
| | | | 7 Hz | | | 14 Hz | | | 7 Hz | | | 14 Hz | | |
		Test	1	2	3	1	2	3	1	2	3	1	2	3
Before Damage	7 Hz	1	1.000	0.883	0.890	0.900	0.888	0.895	0.092	0.080	0.086	0.084	0.087	0.079
		2		1.000	0.891	0.909	0.898	0.905	0.098	0.086	0.090	0.083	0.087	0.086
		3			1.000	0.865	0.864	0.865	0.086	0.078	0.079	0.083	0.096	0.078
	14 Hz	1				1.000	0.989	0.991	0.097	0.095	0.094	0.087	0.091	0.087
		2					1.000	0.990	0.095	0.096	0.093	0.088	0.088	0.085
		3						1.000	0.099	0.098	0.096	0.092	0.088	0.081
After Damage	7 Hz	1							1.000	0.879	0.892	0.813	0.798	0.815
		2								1.000	0.906	0.812	0.809	0.839
		3									1.000	0.805	0.787	0.813
	14 Hz	1										1.000	0.921	0.927
		2											1.000	0.910
		3												1.000

5. Conclusions

This paper presented an initial prototype and the associated methodologies for a sensor package that can be deployed via a drone onto civil structures for structural health monitoring. Alongside this demonstration, Level 1 damage detection (the presence of damage in the structure) was performed using a novel BIRF technique to extract the DSR from data collected by reference accelerometers. Through manual flights, the feasibility of delivering and retrieving a sensor package from the underside of a structure was proven. Similarly, experimental validation showed that the BIRF method is an effective method for determining damage based on data collected from any sensor package.

In future work, the landing mechanism of the UAS system will be fully automated, instead of relying on manual flight controls. The sensitivity of the sensor package will be improved, and wireless communication and data transfer will be added. The use of elastomers will be considered in an effort to reduce the effect of drone vibrations on the measurements. As for the damage detection algorithm, future work will focus on expanding the methodology to include damage quantification and damage localization, which relate to Level 2 and 3 damage detection, respectively.

Author Contributions: The authors confirm the contributions to the paper as follows: study conception and design: N.V., A.D. and D.R.; data collection: S.C., J.S. and S.A.; analysis and interpretation of results: S.C., J.S., S.A., N.V., A.D. and D.R.; draft manuscript preparation: S.C., J.S., S.A., N.V., A.D. and D.R. All authors have read and agreed to the published version of the manuscript.

Funding: This work was partially supported by the University of South Carolina (UofSC) Office of Undergraduate Research (Magellan Awards 80003124 and 80003491), the UofSC College of Engineering and Computing startup fund and the Advanced Railway Technology Group at UofSC.

Institutional Review Board Statement: Not applicable.

Informed Consent Statement: Not applicable.

Conflicts of Interest: The authors declare no conflict of interest.

References

1. Doebling, S.; Farrar, C.; Prime, M.; Shevitz, D. *Damage Identification and Health Monitoring of Structural and Mechanical Systems from Changes in Their Vibration Characteristics: A Literature Review*; Technical Report; Los Alamos National Laboratory (LANL): Los Alamos, NM, USA, 1996; doi:10.2172/249299. [CrossRef]
2. Ho, D.D.; Nguyen, K.D.; Yoon, H.S.; Kim, J.T. Multiscale Acceleration-Dynamic Strain-Impedance Sensor System for Structural Health Monitoring. *Int. J. Distrib. Sens. Netw.* **2012**, *8*, 709208. [CrossRef]

3. Illinois Structural Health Monitoring Project. SHM-H Board High-Sensitivity Accelerometer Sensor Board Datasheet and User's Guide. June 2009. Available online: https://shm.cs.illinois.edu/ (accessed on 1 August 2021).

4. Smart Structure Systems Lab, University of Arizona. Structural Health Monitoring (SHM) Board Series for Imote2 Smart Sensor Platform. Technical Report. 2020. Available online: https://smartstructure.weebly.com/smart-sensor.html (accessed on 1 August 2021).

5. Mohammed, Z.; Elfadel, I.; Rasras, M. Monolithic Multi Degree of Freedom (MDoF) Capacitive MEMS Accelerometers. *Micromachines* **2018**, *9*, 602. [CrossRef] [PubMed]

6. Lent, B. Simple Steps to Selecting the Right Accelerometer. 2018. Available online: https://www.fierceelectronics.com/components/simple-steps-to-selecting-right-accelerometer-0 (accessed on 5 May 2020).

7. Yoder, N.; Adams, D. Commonly used sensors for civil infrastructures and their associated algorithms. In *Woodhead Publishing Series in Electronic and Optical Materials, Sensor Technologies for Civil Infrastructures*; Woodhead Publishing: Cambridge, UK, 2014; pp. 57–85. [CrossRef]

8. Sankarasrinivasan, S.; Balasubramanian, E.; Karthik, K.; Chandrasekar, U.; Gupta, R. Health Monitoring of Civil Structures with Integrated UAV and Image Processing System. *Procedia Comput. Sci.* **2015**, *54*, 508–515. [CrossRef]

9. Sony, S.; Laventure, S.; Sadhu, A. A literature review of next-generation smart sensing technology in structural health monitoring. *Struct. Control Health Monit.* **2019**, *26*, e2321. [CrossRef]

10. Jimenez-Cano, A.; Braga, J.; Heredia, G.; Ollero, A. Aerial manipulator for structure inspection by contact from the underside. In Proceedings of the 2015 IEEE/RSJ International Conference on Intelligent Robots and Systems (IROS), Hamburg, Germany, 28 September–2 October 2015; pp. 1879–1884. [CrossRef]

11. Ruggiero, F.; Lippiello, V.; Ollero, A. Aerial Manipulation: A Literature Review. *IEEE Robot. Autom. Lett.* **2018**, *3*, 1957–1964. [CrossRef]

12. Bodie, K.; Brunner, M.; Pantic, M.; Walser, S.; Pfndler, P.; Angst, U.; Siegwart, R.; Nieto, J. An Omnidirectional Aerial Manipulation Platform for Contact-Based Inspection. *arXiv* **2019**, arXiv:1905.03502.

13. Hamaza, S.; Georgilas, I.; Fernandez, M.; Sanchez, P.; Richardson, T.; Heredia, G.; Ollero, A. Sensor Installation and Retrieval Operations Using an Unmanned Aerial Manipulator. *IEEE Robot. Autom. Lett.* **2019**, *4*, 2793–2800. [CrossRef]

14. Takeuchi, K.; Masuda, A.; Akahori, S.; Higashi, Y.; Miura, N. A close inspection and vibration sensing aerial robot for steel structures with an EPM-based landing device. In Proceedings of the Nondestructive Characterization and Monitoring of Advanced Materials, Aerospace, and Civil Infrastructure 2017, Portland, OR, USA, 26–29 March 2017; Volume 10169, p. 101692U. [CrossRef]

15. Sabamehr, A.; Lim, C.; Bagchi, A. System identification and model updating of highway bridges using ambient vibration tests. *J. Civ. Struct. Health Monit.* **2018**, *8*, 755–771. [CrossRef]

16. Fritzen, C.P.; Kraemer, P. Self-diagnosis of smart structures based on dynamical properties. *Mech. Syst. Signal Process.* **2009**, *23*, 1830–1845. [CrossRef]

17. Liu, J.; Chen, S.; Bergés, M.; Bielak, J.; Garrett, J.H.; Kovačević, J.; Noh, H.Y. Diagnosis algorithms for indirect structural health monitoring of a bridge model via dimensionality reduction. *Mech. Syst. Signal Process.* **2020**, *136*, 106454. [CrossRef]

18. Abdo, M. *Structural Health Monitoring, History, Applications and Future. A Review Book*; Open Science: New York, NY, USA, 2014.

19. Farrar, C.R.; Doebling, S. An Overview Of Modal Based Damage Identification Methods, Engineering Analysis Group. In Proceedings of the Structural Damage Assessment Using Advanced Signal Processing Procedures (DAMAS), Sheffield, UK, 30 June–2 July 1997.

20. Doebling, S.W.; Farrar, C.R.; Prime, M.B. A Summary Review of Vibration-Based Damage Identification Methods. *Shock Vib. Dig.* **1998**, *30*, 91–105. [CrossRef]

21. Civera, M.; Zanotti Fragonara, L.; Surace, C. An experimental study of the feasibility of phase-based video magnification for damage detection and localisation in operational deflection shapes. *Strain* **2020**, *56*, e12336. [CrossRef]

22. Rizos, D.C.; Karabalis, D.L. A time domain BEM for 3-D elastodynamic analysis using the B-spline fundamental solutions. *Comput. Mech.* **1998**, *22*, 108–115. [CrossRef]

23. Rizos, D.C.; Loya, K.G. Dynamic and Seismic Analysis of Foundations based on Free Field B-Spline Characteristic Response Histories. *J. Eng. Mech.* **2002**, *128*, 438–448. [CrossRef]

24. Stehmeyer, E.H.; Rizos, D.C. B-Spline impulse response functions (BIRF) for transient SSI analysis of rigid foundations. *Soil Dyn. Earthq. Eng.* **2006**, *26*, 421–434. [CrossRef]

25. Leon, E.; Rizos, D.C.; Caicedo, J.M. A procedure to develop scalable models for the transient response of sleepers in conventional and high-speed railway lines and implementation to the vertical vibration mode. *Soil Dyn. Earthq. Eng.* **2011**, *31*, 502–511. [CrossRef]

26. Carroll, S.; Kalaitzakis, M.; Vitzilaios, N. UAS Sensor Deployment and Retrieval to the Underside of Structures. In Proceedings of the 2021 International Conference on Unmanned Aircraft Systems (ICUAS), Athens, Greece, 15–18 June 2021; pp. 895–900. [CrossRef]

27. Rizos, D.C. Advanced Time Domain Boundary Element Method for General 3-D Elastodynamic Problems. Ph.D. Thesis, Civil and Environmental Engineering, University of South Carolina, Columbia, SC, USA, 1993.

28. Mulliken, J.; Rizos, D.C. A coupled computational method for multi-solver, multi-domain transient problems in elastodynamics. *Soil Dyn. Earthq. Eng.* **2012**, *34*, 78–88. [CrossRef]

29. Karabalis, D.L.; Rizos, D. Dynamic analysis of 3-D foundations. *Bound. Elem. Tech. Geomech.* **1993**, *6*, 177–208.

30. Alkharousi, S. Identification of Structural Signature Responses from Response Time Histories Using B-Spline Impulse Response Techniques. Master's Thesis, Department of Civil & Environmental Engineering, University of South Carolina, Columbia, SC, USA, 2021.
31. Alkharousi, S.; Rizos, D.C. Identification of Structural Signature Responses from Response Time Histories Using B Spline Impulse Response Techniques. *Struct. Health Monit.* **2022**, in preparation.

Article

Aerial Tele-Manipulation with Passive Tool via Parallel Position/Force Control[†]

Mostafa Mohammadi [1,2], Davide Bicego [3,*], Antonio Franchi [3,4], Davide Barcelli [1] and Domenico Prattichizzo [1,2]

[1]   Department of Information Engineering and Mathematics, University of Siena, 53100 Siena, Italy; mohammadi@dii.unisi.it or mostafa.mohammadi@cariad.technology (M.M.); barcelli@dii.unisi.it (D.B.); prattichizzo@ing.unisi.it (D.P.)
[2]   Department of Advanced Robotics, Italian Institute of Technology, 16163 Genoa, Italy
[3]   Robotics and Mechatronics Group, University of Twente, 7522 NH Enschede, The Netherlands; a.franchi@utwente.nl or antonio.franchi@laas.fr
[4]   LAAS-CNRS, Université de Toulouse, CNRS, 31400 Toulouse, France
*   Correspondence: d.bicego@utwente.nl; Tel.: +31-534-892-626
†   This research was partially supported by the European Union's Horizon 2020 research and innovation program grant agreement ID: 871479 AERIAL-CORE.

Abstract: This paper addresses the problem of unilateral contact interaction by an under-actuated quadrotor UAV equipped with a passive tool in a bilateral teleoperation scheme. To solve the challenging control problem of force regulation in contact interaction while maintaining flight stability and keeping the contact, we use a parallel position/force control method, commensurate to the system dynamics and constraints in which using the compliant structure of the end-effector the rotational degrees of freedom are also utilized to attain a broader range of feasible forces. In a bilateral teleoperation framework, the proposed control method regulates the aerial manipulator position in free flight and the applied force in contact interaction. On the master side, the human operator is provided with force haptic feedback to enhance his/her situational awareness. The validity of the theory and efficacy of the solution are shown by experimental results. This control architecture, integrated with a suitable perception/localization pipeline, could be used to perform outdoor aerial teleoperation tasks in hazardous and/or remote sites of interest.

Keywords: aerial robotics; aerial manipulation; force control; bilateral teleoperation; haptics; quadrotor

Citation: Mohammadi, M.; Bicego, D.; Franchi, A.; Barcelli, D.; Prattichizzo, D. Aerial Tele-Manipulation with Passive Tool via Parallel Position/Force Control. *Appl. Sci.* **2021**, *11*, 8955. https://doi.org/10.3390/app11198955

Academic Editor: Seong-Ik Han

Received: 6 August 2021
Accepted: 18 September 2021
Published: 26 September 2021

Publisher's Note: MDPI stays neutral with regard to jurisdictional claims in published maps and institutional affiliations.

1. Introduction

Aerial robotics has become increasingly popular in research, industry, and for commercial applications. Beyond the traditional visual inspection functionality that made them widely used and appreciated, aerial robots have recently received profound interest for applications which require to seek, establish, and maintain some sort of physical interaction with the environment in order to fulfill a certain task. Relevant examples are epitomized by maintenance operations in the energy sector, for example, oil, gas, refinery, and power plants, in particular, to perform non-destructive tests that require keeping some sensors in touch with objects not easily accessible by a human, due to their installation altitude, and also in hazardous environments [1,2]. Apart from their growing use in industrial and civil sites, these systems are starting to also be employed for the in-contact documentation of historical buildings [3]. Other applications of aerial interaction involve the transportation of cable-suspended payloads [4] and packages for search and rescue missions [5].

Aerial manipulation is the deliberately controlled physical interaction of an aerial manipulator with objects in its environment. For an extensive overview of the works on this topic, the interested reader is referred to [6,7]. By *aerial manipulator* we mean a small size Vertical Take-Off and Landing (VTOL) Unmanned Aerial Vehicle (UAV) equipped with a manipulation tool. This manipulation tool is either an active robotic arm manipulator or a passive tool.

When dexterous manipulation by an aerial manipulator is not required, for example, when the robot is intended to apply desired force vectors to an object in order to push, inspect, or probe its surface, or when the normal grippers are not effective, for example, for an object with a wide flat surface, the use of a lightweight passive tool is preferable to a heavier active arm manipulator, as the smaller payload imposed to the aerial robot results in a more energy-efficient system and longer operational flight time. Moreover, simplicity and low weight of a passive tool allow the usage of a broader range of UAVs.

Despite significant achievements in the fully autonomous control of drones, limited problem-solving capabilities, inadequacy in unexpected environmental conditions, legal restrictions, and imperfect position control [8] often require the presence of human operator(s) in aerial manipulation tasks. In bilateral teleoperation schemes, the human capabilities are enhanced by providing them with tangible interaction information of the remote side in the form of force and motion feedbacks (haptic feedback), besides the traditional visual feedback.

In this paper, we propose an aerial manipulation solution using passive tools with a compliant end-effector. Applying a desired force profile in a uni-lateral contact, and at the same time maintaining the flight stability, that is, position and orientation control, and keeping the contact stable, that is, to avoid losing contact and sliding over the contact surface, represents a challenging control problem, especially when it is performed with an under-actuated aerial manipulator in a bilateral teleoperation loop. Fully-actuated aerial manipulators have been demonstrated to be more effective for this kind of application [8–11] but consume more energy due to internal forces. That is why the use of under-actuated UAVs is investigated in this work. Our aerial manipulator is a quadrotor UAV equipped with a lightweight passive tool rigidly attached to the top of it (Figures 1 and 2). The end-effector has a mechanical damper on its surface to smooth free flight to contact transition, and a passive compliant spherical joint to keep the contact while changing the orientation. This compliant mechanism, along with appropriate control policy conforming with the system constraints, allows involving all the robot's degrees of freedom to generate the desired force vector. The desired motion and force of the aerial manipulator are attained using a *parallel position/force control* scheme within a bilateral teleoperation control framework.

The proposed control scheme regulates the aerial manipulator pose in free flight and the applied force in contact conditions. A human operator using a haptic device (with a limited workspace) commands the aerial manipulator pose (with virtually unlimited workspace) in free flight. When the aerial manipulator's end-effector comes in contact with the environment, the haptic device movement is interpreted as desired force. Position control in free flight and force control in contact are achieved by utilizing a cascaded parallel position/force controller. The reference pose is composed by the operator's pose command (free flight pose command) and the output of a force controller in an outer loop. To avoid losing the contact, the desired force is always kept in a feasible range that satisfies the friction and compliance constraints. On the master side, the human operator is provided with force feedback proportional to the robot's velocity in free flight and the applied force in contact.

The aerial tele-manipulation system presented in this paper may contribute to addressing and solving a broad class of relevant use-case applications where a UAV is remotely operated in hardly accessible and life-threatening sites, for example, at high altitudes, by a human safely located in a protected place, thus relieving him/her from potentially dangerous tasks. A conceptual example is depicted in Figure 1. Apart from the aforementioned applications of remote sensor placement, contact holding and remote button pushing, another idea that could be envisioned is to employ such a system to push boxes located on a shelf onto a conveyor belt, in an industrial warehouse scenario. Thanks to the enhanced situational awareness guaranteed by the haptic feedback, the operator could easily regulate the force applied to the load. Furthermore, the designed control law would ensure that the contact is maintained throughout the manipulation. As should be appreciated, many other relevant applications involving the aforementioned conditions can be easily conceived.

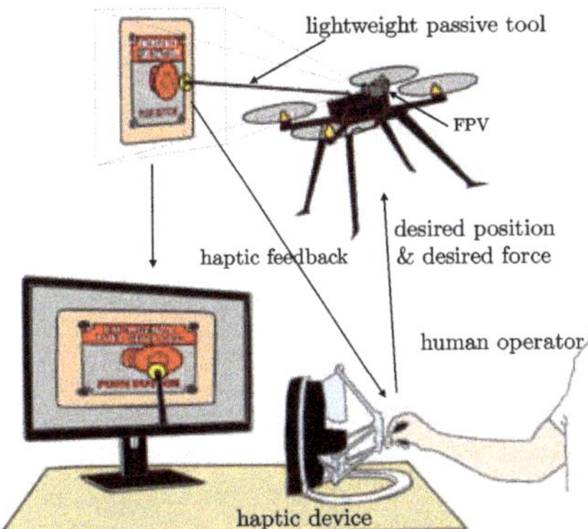

Figure 1. A conceptual example of aerial tele-manipulation with a passive tool: an aerial robot equipped with a lightweight passive tool tasked to press an emergency shut down push button.

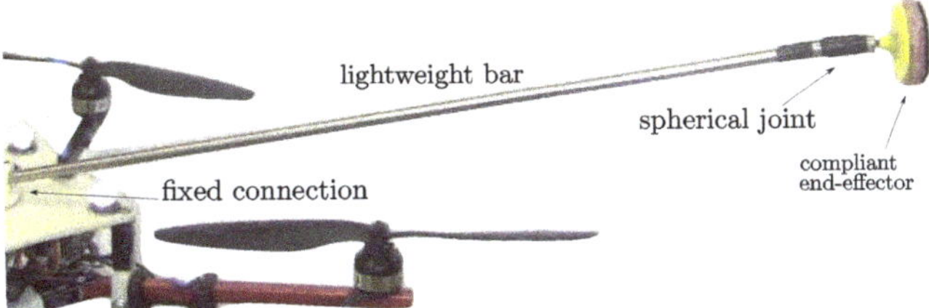

Figure 2. The passive lightweight tool with compliant end-effector used to convert a normal VTOL UAV to an aerial manipulator.

The rest of the paper is organized as follows. The next subsection reviews the related works and the contribution of the paper. Section 2 explains the platform and its dynamic model in the teleoperation system. Section 3 presents the proposed control approach, the stability analysis of which is presented in Section 4. Experimental results are presented in Section 5, while concluding remarks and hints about intended future works are outlined in Section 6.

Related Works and Contribution

There has been a growing interest in aerial robots with physical interaction in the past few years, and many research projects such as [12] have focused on this context. In the following, we try to concisely provide a general overview of the state of the art of this broad topic, focusing then on the works more closely related to the one presented in this paper. Aerial physical interaction with the environment can be macro-categorized as:

(i)   *using active manipulators;* and
(ii) *using passive tools;*

and naturally, different mechanical solutions demand different controllers.

To mention some examples belonging to the first group, in [13] the authors designed and installed a small parallel manipulator on one side of a VTOL, while the use of one and two serial manipulators to grasp objects was proposed in [14] and in [15], respectively. In these works, the authors implemented and validated different instances of hybrid force control. A visual servoing approach to control a quadrotor equipped with a serial manipulator is suggested by [16], while a passivity-based adaptive controller, which can be applied to both position and velocity control to guide a quadrotor aerial manipulator, is utilized by [17]. Furthermore, the behavioral control of an aerial manipulator is presented in [18]. In the multi-robot scenario, a team of quadrotors, equipped with serial manipulators, controlled by a visual servoing technique, is demonstrated in [19]. Differently from these works, we tackle the scenario of aerial tele-manipulation with haptic feedback of an object of interest via parallel/force control, using a passive tool.

In the second group, which encompasses the research presented in this paper, different works have focused on solutions to applications requiring less dexterous manipulation capabilities, but with the benefits of being more cost-effective, versatile, and lightweight, thus allowing for longer operational flight time. In [20], a quadrotor UAV equipped with a rigid tool, controlled based on a mapping between the desired vehicle attitude and the commanded force, is used to establish contacts with the environment. The control strategy therein is designed as a variation of near-hovering control, and does not take into account the friction cone constraints that allow maintenance of the contact between the robot tool-tip and the environment, while instead our strategy does. The authors of [21] propose instead an interesting combination of both mechanical design and control strategy to handle collisions and interaction in a more compliant way, without focusing on direct force control. A hybrid position/force control framework for a quadrotor is presented in [22], which allows the exertion of forces with the quadrotor airframe, without any tool, on the environment. A similar control approach is also adopted in [23] for the very relevant task of tool operation with quadrotors. Planning and control for an aerial robot in contact with its environment, based again on a hybrid position/force switching controller, is presented in [24], where obstacle avoidance is also performed. The control schemes of [22–24] are based on the decoupling of axes of the applied force and motion; that is, force is applied in the motion constrained axes while on the other axes the motion is controlled. Differently from [22–24] and other similar approaches, in this work, we do consider friction constraints in a compliant uni-lateral contact to avoid slipping of the tool in the non-constrained axes of motion and to maintain the contact, which allows us to generate 3D force vectors. For this reason, we use a parallel (and not hybrid) position/force control approach, which also deploys rotational degrees of freedom to attain a broader feasible range of forces. It is worth noting that the traditional parallel position/force controller applied to generic six-DoF grounded manipulators is not directly applicable to our system, as it is an under-actuated floating robot. All the wrench components of the contact interaction are transmitted to the robot CoM and affect its orientation, by which the position is controlled, which makes the problem more challenging, especially in the transition from free flight to contact interaction and vice versa.

Furthermore, in all the aforementioned papers, the use of haptic feedback is not envisioned. Haptic teleoperation of UAVs is mainly used for obstacle avoidance in free flight, aiming to improve the situational awareness of the human operator using haptic feedback, such as the generic hierarchical passive teleoperation control architecture presented in [25]. We use haptic feedback not only to improve the performance of position tracking in free flight, but also to reflect the applied force, in order to let the user feel the force in the contact interaction, which eventually leads to a more accurate tele-manipulation. The stability analysis and experimental results validate the proposed bilateral teleoperation scheme for aerial manipulation using passive tools.

To the best of our knowledge, the problem of teleoperating VTOL UAVs with haptic feedback to establish contact and apply forces on objects of interest while ensuring the compliance with friction constraints and avoiding slipping has not yet been deeply investigated by the community researching aerial physical interaction. We introduced the aerial haptic tele-manipulation idea in [26]. The present paper completes, improves, and extends this concept in the following ways:

(1)   the force control in [26] is based on a mapping that calculates the appropriate robot's desired orientation to generate the desired force; in this work, we use a more efficient sensor-based closed-loop force control;

(2)   the force controller considers the limited friction of the end-effector and object surfaces, and utilizes the independently controlled yaw motion, which allows a

wider range of feasible force commands by using the passive compliant spherical joint mechanism of the end-effector, enforcing the contact maintenance;

(3) the stability analysis of the system is presented;

(4) this work presents experimental results of tracking 3D force vectors applied to both stationary and moving objects.

2. Dynamic Model

The bilateral teleoperation system consists of: a human operator, a haptic device (master), an aerial manipulator (slave), and the remote environment (cf. Section 3). This section presents the dynamic model of the aerial manipulator (a quadrotor UAV equipped with a passive lightweight tool) in contact with the environment. The tool is rigidly connected to the top of the quadrotor, and a lightweight rigid link ensures enough room between the propellers and the end-effector to allow safe contact with the environment. A compliant spherical joint connects the lightweight rigid link to the end-effector, and an elastic shock absorber damper along with the compliant joint help to establish smoother contacts (Figure 2). In the following we assume that the spherical joint rotation is small, so the springs remain in their linear region, the weight of the tool is negligible, and the tool link is rigid.

Let us define the world frame $\mathcal{W} : \{O_\mathcal{W}, x_w, y_w, z_w\}$, the body frame $\mathcal{B} : \{O_\mathcal{B}, x_b, y_b, z_b\}$ for the robot, and the contact frame $\mathcal{C} : \{O_\mathcal{C}, o, t, n\}$ placed at the contact point (Figure 3). The position of $O_\mathcal{B}$ in $\mathcal{W}$ is indicated with $p = [x\,y\,z]^\top \in \mathbb{R}^3$, $R \in SO(3)$ is the rotation matrix representing the orientation of $\mathcal{B}$ in $\mathcal{W}$. We consider the RPY parameterization of R, that is, $\eta := [\phi\,\theta\,\psi]^\top \in \mathbb{R}^3$, where ϕ, θ, ψ are the roll, pitch and yaw angles, respectively, and are bounded as $\frac{-\pi}{2} < \phi < \frac{\pi}{2}$, $\frac{-\pi}{2} < \theta < \frac{\pi}{2}$, and $-\pi < \psi \leq \pi$. The end-effector position expressed in $\mathcal{W}$ is $p_e = p + Rd$, where d is the end-effector position vector in $\mathcal{B}$. The angular velocity of $\mathcal{B}$, denoted by $\omega \in \mathbb{R}^3$, is related to the derivative of Euler angles $\dot{\eta}$ by $\omega = E(\eta)\dot{\eta}$ where $E(\eta) \in \mathbb{R}^{3\times3}$ is defined according to R. The robot dynamics can be expressed in terms of robot pose $x = [p^\top\,\eta^\top]^\top \in \mathbb{R}^6$, in $\mathcal{W}$, as follows.

$$M_s(x)\ddot{x} + C_s(x,\dot{x})\dot{x} + g(x) = w + w_e, \tag{1}$$

where $M_s = diag\{mI_{3\times3}, M\}$, with $M = E^\top JE$, is the inertia matrix in which $m \in \mathbb{R}^+$, $J \in \mathbb{R}^{3\times3}$ are the robot mass and moment of inertia matrix, $C_s = diag\{0_{3\times3}, C\}$, with $C = E^\top(J\dot{E} + S(E\dot{\eta})JE)$ includes the Coriolis/centripetal dynamics in which $S(a)$ is the skew-symmetric matrix of a generic vector a; $g = [mgz_w^\top, 0_3]^\top$ is the gravity vector with g being the gravity acceleration constant. $w = [(u_zRz_b)^\top, u_\eta^\top]^\top$ is the robot control wrench in which the magnitude of the total thrust acting along the z_b direction is denoted with $u_z \in \mathbb{R}^+$, and $u_\eta \in \mathbb{R}^3$ is the rotational control moment. $w_e = [f_t^\top\ (S(d)f_t + \tau_r)^\top]^\top$ is the external wrench applied to the system. The wrench applied to the end-effector is modeled as spring wrench (f_t, τ_r), expressed in $\mathcal{W}$ as $f_t = -R_c^w K_t \delta p^c$ and $\tau_r = -R_c^w K_r \delta \eta^c$, where R_c^w is the rotation matrix representing the orientation of $\mathcal{C}$ w.r.t. $\mathcal{W}$, and $K_t = diag\{k_{to}, k_{tt}, k_{tn}\}$, $K_r = diag\{k_{ro}, k_{rt}, k_{rn}\}$ are the diagonal stiffness matrices, $\delta p^c = [\delta o\ \delta t\ \delta n]^\top$ is the compression of the linear spring, and $\delta \eta^c = [\delta \eta_o\ \delta \eta_t\ \delta \eta_n]^\top$ is the compression of the angular spring, expressed in $\mathcal{C}$, respectively.

We consider the *soft finger model* for the contact in which all three components of force and the normal component of the torque are transmitted in the contact independently [27]. Considering this model and the force torque balance for the end-effector, the wrench transmitted by the end-effector, *that is,* applied force $f = [f_o\,f_t\,f_n]^\top$ and torque $\tau = [0\,0\,\tau_n]^\top$, in $\mathcal{C}$, can be obtained as:

$$\begin{cases} f = -f_t^c - \frac{1}{\|d\|^2}S(d^c)\tau_r^c = K_t\delta p^c + \frac{1}{\|d\|^2}S(d^c)K_r\delta\eta^c \\ \tau_n = k_{rn}\delta\eta_n. \end{cases} \tag{2}$$

Equation (2) is used in the next section to define the control inputs of the force controller. The constraints of the force and torque to keep the contact are as follows:

$$
\begin{cases}
0 \leq f_n \\
\sqrt{f_o^2 + f_t^2}/\mu_s \leq f_n \\
|\tau_n|/\mu_t \leq f_n \\
\sqrt{\tau_o^2 + \tau_t^2}/r_d \leq f_n,
\end{cases}
\tag{3}
$$

where μ_s and μ_t are the linear and angular friction coefficients of the end-effector surface with the object surface, and r_d is the end-effector surface disk radius. The first constraint is the unilateral condition, the second one is to avoid translation slippage, the third one is to avoid rotational slippage, and the last one is to prevent the disk lifting up. The constraint (3) is used in the next section to modify the desired force vector in order to keep the contact.

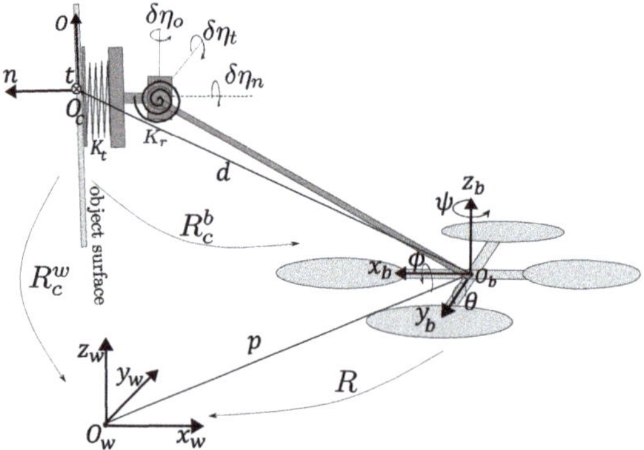

Figure 3. The coordinate frames, and components in the dynamic model of the aerial manipulator in contact with an object.

The dynamic model of the haptic interface, that is, the master robot, considering an inverse dynamic controller with gravity and nonlinear compensation [28], in the operational workspace, can be described as

$$
M_m \ddot{x}_m + K_{mD} \dot{x}_m = K_{mP} \tilde{x}_m + f_h - f_c,
\tag{4}
$$

where $x_m \in \mathbb{R}^{n_m}$ (n_m is the number of master robot's actuated DOFs is the master device pose, $\tilde{x}_m$ is the pose error, M_m is the diagonalized inertia matrix, K_{mP}, K_{mD} are the PD-controller gains regulating the desired master robot pose, f_h is the force applied by the human operator and f_c is the reflected teleoperation force.

3. Control System

The proposed bilateral teleoperation scheme controls the robot's position in the free flight, in an unlimited workspace using a limited workspace haptic device, and regulates the force tracking during physical interaction, keeping the contact based on the human operator's commanded force. The human operator is provided with force feedback proportional to the velocity in free flight and the applied force in contact. The overall teleoperation scheme is depicted in Figure 4, and the parallel position/force controller is shown in Figure 5.

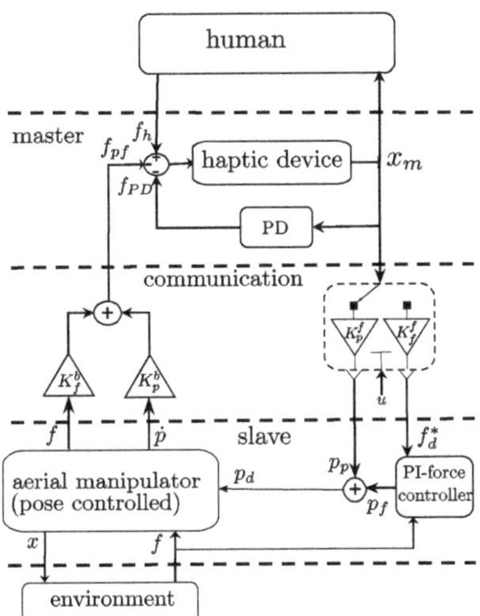

Figure 4. The bilateral teleoperation scheme with a parallel position/force controller in the slave side to control an aerial manipulator.

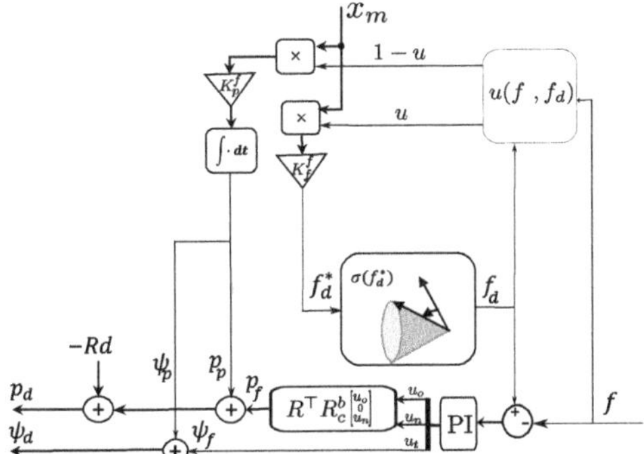

Figure 5. Parallel position/force controller for an aerial manipulator, in the remote side of the teleoperation scheme.

3.1. Position Control

The position control of the mechanically under-actuated quadrotor is implemented using a two layer cascade controller. The orientation is controlled using PID in the inner loop, and the outer loop provides the inner loop with reference roll and pitch (ϕ_d, θ_d) to control the robot planar motion using a gravity-compensated-PD. The yaw motion is controlled independently.

$$
\begin{cases}
u_z = \dfrac{m}{\cos(\phi)\cos(\theta)}\left(g - k_{pz}\tilde{z} - k_{dz}\dot{z}\right) \\[2mm]
u_x = -k_{px}\tilde{x} - k_{dx}\dot{x} \\[2mm]
u_y = -k_{py}\tilde{y} - k_{dy}\dot{y} \\[2mm]
\begin{bmatrix} \sin(\phi_d) \\ \sin(\theta_d) \end{bmatrix} = \dfrac{m}{u_z}\begin{bmatrix} \sin(\psi) & -\cos(\psi) \\ \dfrac{\cos(\psi)}{\cos(\phi)} & \dfrac{\sin(\psi)}{\cos(\phi)} \end{bmatrix}\begin{bmatrix} u_y \\ u_x \end{bmatrix} \\[4mm]
u_\eta = -K_D\dot{\eta} - K_P\tilde{\eta} - K_I \int_0^t \tilde{\eta}(s)\,ds,
\end{cases}
\tag{5}
$$

where $p - p_d = [\tilde{x}\ \tilde{y}\ \tilde{z}]^\top$ is the position error with $p_d \in \mathbb{R}^3$ being the desired position, $k_{px}, k_{dx}, k_{py}, k_{dy}, k_{pz}, k_{dz}, \in \mathbb{R}^+$ are proportional and derivative gains, $\tilde{\eta} = \eta - \eta_d$ is the orientation error with $\eta_d = [\phi_d, \theta_d, \psi_d]^\top$ being the desired orientation, and K_P, K_D, K_I are the proportional, derivative, and integral diagonal gain matrices to regulate the attitude.

3.2. Force Control

The force is regulated by providing the internal position control loop with an appropriate reference. The robot's desired position p_d and desired yaw ψ_d are four commanded states of the system, noting that the system is mechanically under-actuated. As depicted in Figure 5, the outer force regulating the feedback loop generates additional terms (p_f, ψ_f), depending on the force error, that are added to the previously commanded pose (p_p, ψ_p) in the free flight. As we aimed at contact interaction with the environment using the end-effector, we command the end-effector pose; therefore, we convert the end-effector pose to the COM pose by including the term $-Rd$ in the desired pose. Thus, p_d and ψ_d are expressed as:

$$\begin{cases} p_d = p_p + p_f - Rd \\ \psi_d = \psi_p + \psi_f. \end{cases} \tag{6}$$

In the sequel, we see how p_p, ψ_p, p_f, and ψ_f are generated from the user command, given by x_m, based on the aerial manipulator contact condition.

When the robot is in free flight condition, the master robot position (x_m) is interpreted as a position command; while when the end-effector comes in contact x_m is interpreted as the desired force. In order to distinguish the two conditions, let us introduce the contact function $u(f, f_d)$ as:

$$u(t_k) = \begin{cases} 1 & \text{if } f_n(t_k) > 0 \ \& \ f_n(t_{k-1}) > \epsilon \\ 0 & \text{if } f_n(t_k) = 0 \ \& \ f_{d,n}(t_k) < -\epsilon \\ u(t_{k-1}) & \text{otherwise.} \end{cases} \tag{7}$$

This hysteresis-like function is intended to prevent the chattering phenomenon in the attachment and detachment phases. In Equation (7), $f_n(t_k)$ is the normal component of the measured contact force at t_k instance and $f_n(t_{k-1})$ is the previous sample of the same measurement; $f_{d,n}(t_k)$ is the normal component of the commanded force, $u(t_{k-1})$ is the previous output of the function $(u(0) = 0)$, and $\epsilon \in \mathbb{R}^+$ is a small positive value. In the contact condition, if the commanded normal component of the force is negative $(f_{d,n}(t_k) < -\epsilon)$ and the normal component of the applied force is zero, the detachment takes place. During free flight $u(t) = 0$ and the component of the desired position that is intended to control the aerial manipulator position in free flight is generated by integrating the master position as:

$$p_p = p_d(t_0) + \int_{t_0}^{t} (1 - u(t)) K_p^f x_m \, dt, \tag{8}$$

where K_p^f is a 3×3 matrix that rotates and scales the master robot motion appropriately. The aerial manipulator heading ψ_p in free flight is directly commanded by the user, through mapping one of the master robot motions, similar to p_p.

When the end-effector establishes a contact $u(t) = 1$, and thus $(1 - u(t)) = 0$, in this case the integral stores the end-effector position at the contact moment, and the motion of the master robot during the contact interaction is used to command the desired force. The force commanded by the human operator, $f_d^* \in \mathbb{R}^3$, expressed in $\mathcal{C}$, is generated as:

$$f_d^* = u(t) K_f^f x_m, \tag{9}$$

where K_f^f is a 3×3 matrix that rotates and scales the master robot motion appropriately.

To prevent the desired force violating the contact constraint (3), the commanded force is modified as follows:

$$f_d = \begin{cases} f_d^* & \text{if } \frac{1}{\mu_s}\sqrt{(f_o^*)^2 + (f_t^*)^2} + \frac{1}{\sqrt{r_d^2 + \mu_t^2}}|\tau| \leq f_n^* \\ \sigma(f_d^*) & \text{otherwise,} \end{cases} \tag{10}$$

where $f_d^* = [f_o^*, f_t^*, f_n^*]^\top$, and $\sigma(f_d^*)$ is a function projecting the desired commanded force f_d^* to the surface of the contact constraints (3) as follows:

$$\sigma(f_d^*) = \cos(\alpha)f_d^* + \sin(\alpha)(S(a)f_d^*) + (1 - \cos(\alpha))(a^\top f_d^*)a + \frac{1}{\sqrt{r_d^2 + \mu_t^2}}|\tau|n, \tag{11}$$

where $a = (S(f_d^*)n)/\|(S(f_d^*)n)\|$ is the unit axis of rotation, $\alpha = \beta - \gamma$ is the required angle to rotate f_d^* around a to project it on the surface of the friction cone, β is the angle between f_d^* and n, and $\gamma = \tan^{-1}(\mu_s)$ is the translational friction cone angle. This function minimally increases the normal component (by adding $(r_d^2 + \mu_t^2)^{-0.5}|\tau|n$) to avoid torsional slippage and lifting up the end-effector disk, and applies the minimum rotation to its direction to keep it within the feasible force range.

The next step to generate p_f and ψ_f is to feed the force error $\tilde{f} = f - f_d = [\tilde{f}_o\ \tilde{f}_t\ \tilde{f}_n]^\top$ to a PI-controller as:

$$u_f = -K_{Pf}\tilde{f} - K_{If}\int_0^t \tilde{f}dt, \tag{12}$$

where $K_{Pf}, K_{If} \in \mathbb{R}^{3\times3}$ are proportional and integral diagonal gain matrices, respectively. The PI-output $u_f = [u_o\ u_t\ u_n]^\top$—after appropriate transformations—generates p_f and ψ_f.

At the contact moment, in which the springs are in rest position, we define the contact frame $\mathcal{C}$ which is its orientation w.r.t. $\mathcal{B}$ represented by rotation matrix R_c^b. Let $d = [d_x, 0, d_z]^\top$; we can always define $\mathcal{C}$ with a constant R_c^b such that $d^c = [d_o, 0, d_n]^\top$. Expanding (2) we get:

$$\begin{cases} f_o(\delta o, \delta\eta_t) = k_{to}\delta o - \frac{d_n k_{rt}}{\|d\|^2}\delta\eta_t \\ f_t(\delta t, \delta\eta_o, \delta\eta_n) = \frac{d_n k_{ro}}{\|d\|^2}\delta\eta_o + k_{tt}\delta t - \frac{d_o k_{rn}}{\|d\|^2}\delta\eta_n \\ f_n(\delta n, \delta\eta_t) = k_{tn}\delta n + \frac{d_o k_{rt}}{\|d\|^2}\delta\eta_t. \end{cases} \tag{13}$$

Therefore, f_n and f_o can be regulated by commanding the motion along n and t, respectively. To control f_t, we choose $\delta\eta_o$, and the reason is: due to the under-actuation of the quadrotor, changing δt requires changing the roll and pitch angles of the aerial manipulator and this results in applying an undesirable moment around the normal axis of the contact frame. The usage of rotation to generate the desired force also leads to a wider range of feasible forces, as it not only relies on the limited linear friction of the end-effector and object surfaces. $\delta\eta_o$ can be considered as changing the yaw angle ψ (see Figure 3). Therefore, p_f and ψ_f can be obtained as follows:

$$\begin{cases} p_f = R^\top R_c^b [u_o\ 0\ u_n]^\top \\ \psi_f = u_t. \end{cases} \tag{14}$$

3.3. Master Control and Haptic Feedback

The input of the master robot (haptic device), as expressed by (4), receives two elements from the teleoperation scheme: the human force f_h and the haptic feedback force f_c, which is itself constituted by two parts f_{PD}, and f_{pf}. The term f_{PD} is a negative proportional derivative term, based on master position x_m, that is intended to bring back the device to its zero position gently when the device is not moved by the operator, so that the quadrotor will not move or apply force when the haptic device handle is released. On the other hand, f_{pf} is the haptic feedback given to the user depending on the velocity in free flight or force error in contact. The haptic feedack f_c is synthesized as:

$$f_c = f_{PD} + f_{pf} = (K_{Ph}x_m + K_{Dh}\dot{x}_m) + (K_f^b f + (1-u)K_p^b \dot{p}), \tag{15}$$

where $K_f^b, K_p^b, K_{Ph}, K_{Dh} \in \mathbb{R}^{3\times3}$ are positive diagonal gains.

4. Stability Analysis

We first show the stability of the rotational dynamics in contact, then the stability of the force controller, and finally the stability of the teleoperation scheme. In order to facilitate the tractability, let C and W coincide, $R_c^w = I_3$, and at the contact instance $R_c^b = I$ (see Figure 3).

4.1. Rotational Stability

The angular dynamics $M(\eta)\ddot{\eta} + C(\eta,\dot{\eta})\dot{\eta} + K_r\eta + \tau_f = u_\eta$, where $\tau_f = -S(d)f_t$ has the following properties:

- M is symmetric positive definite and bounded as $\mu_M \leq \|M\| \leq \gamma_M$; where μ_M, γ_M are its minimum and maximum eigenvalues. Moreover, M is strictly diagonally dominant (SDD), that is $M_{ii} > \Sigma_{i \neq j}|M_{ij}|$;
- $\dot{M} - 2C$ is skew-symmetric, or $\dot{M} = C + C^\top$;
- $\|C(\eta,\dot{\eta})\| \leq \delta_c\|\dot{\eta}\|$, where $\delta_c \leq 2\sigma_{max}(J)$;
- the rotational stiffness matrix is bounded as $\mu_k \leq \|K_r\| \leq \gamma_k$.

Theorem 1. *Applying the rotational part of the control law* (5) *to the aerial manipulator described by* (1)*, its rotational dynamic is locally asymptotically stable such that $\tilde{\eta}, \dot{\eta} \to 0$ and $\int_0^t \tilde{\eta}(s)ds \to -K_I^{-1}(K_r\eta + \tau_f)$ as $t \to \infty$.*

Proof. Let $\zeta = [\Delta\eta^\top \tilde{\eta}^\top \dot{\eta}^\top]^\top$, where $\Delta\eta = \int_0^t \tilde{\eta}(s)ds + K_I^{-1}(K_r\eta + \tau_f)$, and consider the following scalar function:

$$V(\zeta) = \frac{1}{2}\zeta^\top P\zeta, \tag{16}$$

where the symmetric matrix P is defined as follows:

$$P = \begin{bmatrix} K_I & K_{12} & \varepsilon M \\ K_{12} & K_{22} & \varepsilon M \\ \varepsilon M & \varepsilon M & M \end{bmatrix}, \tag{17}$$

with $\varepsilon = k_\varepsilon \frac{\mu_M}{\gamma_M}$ and $0 < k_\varepsilon < 0.5$, $K_{12} = K_I + \varepsilon K_D - K_r$, and $K_{22} = K_P + \varepsilon K_D - K_{12}K_I^{-1}K_r$. Let $K_s = k_s I$ for $s = \{P, I, D\}$ with $k_s > 0$; in a range satisfying inequities,

$$\begin{aligned} 0 < k_D < \frac{\mu_K}{\varepsilon} - \gamma_M, \quad \varepsilon < \frac{\mu_K}{\gamma_M} \\ k_p > k_I + \frac{\gamma_K}{k_I}(\varepsilon k_D - \mu_K) + \varepsilon\gamma_M. \end{aligned} \tag{18}$$

P becomes SDD with positive diagonal elements, and is therefore positive definite. Thus, $V(\zeta)$ is a positive definite function and hence a Lyapunov function candidate, which is radially unbounded and satisfies the Rayleigh–Ritz inequality [29] as:

$$\mu_P\|\zeta\|^2 \leq V(\zeta) \leq \gamma_P\|\zeta\|^2, \tag{19}$$

where μ_P and γ_P are the minimum and maximum eigenvalues of P. The time derivative of the Lyapunov candidate (16) is:

$$\dot{V} = -\zeta^\top Q\zeta + \zeta^\top B[\dot{\tau}_f^\top \dot{\eta}_d^\top]^\top, \tag{20}$$

with

$$B = \begin{bmatrix} I & -K_{12} \\ K_{12} & -K_{22} \\ \varepsilon K_I^{-1} M & -\varepsilon M \end{bmatrix} \tag{21}$$

$$Q = \begin{bmatrix} \varepsilon K_I & \frac{1}{2}((\varepsilon-1)K_I + \varepsilon K_P) & -\frac{1}{2}\varepsilon C^\top \\ \frac{1}{2}((\varepsilon-1)K_I + \varepsilon K_P) & \varepsilon K_P - (K_I + \varepsilon K_D - K_r) & -\frac{1}{2}\varepsilon(C^\top + M) \\ -\frac{1}{2}\varepsilon C^\top & -\frac{1}{2}\varepsilon(C^\top + M) & K_D - \varepsilon M(I + K_I^{-1}K_r), \end{bmatrix} \tag{22}$$

In the following we shall show that:

$$\dot{V}(\zeta) \le -\zeta_1^\top Q_1 \zeta_1 + \zeta^\top B[\tau_f^\top \ \dot{\eta}_d^\top]^\top , \tag{23}$$

where $\zeta_1 = [\|\Delta\eta\| \ \|\tilde{\eta}\| \ \|\dot{\eta}\|]^\top$ and

$$Q_1 = \begin{bmatrix} \varepsilon k_I & \frac{1}{2}((\varepsilon-1)k_I + \varepsilon k_P) & \frac{1}{2}\varepsilon\delta_C\|\dot{\eta}\| \\ \frac{1}{2}((\varepsilon-1)k_I + \varepsilon k_P) & \varepsilon(k_P - k_D) + \mu_K - k_I & \frac{1}{2}\varepsilon(\gamma_M + \delta_C\|\dot{\eta}\|) \\ \frac{1}{2}\varepsilon\delta_C\|\dot{\eta}\| & \frac{1}{2}\varepsilon(\gamma_M + \delta_C\|\dot{\eta}\|) & k_D - \varepsilon\gamma_M(1 + \frac{\gamma_K}{k_I}) \end{bmatrix}. \tag{24}$$

To obtain (23), we choose k_P, k_I, k_D such that:

$$\begin{aligned} \varepsilon(2\mu_K + \gamma_M) &< k_I < \mu_K \\ (\tfrac{1}{\varepsilon} - \varepsilon)k_I &< k_P < 2\mu_K + \gamma_M \\ \varepsilon\gamma_M(\tfrac{3}{2} + \tfrac{\gamma_K}{k_I}) &< k_D < \mu_K(1 - \varepsilon) \end{aligned} . \tag{25}$$

Consequently, ε must be chosen such that

$$\varepsilon < min\left\{ \frac{\mu_K - k_I}{\mu_K - \frac{k_P}{2} + \frac{\gamma_M}{2}}, \frac{2}{3}\frac{\mu_K(2\mu_K + \gamma_M) - \gamma_K\gamma_M}{\gamma_M(2\mu_K + \gamma_M)}, \frac{\mu_M}{2\gamma_M}, \frac{\mu_K}{\gamma_M} \right\}. \tag{26}$$

Then, it is also assumed that the spherical joint mechanical stiffness is chosen properly in accordance with the system moment of inertia such that $\mu_K > \frac{\gamma_M}{2}(1 + \frac{\gamma_M}{\mu_K})$, which can be simplified as $\sigma_{min}(K_r) > \gamma_M$, which simply means that the higher the inertia, the stiffer the spring that must be chosen. Let $\mu_v = min\{k_I - \varepsilon k_P, \ \varepsilon k_P + 2(\mu_K - \varepsilon k_D) - (\varepsilon + 1)k_I - \varepsilon\gamma_M, \ k_D - \frac{3}{2}\varepsilon\gamma_M - \frac{\varepsilon\gamma_M\gamma_K}{k_I}\}$, for trajectories bounded by:

$$\|\dot{\eta}\| \le \frac{1}{\delta_C}\mu_v. \tag{27}$$

Q_1 will be SDD with positive diagonal elements, and therefore positive definite. Thus, we can conclude:

$$\dot{V}(t, \zeta) \le -\mu\|\zeta\|^2 + \kappa\|\zeta\| , \tag{28}$$

where $\mu = \sigma_{min}(Q_1)$ and $\kappa = \delta_{B1}\delta_f + \delta_{B2}\delta_v$ with $\delta_{B1} = max\{1, \ k_I + \varepsilon k_D - \mu_K, \ k_P + \varepsilon\gamma_M/k_I\}$, $\delta_{B2} = max\{k_I + \varepsilon k_D - \mu_K, \ k_P + \varepsilon k_D - \mu_K(k_I + \varepsilon k_D - \mu_K)/k_I, \ \varepsilon\gamma_M\}$, and $\|\dot{\eta}_d\| \le \delta_v$ is the upper bound on the norm of $\dot{\eta}_d$. $\|\tau_f\| \le \delta_f$ is the upper bound on the norm of the τ_f, which is a reasonable assumption considering that the rotational dynamics is faster than translational [30]. Let W be the positive root of $V = W^2$, as in [31], considering (19) we can state $\dot{W} \le -\frac{\mu}{2\gamma_P}W + \frac{\kappa}{2\sqrt{\mu_P}}$, which means:

$$W(t) \le W(0)e^{-\frac{\mu}{2\gamma_P}t} - \frac{\kappa}{2\sqrt{\mu_P}}\int_0^t e^{-\frac{\mu}{2\gamma_P}(t-s)}ds \le W(0)e^{-\frac{\mu}{2\gamma_P}t} + \frac{\kappa\gamma_P}{\sqrt{\mu_P}\mu}. \tag{29}$$

Considering (19), we can ensure that (27) is satisfied if $W(0) + \frac{\kappa\gamma_p}{\sqrt{\mu_p\mu}} \leq \frac{1}{\delta_C}\mu_v$. Therefore, choosing sufficiently large k_P, and sufficiently small k_I and k_D such that (18) and (25) are satisfied, the solution of the error system converges to zero asymptotically. $\square$

4.2. Force Control Stability

The stability of the force dynamic in the closed-loop force controlled system is investigated in a decentralized manner, that is, each component of the force is analyzed independently by considering the effect of other state variables as disturbances.

4.2.1. Force Along n-Axis

The forces along n-axis is controlled by translational pose command along n, from (13) we have

$$f_n = k_{tn}n + k_j\eta_t \, , \tag{30}$$

where $k_j = \frac{d_o k_{rt}}{\|d\|^2}$. If we substitute the pose controller in (30), assuming the gravity term compensated by feedback linearizing term, the dynamics will be:

$$m_n\ddot{n} + k_{tn}n = -k_p\tilde{n} - k_d\dot{n} \, , \tag{31}$$

where $m_n = m$ is derived from (1). Substituting n and its derivatives, in the left hand side of (31), with f_n from (30), and control terms, in the right hand side, with (12) we obtain:

$$m_f\ddot{f}_n + b_f\dot{f}_n + k_f f_n = -k_1 \int_0^t \tilde{f}_n - k_2\tilde{f}_n + h_n(\ddot{\eta}_t, \dot{\eta}_t, \eta_t) \, , \tag{32}$$

where $m_f = m_n k_{tn}^{-1}$, $b_f = k_d k_{tn}^{-1}$, $k_f = (1 + k_p k_{tn}^{-1})$, $h_n = (m_n k_{tn}^{-1}k_j\ddot{\eta}_t + k_d k_{tn}^{-1}k_j\dot{\eta}_t + (1 + k_p)k_{tn}^{-1}k_j\eta_j)$, and $k_1 = k_p k_{If}$, $k_2 = k_p k_{Pf}$.

One can express (32) in the frequency domain by introducing the controller transfer function $C(s) = k_2 + \frac{k_1}{s}$ and plant transfer function $G(s) = 1/(m_f s^2 + b_f s + k_f)$, where $(s = \sigma + j\omega)$. The system output $f(s)$ is then obtained as:

$$f(s) = \frac{G(s)C(s)}{1 + G(s)C(s)}f_d(s) + \frac{1}{1 + G(s)C(s)}h(s) \, . \tag{33}$$

For the stability of the system, the characteristic polynomial of the system, that is, $s(m_f s^2 + b_f s + k_f) + (k_2 s + k_1)$, must have all roots with a real negative part and, to achieve this, according to the Routh–Hurwitz stability criterion it is required that:

$$k_{Pf} + \frac{1}{k_P} + \frac{1}{k_{tn}} > \frac{m}{k_D}k_{If} \, . \tag{34}$$

Choosing sufficiently high PD gains and an appropriately low I-gain for slowly-varying force commands, that is, $s \to 0$, we obtain $\frac{G(s)C(s)}{1+G(s)C(s)} \to 1$, $\frac{1}{1+G(s)C(s)} \to 0$, thus $f \to f_d$.

4.2.2. Force Along o-Axis

The forces along o-axis are controlled by the translational pose command along o and the stability analysis is the same as the force along the $n - axis$.

4.2.3. Force Along t-Axis

The forces along the t-axis are controlled by the rotational pose command around o, that is, $\eta_o = \psi$ considering the frame convention. From (14) one can write:

$$f_t = k_o\psi + [k_{tt} \ k_{rn}][t \ \eta_n]^\top \, , \tag{35}$$

where $k_o = \frac{d_n k_{ro}}{\|d\|^2}$ and $k_{rn} = \frac{d_o k_{rn}}{\|d\|^2}$. If we substitute the orientation controller in the dynamic model, considering the results of Theorem 1, we obtain:

$$m_{33}\ddot{\psi} + (c_{33} + k_D)\dot{\psi} = -k_p\tilde{\psi} - (c_{31}\dot{\phi} + c_{32}\dot{\theta})\,, \tag{36}$$

where m_{ij}, c_{ij} are extracted from M and C. It is worth noting that m_{33} and c_{33} are constant with respect to ψ and its derivatives. Substituting ψ and from (35) in (36) we get:

$$\frac{m_{33}}{k_o}\ddot{f}_t + \frac{k_D + c_{33}}{k_o}\dot{f}_t + \frac{k_P}{k_o}f_t = -k_1\int_0^t \tilde{f}_t - k_2\tilde{f}_t + h_t\,, \tag{37}$$

with

$$\begin{aligned}
h_t = \; & -(c_{31}\dot{\phi} + c_{32}\dot{\theta}) + \left(\tfrac{m_{33}k_t}{k_o}\ddot{t} + \tfrac{k_t(k_D+c_{33})}{k_o}\dot{t} + \tfrac{k_t k_P}{k_o}t\right) + \\
& + \left(\tfrac{m_{33}k_n}{k_o}\ddot{\eta}_n + \tfrac{k_n(k_D+c_{33})}{k_o}\dot{\eta}_n + \tfrac{k_n k_P}{k_o}\eta_n\right)\,.
\end{aligned} \tag{38}$$

Following the same procedure of force along the n-axis, the Routh–Hurwitz criterion enforces the controller coefficients to be chosen as:

$$k_{Pf} + \frac{1}{k_o} > \frac{m_{33}}{k_D + c_{33}}k_{If}\,, \tag{39}$$

which is fulfilled by setting appropriately high proportional and derivative gains and sufficiently low integral gain, and for slowly-varying force commands $f_t \to f_d$.

Choosing coefficients according to (34) and (39) makes the system over damped, that is, without overshoot, which is that f is not getting higher than f_d. Thus, if f_d is the output of (10), contact maintenance is ensured.

4.3. Stability of Teleoperator in Contact Interaction

During the contact interaction, from (4) we can define the dynamics of the master robot and its controller in the frequency domain as: $G_r(s) = M_m s^2 + K_{mD}s$, $C_m(s) = K_{mP}$. The haptic feedback in the master side (PD term) could also be expressed as: $C_h(s) = K_{hP} + K_{hD}s$. We have shown that the force interaction dynamics and force controller in the remote side could be expressed as: $G_s(s) = diag(M_f s^2 + B_f s + K_f)^{-1}$, $C_m(s) = K_2 + K_1\frac{1}{s}$. We define the transfer function of the master robot and its controller as: $G_m(s) = C_m G_r(I + C_m G_r)^{-1}$. The internal stability of the system requires that the roots of the denominator of G_m all have real negative parts, for which $K_{mP}, K_{mD} > 0$ suffice. The overall transfer function of the master side, with force input and position output, is defined as:

$$G_1(s) = G_m(I + G_m C_h)^{-1}\,. \tag{40}$$

The characteristic polynomial of G_1 is $(M_m K_{mP} K_{hD})s^3 + (M_m K_{mP}(1 + K_{hP}) + K_{mD}K_{mP}K_{hD})s^2 + (K_{mD}K_{mP}K_{hP} + K_{mP}K_{mD})s + 1$. For internal stability, the controller coefficients must conform with the following constraint:

$$k_{mD}k_{mP}k_{hP} + k_{mP}k_{mD} > \frac{\sigma_{max}(M_m)k_{hP}}{\sigma_{min}(M_m)(1+k_{hP})+k_{mD}k_{hD}}\,. \tag{41}$$

The transfer function of the slave side, with force input and force output, is obtained as: $G_2(s) = G_s C_s(I + G_s C_s)^{-1}$. Its internal stability constraints are expressed by (34) and (39). The teleoperator output f (see Figure 6), can be obtained as:

$$\begin{aligned}
f(s) = \; & G_2 K_f^f G_1(I + G_2 K_f^f G_1 K_f^b)^{-1} f_h(s) \\
& + (I + G_s C_s)^{-1}h(s).
\end{aligned} \tag{42}$$

The constraint (34), (39) and (41) gives all the poles of G_1 and G_2 a negative real part; therefore, G_1 and G_2 are strictly positive real and thus passive systems, and the negative feedback interconnection of a passive system is a passive system [32]; thus, (42), that is, the teleoperator in the contact interaction, is stable.

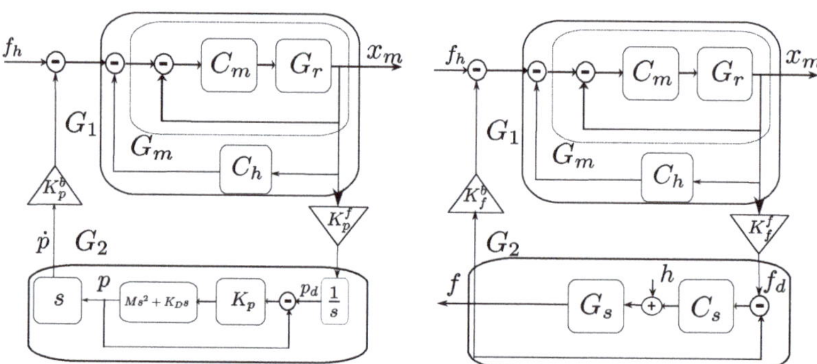

Figure 6. Block diagram of teleoperation system using the aerial manipulator in free flight (**left**) and in contact interaction (**right**).

4.4. Stability of Teleoperator in Free Flight

During the free flight, assuming fast rotational dynamics compared to translational dynamics, and gravity compensation, we may express the slave robot and its controller as $G_s(s) = Ms^2 + K_D s$ and $C_s(s) = K_p$, the input to G_s is the integrated value of scaled x_m, and the force feedback to the master side is the robot velocity. Therefore, G_2 can be expressed as (Figure 6):

$$G_2 = (\frac{1}{s}I)(G_s C_s (I + G_s C_s)^{-1})(Is) = G_s C_s (I + G_s C_s)^{-1}, \tag{43}$$

which is internally stable by choosing pose controller gains $k_P, k_D > 0$. The master side of the teleoperator is the same as in the contact interaction; therefore, G_1 does not change. Considering velocity $\dot{p}$ as the output of the system, it can be obtained as:

$$s\boldsymbol{p}(s) = G_2 K_p^f G_1 (I + G_2 K_p^f G_1 K_p^b)^{-1} f_h. \tag{44}$$

$s\boldsymbol{p}(s)$, that is, velocity in frequency domain, is the system output constituted by the negative feedback interconnection of passive systems G_1 and G_2. Therefore, the teleoperator in free flight is passive and stable.

5. Experimental Results

In order to evaluate the proposed aerial tele-manipulation solution, and to assess the functionalities of the proposed controller in the bilateral teleoperation scheme, two experiments with fixed and movable objects were performed. In the first experiment, a human operator drives the aerial manipulator to establish a contact with a stationary target and applies force to it, receiving force feedback. In the second experiment, the human operator drives the quadrotor to establish a contact with a wheeled cart and pushes it to generate motion, while receiving force feedback.

We encourage the interested reader to watch the video of the experiments in the multimedia attachment to this paper, cf. Supplementary Materials, to better appreciate the presented validation.

5.1. Experimental Setup

Our aerial manipulator was equipped with a lightweight tool (Figure 2) with a total weight of 0.05 kg. It was rigidly connected to the top of a quadrotor UAV with 1.0 kg weight. The quadrotor platform used for the experiments was a Mikrokopter$^{©}$ x4 platform (HiSystems GmbH). The distance vector from the quadrotor COM to the end-effector surface was $d = [0\,m, 0.5\,m$ and $0.2\,m]$. The link lengths were $d_1 = 0.51\,m$, and $d_2 = 0.02\,m$ long. A compliant spherical joint, that connects the lightweight rigid link to the end-effector, and an elastic shock absorber damper on the end-effector helped to establish smooth contacts. The end-effector surface was covered by a high friction material to expand the feasible force vector range.

The robot positioning was performed by a Vicon tracking system (Vicon Capture Systems, London, UK). The quadrotor thrust and rotational controller was implemented on its onboard microcontroller (Atmel AVR 8-bit, ATmega-1284, running at 20 MHz), based on the inertial sensors of the robot, the rest of the control law was implemented on an external PC (Core i7, 16GB RAM, running Ubuntu 14), communicating with the robot using a pair of Zig-Bee transceiver chips. An Omega.3 haptic device (Force Dimension, Nyon, Switzerland) was used as the master device. We used the force/torque ATI sensor (ATI Industrial Automation, Apex, NC, USA) embedded inside the object to measure the applied force by the aerial manipulator. The software was implemented in the ROS, and all control loops ran at a frequency of 100 Hz.

5.2. Results

5.2.1. Stationary Object Experiment

Initially the quadrotor was located at the origin of the global coordinate, while the object, which was a $0.15\,m \times 0.15\,m$ plate, was located at $[1.0\,m, 0.0\,m$ and $-0.85\,m]$ with downward pointing z_w. The human operator drove the quadrotor towards the object, and once the robot reached the object, the driver commanded a variable continuous force vector, by means of the haptics device. Finally, the human operator commanded the robot to leave the object, and brought it back to free flight. Figure 7-top shows the snapshots of different moments of the experiment, while Figures 8-left, 9-left, and 10-left show the results of the experiment.

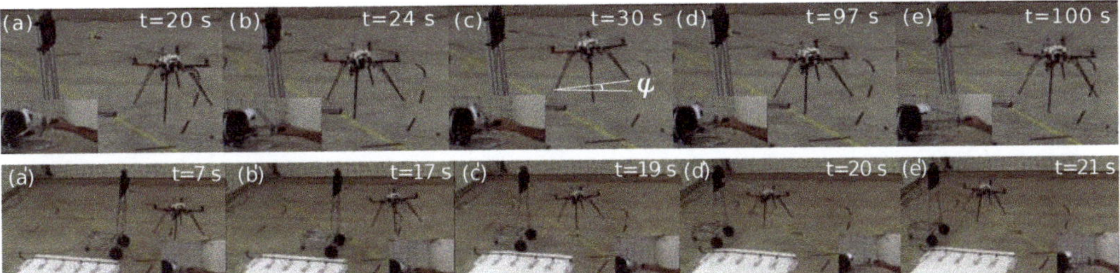

Figure 7. Snapshots of stationary object (**top**) and movable object (**bottom**) experiments. The experiments aimed to dock and apply a commanded time-varying force to the objects. (**a,a′**) In free flight, (**b,b′**) establishing a contact with the object, (**c,c′**) applying force vectors, (**d,d′**) releasing the contact, and (**e,e′**) in free flight again.

Figure 8a shows the position tracking during the experiment, while the position error is explicitly reported in Figure 9-left, for the reader's convenience. As can be seen, before the contact event, and after the quadrotor leaves the object, the position error is low, that is, bounded below 7 cm, while during the contact there is significant position error, that is, ≈35 cm, specifically in x-direction which corresponds to the normal component of force. It is worth underlining the fact that, during the interaction, the position error does not provide relevant information for evaluating the controller's performance, as during that phase the platform is force-controlled, and the position error represents a necessary condition to guarantee the force tracking. As a matter of fact, a higher position error during the contact with stationary object means a higher thrust and a higher pitch angle is demanded, which results in larger applied forces.

Figure 8b shows the contact maintenance function output, which keeps the desired force vector inside the friction cone. It can bee seen that, when the commanded force violates the friction constraints of (10), the normal component of the desired commanded force is increased while the other two tangential components are decreased, and the desired norms of the input (f_d^*) and output (f_d) are equal.

Figure 8c shows the force tracking control result during the experiment. As is evident, the force tracking performance is very good (average absolute error of 0.11 N, which represents less than 10% relative error, and a maximum absolute error of 0.51 N). The contact transition is shown in a separate window inside Figure 8c, which introduces a very smooth contact transition with only a small single bouncing event. Note that in the transition phase of the robot force control with a rigid environment, having a small amount of bouncing and inadvertent losses of contact are common, even for grounded robotic arm manipulators with non-zero reaching velocity [33].

The haptic feedback components are shown in Figure 10-left. The master position (the position of the haptic device's handle with respect to the center of its workspace) is shown in creating the spring-damper force f_{PD} that brings back the master device's handle to the center of its workspace (Figure 10a). The position tracking error of the aerial manipulator in the free flight condition along with the force applied to the object in contact create the second constitutive component of the haptic feedback f_{pf}, which is shown in Figure 10b. The haptic feedback in the contact condition is equal to the measured force from the force sensor with the opposite direction.

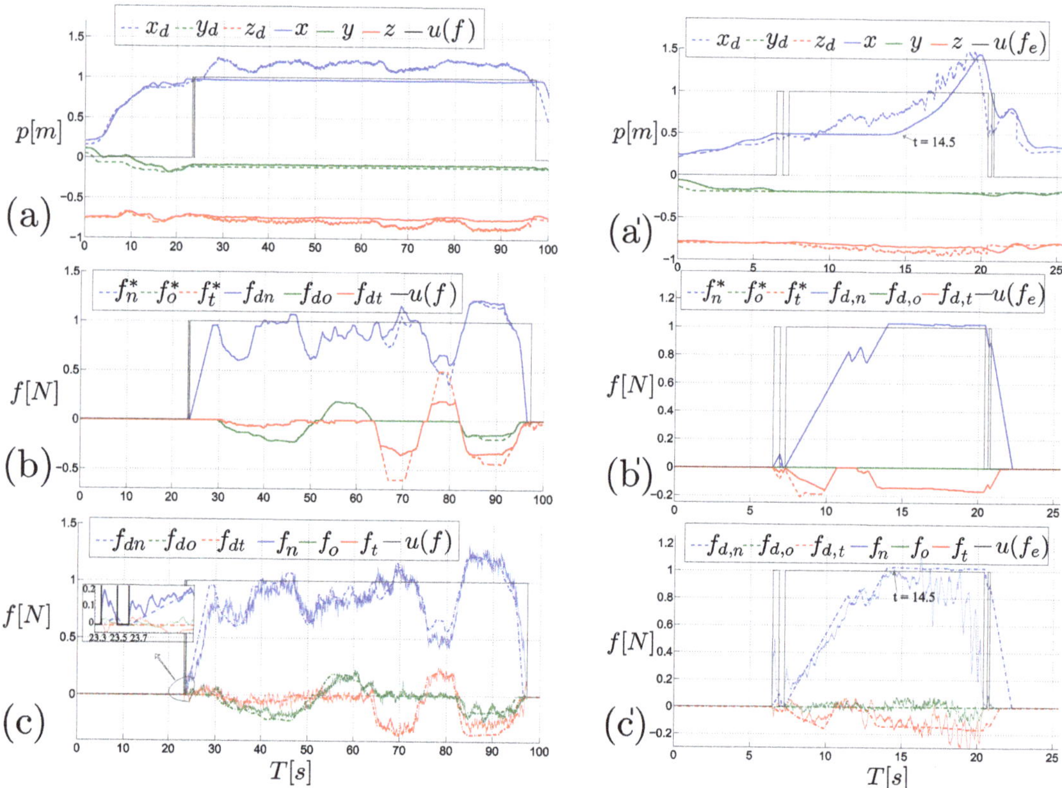

Figure 8. Stationary object experiment (**left**): docking and applying a force commanded by the human operator to a stationary object: (**a**) position tracking, (**b**) contact maintenance, that is, keeping of the desired force within the friction cone, and (**c**) force tracking. Movable object experiment (**right**): docking and applying force to a movable object in order to push it 1 m away: (**a′**) position tracking, (**b′**) contact maintenance, that is, keeping of the desired force within the friction cone, and (**c′**) force tracking.

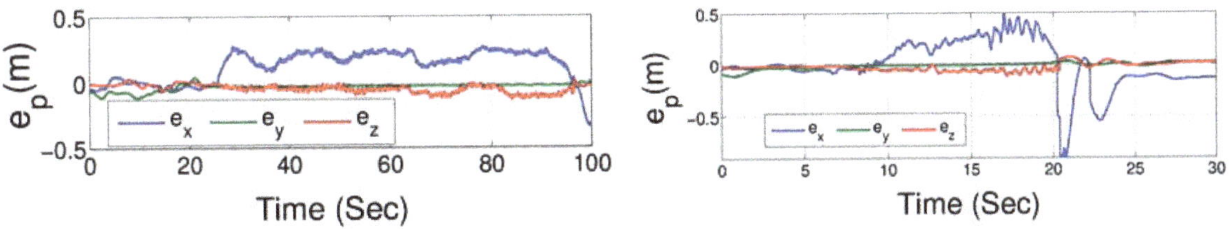

Figure 9. Position error in the stationary object experiment (**left**) and movable object experiment (**right**): as can be seen, the error is bigger during the interaction phase, in both cases, as a consequence of the force tracking. Furthermore, the error is higher in the movable object experiment, in particular during the undocking phase.

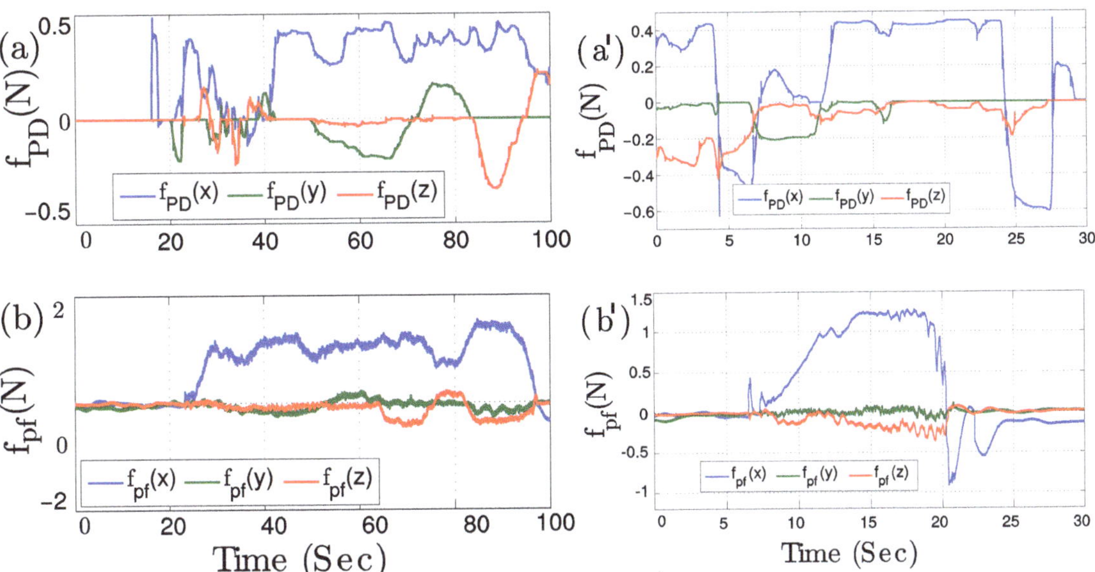

Figure 10. Stationary (**left**) and movable (**right**) object experiments: (**a,a′**) spring–damper force brings back the master device's handle to the center of its workspace, (**b,b′**) the haptic feedback f_{pf} that renders the applied force to the object in contact and position error in free flight.

5.2.2. Movable Object Experiment

In the second experiment, the robot was initially located at the global frame origin. The movable object, which was a cart with plate of $0.15\,\text{m} \times 0.15\,\text{m}$ attached to it, was located at $[0.5\,\text{m}, 0.0\,\text{m}$ and $-0.90\,\text{m}]$, with downward pointing z_w. The human operator was driving the quadrotor towards the cart, and once the robot reached the cart, the driver pushed it until it passed $1.0\,\text{m}$, and eventually the driver commanded the robot to leave the object, and brought the robot back to free flight. Figure 7-bottom shows the snapshots of this experiment, while Figures 8-right, 9-right, and 10-right depict the results of the experiment.

Figure 8a′ shows the position tracking during the experiment. The difference between the new situation (a movable object) compared to the previous experiment (a stationary object) can be seen through this plot, where after establishing the contact, while the quadrotor is applying the force to the object (cart), its position is not changing until the moment (in this experiment at 14.5 s) the applied force overcomes the static friction of the cart wheels with the ground. Then, the cart (and the aerial manipulator too) experiences an accelerated motion. The position error is explicitly reported in Figure 9-right), for the reader's convenience. As in the previous scenario, before the interaction the error is relatively close to zero, that is, bounded below 7 cm, while during the contact there is significant position error, that is, $\approx$50 cm, specifically in x-direction which corresponds to the normal component of force. As already mentioned, this is in accordance with the fact that, in this phase, the robot is tracking the desired force provided by the user, and not the desired position. During the undocking phase, namely when the human operator sees the accelerating cart passing the goal line and pushes back the handle of the haptic device in order to bring the robot back, a peak of $\approx$75 cm is reached due to a small false contact detection, after which the position error converges back to its typical free-flight values. The large error value in the detachment phase can be improved by limiting the integral term of the PI-force controller, and increasing the proportional gain of PI-controller, or also using a more agile controller in the inner position loop. It is worth noting that, in this experiment, the goal was to evaluate the efficiency of the proposed controller in applying force to objects in order to move them in a way the human operator wants, rather than performing precise force tracking.

In relation to this last point, another meaningful note is that the two experiments here discussed were explicitly designed and performed with the main goal of validating the presented aerial robotic solution in indoor laboratory conditions, which imply the use of a motion capture system for state estimation, a cabled connection between the robot and the ground workstation, and the absence of non-negligible external disturbances. Under these conditions, we basically experienced a rate of success of 100%. A more fair and comprehensive validation and testing of the this system in outdoor, mocap-denied, and realistic scenarios in the presence of external disturbances and using only onboard computation, which is not in the scope of this paper, will be the subject of future work.

Figure 8b' shows the contact maintenance function input and output forces. Since the main purpose is to push the cart, the major component of the commanded force is along the normal direction; therefore, most of the time the feasible commanded force is the same as the desired commanded force by the human operator and, as is evident from Figure 8c', the contact is kept during the force interaction.

The force tracking control results, shown in Figure 8c', represent a more difficult task compared to the previous task (force tracking control in contact with a stationary object). As can be seen, while the cart is stationary, that is, when the force applied by the quadrotor end-effector is compensated for by the friction of the cart wheels with the surface, the force tracking has a similar performance to the previous experiment. On the other hand, once the cart starts to move the error increases. However, considering the particular application of this experiment, which aims at pushing a cart forward, as far as the contact is kept and the operator is able to accomplish the task, the experiment is considered successful, although the average of the absolute force error norm is 0.5 N, that is a 20% relative error.

The haptic feedback components are shown in Figure 10-right. f_{PD} is depicted in Figure 10a', and the haptic feedback component related to the slave side, f_{pf}, is shown in Figure 10b'. As is evident, the magnitude of the latter is generally bigger than the magnitude of the former, allowing the human operator to be aware of the aerial manipulator condition in the master side (this is also true for the experiment with the stationary object). Haptic feedback in this experiment is even more important for the human operator to not lose the contact, as f_{pf} allows the human operator to feel the applied force during the contact interaction. When the perceived force is small, it means that the contact is weak, and the human operator can command a higher force to ensure the contact maintenance.

5.3. Discussion

The experimental results validated the capability and efficacy of the aerial manipulator with a passive tool in free flight pose tracking and in contact force tracking while keeping the contact and maintaining the flight stability. After performing several experiments to investigate the properties of the system and the controller, the following results from interpreting observations were obtained. In the free flight phase, the autonomous control of pose is not very precise, and there are position/velocity errors that prevent the autonomous controller from doing very fine tasks, even in the case of knowing the environment perfectly.

This happens due to the existence of aerodynamics disturbances and unmodeled dynamics that are not incorporated in the controller design. However, a human teleoperated aerial system is capable of doing such tasks, thanks to superior human learning and sensory motor capabilities. A similar observation has been made in [8].

The contact establishing phase (docking) involves bouncing; this is a natural behavior because the reaction force of the contact is directly applied to the robot CoM that is a floating (hovering) object, and this results in jumping back; unlike grounded manipulators this force is not transmitted to the ground. To mitigate this effect, velocity slowing-down policy and physical shock absorber on the end-effector surface is used, that results in a smooth transition and reduces the bouncing effect significantly. Moreover, deploying f_{PD} in the haptic feedback prevents the operator from loosely grasping the master robot, which could increase the effect. Note that in establishing a contact with non-zero reaching velocity, having a small amount of bouncing is commonly acceptable, even for grounded robotic arm manipulators [33].

Contact maintenance function, in conjunction with appropriate gain tunings (to avoid overshoots as mentioned in the stability analysis section) plays an important role; without respecting these constraints the end-effector could slide on the surface. Therefore, using this function is necessary if the task involves keeping the contact, unlike the hybrid position-force controllers such as in [2,24,34] in which it is desirable to slide over the surface.

The force error constituted by a non-zero-mean part (DC) and a high frequency signal (AC). The DC part is the effect of the not-force-controlled axes of motion (due to the under-actuation and uni-lateral contact), and the AC part is because of the propeller rotational (aero)dynamics. The DC part can be decreased by choosing appropriately large gains, while increasing these gains increases the AC part. Therefore, care is to be taken in tuning the controller to achieve an acceptable trade-off depending on the task.

The detachment phase also showed some bouncing effects, which is due to the time it takes to deplete the integral term, and imprecise free flight position control in the vicinity of the object. This effect is mitigated by deploying the switch function (7), and by bounding the integral term and choosing a small coefficient for it while choosing appropriately big proportional and derivative terms. A similar phenomenon is also observed in the case of grounded manipulators controlled by the parallel position/force control method in [35] and referred to as the sticky-effect.

The proposed controller can also be utilized in redundant omni-directional or partially omni-directional aerial manipulators, such as [1,8], by keeping their position controllers and providing it with a reference based on the presented force controller that considers contact maintenance and deploying orientation to generate the desired force.

6. Conclusions and Future Works

The topic of aerial physical interaction using a typical under-actuated VTOL UAV equipped with a passive tool was considered in this paper. We use a simple passive light weight tool, instrumented with a compliant end-effector, rigidly attached to the top of an aerial robot. We propose a control method in a bilateral teleoperation scheme to let a human operator drive the aerial manipulator in a remote environment, controlling the position of the robot in free flight and regulating the force applied to the object in contact interaction while maintaining the flight stability and keeping the contact. For this purpose, a parallel position/force controller is utilized, which also uses the rotational dynamic axes, that is, yaw motion, to generate the desired force that allows maintenance of the contact with a wider range of forces. On the one hand, theoretical proof of the stability of controlled system is derived and presented. On the other, experiments on driving the aerial robot toward the desired point, docking, and applying the desired forces to stationary and movable objects represented the feasibility and efficacy of the proposed solution. The human operator is provided with force haptic feedback proportional to the velocity in free flight and the applied force in contact, allowing for an improved situational awareness.

Future work will involve vision based outdoor implementation of the proposed approach, and investigating bandwidth and delay effects in haptic teleoperation based on passivity theories such as [36]. Performing more complex tasks cooperatively with a team of aerial manipulators will also be considered.

Supplementary Materials: The following are available at https://www.mdpi.com/article/10.3390/app11198955/s1, Video S1: The video of the experiments in the multimedia.

Author Contributions: Conceptualization, M.M., A.F., D.B. (Davide Barceli) and D.P.; Funding acquisition, A.F. and D.P.; Investigation, A.F. and D.P.; Methodology, M.M., A.F., D.B. (Davide Barceli) and D.P.; Project administration, A.F. and D.P.; Software, M.M. and D.B. (Davide Barceli); Supervision, A.F. and D.P.; Visualization, D.B. (Davide Bicego); Writing—original draft, M.M., D.B. (Davide Bicego), A.F., D.B. (Davide Barceli) and D.P.; Writing—review & editing, M.M., D.B. (Davide Bicego), A.F. and D.P. All authors have read and agreed to the published version of the manuscript.

Funding: This research was partially supported by the European Union's Horizon 2020 research and innovation program grant agreement ID: 871479 AERIAL-CORE.

Institutional Review Board Statement: Not applicable.

Informed Consent Statement: Not applicable.

Data Availability Statement: Not applicable.

Conflicts of Interest: The authors declare no conflict of interest.

References

1. Tognon, M.; Chavez, H.A.T.; Gasparin, E.; Sablé, Q.; Bicego, D.; Mallet, A.; Lany, M.; Santi, G.; Revaz, B.; Cortés, J.; et al. A Truly Redundant Aerial Manipulator System with Application to Push-and-Slide Inspection in Industrial Plants. *IEEE Robot. Autom. Lett.* **2019**, *4*, 1846–1851. [CrossRef]
2. Trujillo, M.Á.; Martínez-de Dios, J.R.; Martín, C.; Viguria, A.; Ollero, A. Novel Aerial Manipulator for Accurate and Robust Industrial NDT Contact Inspection: A New Tool for the Oil and Gas Inspection Industry. *Sensors* **2019**, *19*, 1305. [CrossRef]
3. Smrcka, D.; Baca, T.; Nascimento, T.; Saska, M. Admittance Force-Based UAV-Wall Stabilization and Press Exertion for Documentation and Inspection of Historical Buildings. In Proceedings of the 2021 International Conference on Unmanned Aircraft Systems (ICUAS), Athens, Greece, 15–18 June 2021; IEEE: Piscataway, NJ, USA, 2021; pp. 552–559.
4. Li, G.; Ge, R.; Loianno, G. Cooperative Transportation of Cable Suspended Payloads With MAVs Using Monocular Vision and Inertial Sensing. *IEEE Robot. Autom. Lett.* **2021**, *6*, 5316–5323.
5. Bernard, M.; Kondak, K.; Maza, I.; Ollero, A. Autonomous transportation and deployment with aerial robots for search and rescue missions. *J. Field Robot.* **2011**, *28*, 914–931. [CrossRef]
6. Ruggiero, F.; Lippiello, V.; Ollero, A. Aerial Manipulation: A Literature Review. *IEEE Robot. Autom. Lett.* **2018**, *3*, 1957–1964. [CrossRef]
7. Ollero, A.; Tognon, M.; Suarez, A.; Lee, D.J.; Franchi, A. Past, Present, and Future of Aerial Robotic Manipulators. *IEEE Trans. Robot.* **2021**, 1–20. [CrossRef]
8. Park, S.; Lee, J.; Ahn, J.; Kim, M.; Her, J.; Yang, G.H.; Lee, D. ODAR: Aerial Manipulation Platform Enabling Omnidirectional Wrench Generation. *IEEE/ASME Trans. Mechatron.* **2018**, *23*, 1907–1918. [CrossRef]
9. Ryll, M.; Muscio, G.; Pierri, F.; Cataldi, E.; Antonelli, G.; Caccavale, F.; Bicego, D.; Franchi, A. 6D Interaction Control with Aerial Robots: The Flying End-Effector Paradigm. *Int. J. Robot. Res.* **2019**, *38*, 1045–1062. [CrossRef]
10. Staub, N.; Bicego, D.; Sablé, Q.; Arellano-Quintana, V.; Mishra, S.; Franchi, A. Towards a Flying Assistant Paradigm: The OTHex. In Proceedings of the 2018 IEEE International Conference on Robotics and Automation, Brisbane, QLD, Australia, 21–25 May 2018; pp. 6997–7002.
11. Rashad, R.; Bicego, D.; Jiao, R.; Sanchez-Escalonilla, S.; Stramigioli, S. Towards Vision-Based Impedance Control for the Contact Inspection of Unknown Generically-Shaped Surfaces with a Fully-Actuated UAV. In Proceedings of the 2020 IEEE/RSJ International Conference on Intelligent Robots and Systems (IROS), Las Vegas, NV, USA, 24 October–24 January 2021; IEEE: Piscataway, NJ, USA, 2020; pp. 1605–1612.
12. Aerial-Core. EU Coll. Proj. H2020-ICT-2019-2. Available online: https://aerial-core.eu/ (accessed on 1 September 2021).
13. Scholten, J.L.; Fumagalli, M.; Stramigioli, S.; Carloni, R. Interaction control of an UAV endowed with a manipulator. In Proceedings of the Robotics and Automation (ICRA), 2013 IEEE International Conference on Karlsruhe, Karlsruhe, Germany, 6–10 May 2013; IEEE: Piscataway, NJ, USA, 2013; pp. 4910–4915.
14. Fanni, M.; Khalifa, A. A new 6-DOF quadrotor manipulation system: Design, kinematics, dynamics, and control. *IEEE/ASME Trans. Mechatron.* **2017**, *22*, 1315–1326. [CrossRef]

15. Orsag, M.; Korpela, C.M.; Bogdan, S.; Oh, P.Y. Hybrid adaptive control for aerial manipulation. *J. Intell. Robot. Syst.* **2014**, *73*, 693–707. [CrossRef]

16. Mebarki, R.; Lippiello, V.; Siciliano, B. Image-based control for dynamically cross-coupled aerial manipulation. In Proceedings of the 2014 IEEE/RSJ International Conference on Intelligent Robots and Systems, Chicago, IL, USA, 14–18 September 2014; IEEE: Piscataway, NJ, USA, 2014; pp. 4827–4833.

17. Kim, S.; Seo, H.; Choi, S.; Kim, H.J. Vision-guided aerial manipulation using a multirotor with a robotic arm. *IEEE/ASME Trans. Mechatron.* **2016**, *21*, 1912–1923. [CrossRef]

18. Baizid, K.; Giglio, G.; Pierri, F.; Trujillo, M.A.; Antonelli, G.; Caccavale, F.; Viguria, A.; Chiaverini, S.; Ollero, A. Behavioral control of unmanned aerial vehicle manipulator systems. *Auton. Robot.* **2017**, *41*, 1203–1220. [CrossRef]

19. Mebarki, R.; Lippiello, V.; Siciliano, B. Toward image-based visual servoing for cooperative aerial manipulation. In Proceedings of the 2015 IEEE International Conference on Robotics and Automation (ICRA), Seattle, WA, USA, 26–30 May 2015; IEEE: Piscataway, NJ, USA, 2015; pp. 6074–6080.

20. Gioioso, G.; Ryll, M.; Prattichizzo, D.; Bülthoff, H.H.; Franchi, A. Turning a near-hovering controlled quadrotor into a 3D force effector. In Proceedings of the 2014 IEEE International Conference on Robotics and Automation (ICRA), Hong Kong, China, 31 May–7 June 2014; IEEE: Piscataway, NJ, USA, 2014; pp. 6278–6284.

21. Bartelds, T.; Capra, A.; Hamaza, S.; Stramigioli, S.; Fumagalli, M. Compliant Aerial Manipulators: Toward a New Generation of Aerial Robotic Workers. *IEEE Robot. Autom. Lett.* **2016**, *1*, 477–483. [CrossRef]

22. Bellens, S.; De Schutter, J.; Bruyninckx, H. A hybrid pose/wrench control framework for quadrotor helicopters. In Proceedings of the Robotics and Automation (ICRA), 2012 IEEE International Conference on, Saint Paul, MN, USA, 14–18 May 2012; IEEE: Piscataway, NJ, USA, 2012; pp. 2269–2274.

23. Nguyen, H.N.; Lee, D. Hybrid force/motion control and internal dynamics of quadrotors for tool operation. In Proceedings of the 2013 IEEE/RSJ International Conference on Intelligent Robots and Systems, Tokyo, Japan, 3–7 November 2013; IEEE: Piscataway, NJ, USA, 2013; pp. 3458–3464.

24. Alexis, K.; Darivianakis, G.; Burri, M.; Siegwart, R. Aerial robotic contact-based inspection: Planning and control. *Auton. Robot.* **2016**, *40*, 631–655. [CrossRef]

25. Mersha, A.Y.; Stramigioli, S.; Carloni, R. On bilateral teleoperation of aerial robots. *IEEE Trans. Robot.* **2014**, *30*, 258–274. [CrossRef]

26. Gioioso, G.; Mohammadi, M.; Franchi, A.; Prattichizzo, D. A force-based bilateral teleoperation framework for aerial robots in contact with the environment. In Proceedings of the 2015 IEEE International Conference on Robotics and Automation (ICRA), Seattle, WA, USA, 26–30 May 2015; IEEE: Piscataway, NJ, USA, 2015; pp. 318–324.

27. Prattichizzo, D.; Trinkle, J.C. *Grasping, Springer Handbook of Robotics*; Springer: Berlin/Heidelberg, Germany, 2008; pp. 671–700.

28. Siciliano, B.; Sciavicco, L.; Villani, L.; Oriolo, G. *Robotics: Modelling, Planning and Control*; Springer Science & Business Media: Berlin/Heidelberg, Germany, 2010.

29. Marquez, H.J. *Nonlinear Control Systems: Analysis and Design*; Wiley-Interscience: Hoboken, NJ, USA, 2003.

30. Antonelli, G.; Cataldi, E.; Arrichiello, F.; Giordano, P.R.; Chiaverini, S.; Franchi, A. Adaptive trajectory tracking for quadrotor MAVs in presence of parameter uncertainties and external disturbances. *IEEE Trans. Control. Syst. Technol.* **2018**, *26*, 248–254. [CrossRef]

31. Wen, J.T.; Murphy, S.H. *PID Control for Robot Manipulators*; Rensselaer Polytechnic Institute: Troy, NY, USA, 1990.

32. Lawrence, D.A. Stability and transparency in bilateral teleoperation. *IEEE Trans. Robot. Autom.* **1993**, *9*, 624–637. [CrossRef]

33. Tarn, T.J.; Wu, Y.; Xi, N.; Isidori, A. Force regulation and contact transition control. *IEEE Control Syst.* **1996**, *16*, 32–40.

34. Yüksel, B.; Secchi, C.; Bülthoff, H.H.; Franchi, A. Aerial physical interaction via IDA-PBC. *Int. J. Robot. Res.* **2019**, *38*, 403–421. [CrossRef]

35. Hashtrudi-Zaad, K.; Salcudean, S.E. Bilateral parallel force/position teleoperation control. *J. Robot. Syst.* **2002**, *19*, 155–167. [CrossRef]

36. Lee, D.; Spong, M.W. Passive bilateral teleoperation with constant time delay. *IEEE Trans. Robot.* **2006**, *22*, 269–281. [CrossRef]

applied sciences

Article

Release of Sterile Mosquitoes with Drones in Urban and Rural Environments under the European Drone Regulation

Manuel Garcia [1], Ivan Maza [1,*], Anibal Ollero [1], Daniel Gutierrez [2], Idoia Aguirre [3] and Antidio Viguria [4]

[1] GRVC Robotics Laboratory, University of Seville, Avda. de los Descubrimientos, s/n, 41092 Seville, Spain; manuel.garcia.rivero@gmail.com (M.G.); aollero@us.es (A.O.)
[2] Instituto Tecnológico de Galicia, c/Cantón Grande 9, Planta 3, 15003 La Coruña, Spain; dgutierrez@itg.es
[3] Centro Tecnológico de Automoción y Mecatrónica, c/Cantón Grande 9, Planta 3, Plaza Cein 4, 31110 Noain, Spain; info@naitec.es
[4] Advanced Center for Aerospace Technologies, c/Wilbur y Orville Wright 19, La Rinconada, 41300 Seville, Spain; aviguria@catec.aero
* Correspondence: imaza@us.es

Featured Application: Drones can be a very useful tool to help mosquito control tasks in areas where these insects are an important vector of disease transmission. Storing sterile mosquitoes in drones and releasing them in targeted areas where previous suppression of wild populations has been performed can be a major advance in this methodology. This is part of the sterile insect techniques (SIT) which have been demonstrated to be an environmentally friendly solution for the control of insect pests. However, the use of drones in the areas where these operations are foreseen is limited by the fact that we have a regulation that advances at a slower pace than technology. Of particular interest is the case of the suppression of mosquito populations in urban areas below the transmission threshold, where drones and their operating conditions must meet demanding safety requirements. This article presents the current regulatory situation in Europe that affects drone operations and its applicability to the case of the release of sterile mosquitoes to control the population of these insects in two different scenarios: urban and rural areas.

Abstract: In recent years, several countries have developed the use of sterile insect techniques (SIT) to fight against mosquitoes that transmit diseases. From a technical and economic point of view, the use of drones in the aerial release of sterile mosquitoes leads to important improvements in aerial coverage and savings in operational costs due to the requirement of fewer release sites and field staff. However, these operations are under the European drone regulation, one of the most advanced in the world. The main contribution and novelty of this paper with respect to previous work is the analysis of the SIT application with drones under the European risk-based regulation in two scenarios: urban and rural areas. The specific operations risk assessment (SORA) methodology has been applied to assess the risk of drone operations in these scenarios. The paper presents the operational requirements for aerial release of mosquitoes with drones along with the regulatory considerations that must be applied. Finally, an overview of the conditions in operation that could relax risks and mitigation measures is also discussed.

Keywords: mosquitoes' control; drones; drone regulation; unmanned aircraft systems (UAS); U-space; SORA methodology; sterile insect technique (SIT)

Citation: Garcia, M.; Maza, I.; Ollero, A.; Gutierrez, D.; Aguirre, I.; Viguria, A. Release of Sterile Mosquitoes with Drones in Urban and Rural Environments under the European Drone Regulation. *Appl. Sci.* **2022**, *12*, 1250. https://doi.org/10.3390/app12031250

Academic Editor: Augusto Ferrante

Received: 6 December 2021
Accepted: 14 January 2022
Published: 25 January 2022

Publisher's Note: MDPI stays neutral with regard to jurisdictional claims in published maps and institutional affiliations.

1. Introduction

Mosquitoes are considered one of the deadliest reservoirs of vector borne disease in the world. According to the World Health Organisation (WHO), several million of deaths and hundreds of millions of cases occur each year from mosquito-borne diseases. Mosquito control is essential to reduce the transmission of diseases such as malaria, dengue fever, and Zika. In [1] the main, mosquito-borne diseases are shown with a summary of statistics and places where mosquito pest control is essential.

Different techniques have been used in recent years to control mosquitoes that transmit human disease. Most mosquito control techniques require the use of insecticides with high costs associated in terms of personnel and time. Furthermore, the presence of these toxic products has a large impact not only on human health but also on the environment. Concerns about resistance to pyrethroids in recent years [2] have led researchers to explore alternative solutions, such as the sterile insect technique (SIT) for mosquito control. The release of sterile insects to contain and suppress [3] mosquito populations was proposed by Dahmana and Mediannikov (2020) [4]. This technology has been used for a long time, for example, in 1980 in a program in El Salvador [5], but the required logistics have limited its applicability in practice.

The use of drones is seen as a promising tool for the release of sterile insect populations into targeted environments. Furthermore, these releases are carried out repeatedly over a certain period. The ease of use of drones and their ability to access any environment from the air make them a viable system in this mosquito control technique. Drones may provide a means of releasing sterile mosquitoes over large areas due to the ability of drones to travel long distances in short periods of time. A standard DJI Phantom 4 type light drone can fly for 25 min, being able to travel up to 25 km. Furthermore, one of the main advantages of using drones is that they can reach inaccessible areas where humans cannot.

There are some interesting works on this aspect. In [6] a fully automated system for releasing adult mosquitoes with a DJI M600 drone in a region of Brazil was reported. The system enabled a homogeneous dispersal of sterile male *Aedes aegypti* while maintaining their quality, leading to a homogeneous sterile-to-wild male ratio due to their aggregation at the same sites. This article concluded that the use of drones for the release of sterile mosquitoes leads to important improvements in aerial coverage and savings in operational costs due to the requirement of fewer release sites and field staff. One of the first works in this direction, dating back to 2017, can be found in [7] when the WeRobotics organization modified a DJI Matrice 600 drone to integrate a sterile mosquito release mechanism.

The effect of storage conditions on the survival of male Aedes aegypti mosquitoes during transport was studied in [8]. During transport from the rearing facility to the release site and during actual release in the field, damage to male mosquitoes should be minimized to preserve their reproductive competitiveness. The short flight range of male Aedes aegypti requires elaborate release strategies, such as release via drones. The authors conducted shipping in a 'real-life' setting to determine a good storage temperature and a compaction rate for the survival of the mosquitoes. In [9], optimized chilling conditions for handling male adults of *Aedes albopictus* prior to release were analyzed. The authors claimed that further studies are required to develop drone release systems specific for chilled mosquitoes to improve release efficiency, as well as to compare the population suppression efficiency between release of postchilled and nonchilled males in the field. In any case, the weight of the payload for the UAS was not high and some low-cost solutions [10,11] could be applied.

The sterile insect technique is also used to control the codling moth pests and the authors of [12] evaluated the use of small uncrewed aircraft systems (UAS) for the release of sterile codling moths. Sterile codling moths released from higher altitudes were more widely distributed and drifted more in strong winds, compared to those released from lower altitudes. Most of the released insects were recaptured in a 50 m wide swath under the release route. Recapture rates for aerially released insects were 40–70% higher compared to those released from the ground. The authors claimed that drones provide a promising alternative to ground release and conventional aircraft for the release of sterile codling moths. For the same pest, in [13] the authors compared the recapture rate of sterile moths following their release by four methods, and the efficiency of each system. The methods were the following: a fixed-wing unmanned plane flying ~40–45 m high at 70 km/h, an unmanned hexacopter travelling 20 m high at 25 km/h, and manually from the ground via bicycles or motor vehicles. The highest recapture rate followed delivery by hexacopter, then bicycle, vehicle, and plane, whereas the methods in ascending order of time per hectare for delivery were the following: plane and vehicle, hexacopter, then bicycle.

From an analysis of the state of the art, drone technology is mature enough from a technical point of view to perform the SIT application. However, the use of drones for the release of sterile insects is affected by the new regulation that has been approved at the European level [14] and came into force in January 2021. Then, the main contribution and novelty of this paper with respect to previous work is the risk analysis needed to put this application in practice with drones in rural and urban areas under the European drone regulation.

The European Aviation Safety Agency (EASA) is responsible for implementing, maintaining and monitoring compliance with newly established rules. Historically, each European Union (EU) member state maintained its own drone regulation at national level. Recently, a new EU regulation was approved that affects all member states equally. This new regulation was developed in the spirit of harmonizing rules and promoting the growth of the drone sector. The European regulation applies to any drone regardless of its mass and use. Drone operations are classified according to risk and are broken down into three operational categories. Table 1 shows this categorization.

Table 1. Categorization of drone operations in the new European regulation.

Category	Risk
Open	Low-risk operation
Specific	Medium-risk operation
Certified	High-risk operation, similar to manned aviation operations

In addition to this, EASA has published the acceptable means of compliance (AMC) and guidance material (GM) [15], which complement the regulation and explains in detail the different categories set out above. The document entitled Specific Operations Risk Assessment (SORA) details the methodology for assessing the risk of drone operations. In this regard, several papers about risk analysis and SORA can be found in the literature. Reference [16] analyzes the application of SORA for a multi-UAS airframe inspection (AFI) operation, which involves the deployment of multiple UAS with autonomous features within an airport. In [17], the authors present the most important risks related to conducting operations with the use of UAS by first responders (FRs), while reference [18] presents the application of the SORA methodology for media production with a small UAS team. In addition, [19] describes the application of SORA to the flight of large remotely piloted aircraft systems (RPAS) in Australian airspace highlighting its distinguishing factors.

The paper is structured as follows. First, the current European drone regulation will be explained in detail in Section 2. Materials and methods are described in Section 3, where the SORA methodology will be shown as a basis to assess the risk of drone operation and help categorize the operation. Then, the concept of U-Space will be introduced in Section 4 since it is the framework that will help integrate drone operations into the airspace. Section 5 explains the results of the application of the SORA methodology to some typical mosquito-related operations. Finally, Section 6 closes the paper with a discussion of the analysis.

2. Current European Drone Regulation

As mentioned above, the European Commission adopted a new set of provisions for the use of drones within the Single European Sky strategy, which is an initiative of the European Commission aiming to reform the current air traffic management system in most of Europe. The aim is not only to guarantee standards on the safety, efficiency, and environmental impact of the air traffic, but also to integrate drones safely into airspace.

EU legislation has been radically amended in recent years. In July 2018, European lawmakers passed the new Regulation (EU) 2018/1139 on common rules in the field of civil aviation, which included a new mandate for the EU EASA on drones and urban air mobility [14]. This regulation adopts a new comprehensive legal strategy for the drone sector and repeals Reg. (EC) 2008/216 [20], which only concerned drones with a take-off weight of more than 150 kg, while drones with a maximum take-off weight of less than 150 kg were within the jurisdiction of the member states.

Since neither the EU Parliament nor the EU Council had any objections, both the Implementing and Delegated Acts (Commission Delegated Regulation (EU) 2019/945 [21] and Commission Implementing Regulation (EU) 2019/947 [22]) were published in June 2019 and entered into force on 1 July 2019. Drone operators were required to register either in the member state of their residence or where the drone operator registered their main place of business by July 2020. These regulations were gradually enforced over the course of a one-year transition period from their date of publication. This transition period provided Member States and drone operators with time to prepare before full regulatory enforcement in 2022.

The new EU regulatory framework covers all types of existing and future drone operations, enabling operators—once authorized in their state of registration—to freely circulate between Member States. The purpose of introducing these new regulations is to ensure the safety of drone operations, as well as to protect the privacy of EU citizens, with respect to personal data protection and the environment while allowing free access to airspace. The new regulations establish technical and operational requirements, provisions for drone operations, and personnel (minimum requirements and operator training), including both pilots and organizations. The EU regulatory framework defines drone capabilities, types of operation and labels these into three broad risk-based categories (open, specific, and certified) following the distinction suggested by EASA in the Opinion 01/2018 [23]. These three categories of operations are based on the levels of risk involved per drone flight, and each adopts a varied regulatory approach, with drone flight operational limitations decreasing with the requirement for greater authorization from a member state's national aviation authority.

Regulation 2019/947 presents a comprehensive system of unified legal regulations that classifies drone operations into the above-mentioned three categories based on different criteria:

- Open (Article 4 of Regulation 2019/947). Operations in this category shall not be subject to any prior operational authorization, nor to an operational declaration by the drone operator before the operation if the following conditions are met. The drone belongs to one of the classes set out in the Delegated Regulation (EU) 2019/945 or is privately built or meets the conditions defined in Article 20 of Regulation 2019/947. The unmanned aircraft has a maximum take-off mass of less than 25 kg, and the remote pilot keeps the unmanned aircraft in visual line of sight (VLOS) conditions at all times except when flying in follow-me mode or when using an unmanned aircraft observer. During open operations, the remote pilot ensures that the unmanned aircraft is maintained within 120 m from the closest point on the surface of the Earth (except when it overflies an obstacle upon request to its owner) and at a safe distance from people (never flying over crowds). The unmanned aircraft cannot carry dangerous goods and does not drop any material. Open operations are further divided into three subcategories: A1 (fly over people), A2 (fly close to people) and A3 (fly far from people).

- Specific (Article 5 of Regulation 2019/947). Operations fall into this category as long as the concept of operation exceeds the limitations defined in the open category. The drone operator shall apply to obtain an operational authorization from the competent authority in the member state where it is registered, submitting a risk assessment including adequate mitigating measures. This risk assessment approach allows handling new technologies and operations such as beyond visual line of sight (BVLOS), fully autonomous drones, urban areas, etc. However, if the operation complies with one of the standard scenarios (STS) defined by EASA, the drone operator shall not be required to obtain the above-mentioned operational authorization, and a declaration (responsible) by the drone operator will be sufficient. EASA has defined two STS:

- o   STS01: VLOS operation in a controlled ground area in an urban environment;
- o   STS02: BVLOS operation in a controlled ground area in a sparsely populated environment.

Apart from that, an operational authorization or a declaration shall not be required for drone operators holding a light UAS operator certificate (LUC) with appropriate privileges, which is valid in all EU member states without additional demonstrations.

- Certified (Article 6 of Regulation 2019/947). An operation is classified as being in the certified category when, according to the risk assessment, the operation cannot take place without a certificate for the operator, a certificate for the airworthiness of the drone, and a license for the remote pilot (unless fully autonomous). In any case, the following operations are within the certified category: operations over assemblies of people with an aircraft of characteristic dimensions of 3 m or more, transportation of people, and transportation of dangerous goods if, in case of accident, they pose a high risk for third parties.

In October 2019, EASA published its guidance material (GM) and a description of the means of complying with the regulation (acceptable means of compliance (AMC)). The AMC shed light on how to carry out the SORA, which is required for operation, depending on the 'specific' category under scrutiny. Along the same lines, EASA has published some predefined risk assessments as AMC to Article 11, to cover most common drone operations. The intent is to simplify the burden for drone operators, paving the way for the full implementation of the new legal framework.

Finally, there are some points to be remarked: there is no distinction between professional and leisure activities with drones or between experimental flights and aerial works.

3. Materials and Methods: The SORA Risk Analysis Methodology

In the current regulatory framework for drones or UAS in Europe, a risk-based approach is used to assess the safety of drone operations. The SORA methodology has been developed by the Joint Authorities for Rulemaking of Unmanned Systems (JARUS) and focuses primarily on the specific category.

The SORA is a multistage process of risk assessment aiming at risk analysis of certain unmanned aircraft operations, as well as defining necessary mitigations and levels of robustness. The application of this methodology is an acceptable means of evaluating the risks associated with the operation of a drone within the specific category and to determine the acceptability of the proposed operation. In this section, a general description of the SORA methodology is presented.

3.1. Introduction to SORA

JARUS pursues a consensus from various national aviation authorities and stakeholders on a common procedure to identify and qualitatively assess safety risks for drone operations. In particular, JARUS developed SORA [24] in 2019.

Any risk assessment methodology uses a schema based on the ISO 31000 standard, providing a reference framework and guiding the general risk management process (see Figure 1).

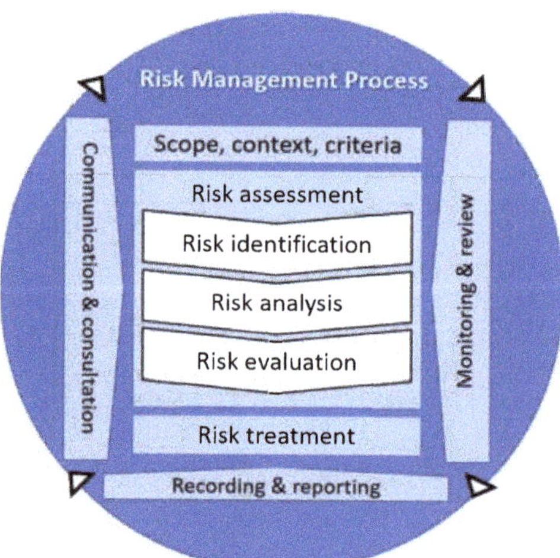

Figure 1. General aspects of risk management from ISO 31000. The risk management process involves the systematic application of policies, procedures and practices to the activities of communicating and consulting, establishing the context and assessing, treating, monitoring, reviewing, recording and reporting risk.

SORA is a method based on a holistic risk model that provides a generic framework to identify possible hazards and threats, as well as relevant harm and threat barriers applicable to drone operation. The aim is to establish a sufficient level of confidence that a specific operation can be carried out safely. To achieve this, SORA requires first the applicant to collect and provide the relevant technical, operational, and system information needed to assess the risk associated with the intended operation of the drone.

Some key aspects related to the applicability of the methodology are the following.

- It aims to assess the safety risks involved in the operation of drones of any class, size, and type of operation and particularly suited, but not limited to 'specific' operations for which a risk and hazard assessment is required.
- The safety risks associated with collisions between drones and manned aircraft are within the scope of the methodology.
- Security aspects are excluded when not limited to those confined by the airworthiness of the systems (e.g., aspects relevant to the protection from unlawful electromagnetic interference).
- Privacy aspects are excluded from the applicability of this methodology.

3.2. The SORA Process

The SORA methodology provides a logical process for analyzing the proposed concept of operations (ConOps) and establishing an adequate level of confidence that the operation can be conducted with an acceptable level of risk. There are ten steps that support the SORA methodology, as shown in Figure 2. The process begins with the ConOps description, which provides the relevant technical, operational, and system information needed to assess the risk associated with the intended operation of the drone, both ground risk and air risk.

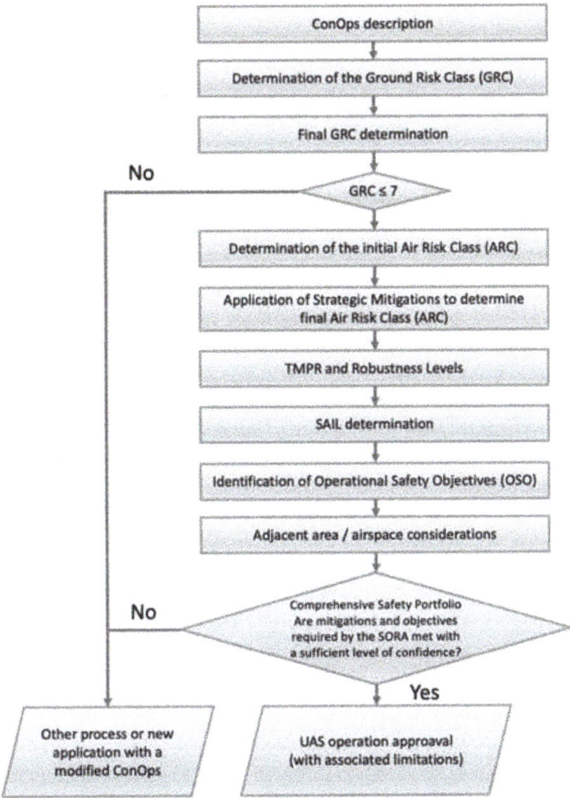

Figure 2. The SORA process adapted from [24] shows the ten steps that support the SORA methodology.

The ground risk class (GRC) is the intrinsic ground risk of the drone related to the risk of a person being struck by the drone (in the case of loss of drone control with a reasonable assumption of safety). To compute it, the applicant needs to identify the maximum drone characteristic dimension (e.g., wingspan for fixed wing, blade diameter for rotorcraft, maximum dimension for multicopters, etc.) and the knowledge of the intended operational scenario (VLOS or BVLOS; population density of the overflown areas). Intrinsic risks can be controlled and reduced by mitigations. The mitigations used to modify the intrinsic GRC have a direct effect on the safety objectives associated with a particular operation and, therefore, are important to ensure its robustness. The final determination of the GRC is based on the availability of these mitigations for operation.

The air risk class (ARC), understood as the intrinsic risk of mid-air collision, determines the air risk category. Identification of the ARC must take into account the impact on other air traffic and air traffic management (the altitude of the operation; controlled versus uncontrolled airspace; aerodrome versus non-aerodrome environment; airspace over urban versus rural environment). The ARC may be lowered by applying strategic and tactical mitigation means (detect and avoid systems or alternate means) resulting in the residual ARC. Strategic conflict strategies will deal with the planning at a global level of the route that the drone must follow to execute the mission, while tactical conflict strategies will deal with reactive local maneuvers that are executed during the flight to fulfill certain functionalities, such as avoiding possible encounters with other aircraft.

After determining the final GRC and residual ARC, it is now possible to derive the specific assurance and integrity levels (SAIL) associated with the proposed ConOps. The

SAIL represents the level of confidence that the operation will remain under control. On the other hand, the SAIL is also used to evaluate the defenses within the operation in the form of operational safety objectives (OSO) and to determine the associated level of robustness. These OSOs appear in the SORA as a list of objectives that have historically been used to ensure safe drone operations regarding technical issues, external systems supporting the operation, the effects of human errors, and the effects of adverse operating conditions.

In this paper we address two different scenarios that are highly representative of mosquitoes-related operations. First, it is considered the case of an urban environment targeted for a SIT mosquito program. Usually, large cities and their surroundings are immersed within the airspace associated with airports. Therefore, the operation is considered to be carried out in controlled airspace. The following conditions for the urban scenario constitute the ConOps for this case:

- VLOS: The operation takes place within the visual line of sight of the pilot;
- Moderately populated environment;
- Inside of controlled airspace;
- VLL: very low level operation, flights below 150 m of altitude.

The other case is in rural environments, without people around, which are aligned with population control in areas where this type of insect reproduces: lakes, humid areas, areas with dense vegetation, etc. In this case, the operational conditions that are considered as inputs for the ConOps are the following:

- BVLOS: the operation takes place beyond the visual line of sight of the pilot, without observers who can help the pilot;
- Sparsely populated environment;
- Out of controlled airspace;
- VLL: very low level operation, flights below 150 m of altitude.

4. Integration into Airspace (U-Space)

As indicated in [25], the expected number of drone operations in the European market is expected to reach 10 billion euros per year by 2035 and more than 15 billion euros per year by 2050. This volume of operations will pose safety, security and airspace integration issues in European airspace, especially at low flight levels (which are currently defined as altitudes below 150 m) with drones belonging mostly to open and specific categories.

Many organizations are iteratively maturing their approach to address this volume of drone operations, as outlined in various concepts developed by bodies such as the Single European Sky Airspace Research (SESAR) Joint Undertaking (JU) through the CORUS project [26], the Federal Aviation Administration (FAA) [27], and global standardization bodies such as the Global UTM Association (GUTMA) [28].

In Europe, safe drone traffic management and its safe operation within the existing air traffic environment in a harmonized manner across the European airspace are ensured by U-space (outside Europe, this concept is commonly referred to as unmanned traffic management or UTM). In [29], SESAR JU defined U-space in a blueprint as a set of services based on high levels of digitization, as well as automation of functions and procedures designed to guarantee access to airspace for a large number of drones in a safe and efficient way, with an initial focus on operations at very low levels (VLL), with a maximum height above ground level of 150 m. Then, U-space is a technological framework designed to facilitate any type of operation, in all classes of airspace and in any environment, even the most congested, while providing an appropriate interface for manned aviation and air traffic control.

The ConOps (concept of operations) for U-space was developed in the CORUS project and published on 30 September 2019 [26]. CORUS was focused on drones belonging to open or specific categories that operate in VLL, which are split into three types according to the services provided (see Figure 3):

- X: No conflict resolution service is offered;

- Y: Only preflight conflict resolution is offered;
- Z: Preflight conflict resolution and in-flight separation are offered.

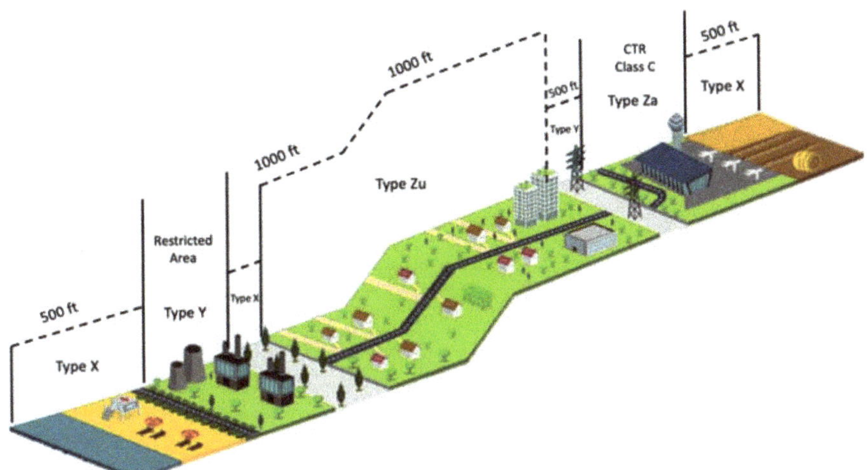

Figure 3. Classification of airspace adapted from the CORUS project [26]. The VLL airspace is divided into X, Y, Z volumes: X (low risk), Y (higher risk—access only with approved operation plan and specific technical requirements per volume) and Z (highest risk—access only with approved operation plan being Zu under U-space and Za the ATC controlled airspace).

Type Y airspace will be available from U2 and will facilitate VLOS, EVLOS (extended visual line of sight) and BVLOS flight. The risk mitigations provided by U-space mean that the Y airspace is more amenable to other flight modes than X.

Type Z airspace may be subdivided into Zu and Za, controlled by UTM and air traffic management (ATM), respectively. Za is simply a normal controlled airspace and is therefore immediately available. Zu airspace will be available from U3.

Because U-space provides more risk mitigations for Z type, it is more amenable to other flight modes, and allows higher density operations than Y airspace. Z allows VLOS and EVLOS and facilitates BVLOS and automatic drone flight.

Finally, EASA issued Opinion No 01/2020 [30] on 13 March 2020, which proposed a draft of the high-level regulatory framework for U space, closely linked to the two existing drone regulations [21,22]. The opinion proposed a first set of what were considered by EASA as the minimum necessary rules, which are to be complemented later with further provisions enabling a more mature state of airspace integration. The objective of the opinion was to create and harmonize the necessary conditions for manned and unmanned aircraft to operate safely in the U-space airspace focusing on strategic and pretactical traffic management techniques in order to mitigate the risk of collisions by requiring adapted services and sharing essential traffic information. In this sense, the opinion acknowledged that ensuring that U-space participants are cooperative is required until further development of detect and avoid (DAA) systems to ensure safe operations, in particular for BVLOS operations.

The airspace in which aircrafts fly can be classified as controlled airspace in which air traffic control (ATC) services such as clearance and traffic information. However, in uncontrolled airspace, there are no such services for manned aircraft. With the integration of U-Space in the future, when the member states designate a volume of airspace as U-space airspace, there will be a restriction: both for drone operators, to use U-space services to fly in that airspace; and for manned aircraft operators, to make available their position at regular intervals.

Based on the evaluation conducted by EASA of existing U-space services and their maturity, the following U-space services are considered necessary and mandatory to ensure safe and efficient operations in each U-space airspace implementation: network identification, geoawareness, traffic information, and drone flight authorization. In addition, member states may decide that additional U-space services are needed to support safe and efficient drone operations in specific volumes of U-space airspace implementation. They can decide to mandate them based on their risk assessment.

5. Results: Application of the European Drone Regulation to Mosquito Release Operations

This section will show the results of the application of the SORA methodology to the scenarios previously presented in Section 3.2: urban and rural scenarios. One of the main parameters for the application of the SORA analysis is the size of the drone. In both scenarios, a very small size drone is considered, with a characteristic dimension smaller than 1m or typical kinetic energy expected less than 700 J. In this category, drones with up to approximately 2 kg of weight can be included, which can have enough payload capacity and high endurance to perform mosquito release operations from the air according to the state of the art in Section 1. The release of mosquitoes from the drone does not require heavy systems to be integrated on the aircraft, nor do mosquitoes weigh too much, thus very small size drones are suitable. In the following, it should be noticed that for the characteristics of the drone and other parameters we consider ranges instead of particular values. The goal is to make the analysis broader and more useful to the drone community.

The SORA methodology is performed for each scenario, and the following subsections show the main results. In addition to that, a final overview of the conditions in the operation that could relax risks and mitigation measures is presented. Instead of showing the detailed steps of the SORA methodology, an added value of this paper is to summarize the main points. To this end, a division of the SORA into three fundamental parts is considered: evaluation of the GRC, evaluation of the ARC and the TMPR (tactical mitigation performance requirements), and final evaluation of the SAIL and identification of the OSOs.

5.1. Urban Scenario

Once the ConOps was defined in Section 3.2, it is possible to start with the application of the SORA methodology to analyze the level of risk of the operation, as well as to identify the mitigating measures to reach an acceptable risk in order to carry out the operation safely.

For the GRC evaluation, the drone is considered to fly over a controlled ground area. This means the ground area where the drone is operated and within which the operator can ensure that only the involved persons are present. In this way, taking into account the size considered for the drone, the resulting GRC is one. This is the lowest value of the GRC in the final evaluation table, so it is considered acceptable in this case. When controlling the ground area is not feasible, the GRC increases a lot since the case would turn to VLOS in a populated environment, resulting in a GRC of four. In this case, some mitigation measures should be applied to reduce this GRC, such as the integration of a parachute and the creation of ground risk buffers, which should both be approved by the authority through analysis and results.

Regarding the risk assessment in air, considering the type of VLL operation and in controlled airspace, an ARC-c is obtained. It is possible to apply both strategic and tactical mitigations to try to reduce that ARC level. First, strategic mitigations can be applied due to operational restrictions that may imply the geographical limitation of the volume in which the operation takes place or the temporal limitation to establish specific terms in which the operation is executed. It is also possible to establish strategic mitigation measures by establishing common structures and rules for all aircraft that will share such airspace. The last possibility is to apply tactical mitigations, which correspond to the measures that are applied once the aircraft is in flight, to reduce the risk of an encounter with another aircraft. This includes measures such as situational awareness through VLOS or alternative detect and avoid (DAA) systems.

According to [15], if the applicant considers that the assigned initial ARC is correct, which makes sense in this scenario, then that initial ARC becomes the final one, resulting in an ARC-c. This has implications at the level of the TMPR. In this scenario, the fact that the operation is VLOS is considered as an acceptable tactical mitigation for all levels of ARC. Despite this, the operator is advised to consider additional means to increase situational awareness of the air traffic operating in the vicinity. Additionally, the operator is required to have a conflict resolution scheme in which the applicant explains the methods used for detection and defines the criteria and decision-making in the case of possible encounters.

Once the GRC and ARC have been defined, it is possible to determine the SAIL associated with the operation. Taking into account the previous consideration, the result obtained is SAIL IV. This implies that the robustness levels required for the different OSOs are between medium (M) and high (H), meaning that the requirements that must be met in relation to the technical aspects of the drone and operation, operational procedures, pilot competencies, design, dependence on external systems such as GPS, human error, and operational conditions are moderately demanding and, in some cases, very demanding. This high level of SAIL is mainly because most cities have airports in their proximity and therefore are immersed in controlled airspace, which implies a high ARC and, therefore, a high SAIL. This also occurs for all levels of GRC less than or equal to 2, that is, for aircraft less than 25 kg, as can be seen in the SAIL allocation table shown in Table 2. Section 5.3 will show other alternatives for this kind of operation that offer more relaxed conditions.

Table 2. The SAIL determination table adapted from [24] is used to obtain the SAIL assigned to a particular ConOps.

Final GRC	Residual ARC			
	a	b	c	d
≤ 2	I	II	IV	VI
3	II	II	IV	VI
4	III	III	IV	VI
5	IV	IV	IV	VI
6	V	V	V	VI
7	VI	VI	VI	VI
>7	Category C operation			

This result comes from the formal application of the SORA to the ConOps described, with the result as analyzed being a fairly high SAIL for the reasons that have been discussed. However, it is possible to consider one of the STS published by EASA for this scenario, under which, as explained in Section 2, the regime is declarative by the operator and it is not necessary to have an operational authorization, which would greatly facilitate the operation in the regulatory aspect. The STS-01 scenario is the one that best fits the ConOps described, with the following characteristics: VLOS operation at a maximum height of 120 m, at a ground speed of less than 5 m/s, over controlled ground areas that can be in a populated environment, using drones with maximum weight up to 25 kg. Hence, the operator must declare that the conditions of its operation comply with the characteristics of STS-01.

The difference from the previous formal application of SORA is that in STS-01 the ARC level considered is ARC-b, which finally results in a much lower final SAIL (SAIL II). In order to consider this ARC-b, the airspace in which operations are intended to be conducted must have a low probability of the drone encountering manned aircraft or other airspace users. Even in urban areas with controlled airspace, as it is the analyzed airspace, this could be achieved by means of limiting the altitude of the operation (for example, fly always below the highest surrounding buildings) or following coordination procedures with ATC of the nearby airport/aerodrome. However, other restrictions imposed by the STS-01 definition are required, such as the control of the ground area from uninvolved people.

For this STS-01 it has been proposed that drones operating under this scenario should carry a C5 class mark. The main requirements that apply to class C5 drones are as follows.

- Be an aircraft other than a fixed-wing aircraft, with less than 25 kg of weight, which is the case considered in our analysis;
- Be equipped with a geo-awareness function;
- Provide to the remote pilot clear information about height of the drone above surface or take-off point;
- Limit the groundspeed to not more than 5 m/s;
- Limit the maximum height above take-off point or surface to 120 m;
- Provide means to the remote pilot to terminate the flight of the drone;
- Provide means to the remote pilot to monitor the quality of the command and control link, providing alerts in case of degradation or loss of communications.

Thus, by complying with the conditions described for STS-01, which fit very well with the ConOps described above, the operator could use this scenario and work under a declarative regime, which greatly speeds up all the bureaucracy associated with the operation in all aspects.

5.2. Rural Scenario

According to the ConOps presented in Section 3.2, for the assessment of the GRC, the drone is considered to fly over a sparsely populated area. This means that the flight is not executed over a populated environment and there is no need to control the presence of uninvolved persons in the area of operation of the drone. In this way, and taking into account the size considered for the drone, the resulting GRC is three. As in the previous case, it is possible to apply mitigating measures to reduce this GRC, such as integrating a parachute into the drone, but considering the SAIL determination table, there is not much difference between GRC 2 and 3 in terms of final SAIL, for levels of ARC-b or higher, so it is considered a nonessential mitigating measure.

As for the evaluation of the ARC in this scenario, considering the type of VLL operation and in noncontrolled airspace, an ARC-b is obtained. As in the previous case, tactical and strategic mitigation measures can be applied to try to reduce the level of ARC obtained. In this case, it is noted that lowering ARC-b to ARC-a is a complicated process, since the ARC-a level corresponds to an atypical or segregated airspace where the probability of encountering another aircraft is practically zero. Therefore, even if measures are applied to mitigate the risk of encounter in the air, the level of ARC-b is considered adequate for this scenario.

In this case, the evaluation of the TMPR is important because it is a BVLOS flight. Because the resulting ARC is ARC-b, the robustness level assignment for TMPR is low, as it is considered an airspace where the probability of encountering another aircraft is low but not negligible. Operations with a low TMPR level are supported by technology designed to help the pilot detect other traffic. This is where detect and avoid (DAA) systems come in, which can be based on ground systems, such as U-Space, or air systems such as the transponder called automatic dependent surveillance—broadcast (ADS-B) or the traffic awareness and collision avoidance system from FLARM (https://flarm.com/ (accessed on 1 December 2021)) for general aviation, light aircrafts, and drones. This ARC imposes low performance requirements for the DAA system; hence, in this case, the DAA system is required to detect approximately 50% of all possible aircrafts present in the operational volume in which an encounter could occur.

Having already obtained the final levels of GRC and ARC, it is possible to proceed with the determination of the SAIL. In this case, the resulting level is SAIL II, so it is an operation with lower risk levels than the case previously considered of the urban environment. This means that the robustness levels required for the different OSOs are low (L), and even many of them are optional (O). Certain objectives with a medium level of robustness (M) are appreciated corresponding mainly to operational procedures. As can be checked in

Table 2, for ARC-b, the SAIL level is II for GRC 3 and GRC 2, so drones with weight less than 25 kg can be considered with the same level of risk in this scenario.

5.3. Alternatives to Relax Risk Levels

Certain operational conditions can help relax risk levels and have less demanding mitigating measures. The environment in which the release of sterile mosquitoes is the most important is the urban scenario. At the same time, this scenario is the most complex due to the conditions that exist in this case: normally they are immersed in the airspaces affected by the nearby airports, they are populated environments with the presence of people on the ground, and there is more possibility of traffic from other aircraft (emergency, medical and police helicopters, etc.). However, it is important to note here that there is a current discussion between the different stakeholders on how to redefine and redistribute airspace. Currently, controlled airspace in urban areas covers most of the urban volume from ground to high altitude due to the proximity of their airports, so any drone that wants to fly in VLL would be in controlled airspace. That is why the drone and airspace community is studying relaxing the airspace distribution and starting from different heights depending on the distance to the airport, establishing a kind of cone to categorize the airspace.

According to the European drone regulation [14], drone operations can be included in the open category if they are considered low risk operations. The main characteristics of this kind of operation are that the drone should have a CE mark, weigh less than 25 kg and be operated in VLOS or assisted by an observer, and that the maximum height is 120 m. Today, the ASD-STAN organization has established a D05/WG08 working group called the UAS unmanned aircraft system, in which standardization activities are being developed for the definition of means of compliance to obtain the CE mark. Furthermore, there is one major restriction regarding the impossibility of carrying dangerous goods or dropping any material. This last aspect is crucial when it comes to discerning with the aviation authority the possibility of including this type of mosquito release operation in the open category. This must be further discussed with the aviation authority to reach an agreement.

By analyzing the different operational alternatives of the open category for mosquito release operations, options that include flights in urban environments will be highlighted. Within the subcategories into which the open category is divided, those particularities that affect the urban environment are as follow.

- Subcategory A1: The drone can fly in urban areas but not over an assembly of people and is trying to reduce overflying uninvolved persons. If uninvolved persons are overflown, the remote pilot must reduce as much as possible the time during which the drone overflies those persons. Apart from that, maximum take off weight (MTOW) of drones in this subcategory can be up to 900 g.
- Subcategory A2: The drone can fly in urban areas, but a safety distance of at least 30 m must be maintained with respect to the uninvolved people in the operation. In this case, the drone can weigh up to 4 kg.

The STS-01 standard scenario analyzed above in Section 5.1 is considered to be an extension of the A2 subcategory of the open category because of the similarities in the conditions and requirements that are posed. However, operations in an urban environment can fall under these two subcategories within the open category as long as they comply with the airspace restrictions imposed by the national civil aviation authority. This is the point that differentiates these subcategories from the STS-01 approach raised in Section 5.1, since most authorities reject the possibility of flying the open category in controlled airspace, in which most cities with nearby airports are immersed. Therefore, there is a current trend of redefinition of airspace in cities, which would allow the dimensions of these controlled spaces to be relaxed by not reaching the ground when there is a certain distance from the airport. This would allow flying in the open category in an urban environment under the criteria previously defined in most cities, as long as there is a certain distance from the aerodromes.

6. Discussion

This article presented the current legal framework in the EU for the use of drones as an innovative tool for the release of sterile mosquitoes within a SIT operational program against human disease transmitting mosquitoes in two different environments: urban and rural. Targeting urban environments for sterile mosquito release is a promising scenario due to the density of human populations. Additionally, drones are characterized by their ease of use and versatility of operation, and can significantly help release these mosquitoes in hard-to-reach areas and distribute them evenly over a given zone.

We analyzed the current European drone regulation and its impact on possible mosquito release operations. This regulation is based on the risk of the operation, and not in the weight of the drone, so it opens the possibility, for example, to fly drones in urban areas in the open category as explained in Section 5.3, which has been really complex until now. In addition, the SORA methodology proposed by the JARUS organization and accepted by the EASA authority within its auxiliary material to the regulations was described. SORA is a tool for analyzing the risks of drone operations, evaluating their level of safety, and identifying possible mitigating measures to achieve the desired level of safety. In addition, the concept of U-Space is introduced as a set of services that will help to manage drone air traffic.

Finally, the European regulation establishes common standards for all member states that will help harmonize and encourage the drone market. This new regulatory framework will affect all drone operations, including the release of sterile mosquitoes. As analyzed in this article, there are different alternatives for this operation that, depending on the risk of the operation, will fundamentally determine its classification in an open or specific category. Among the key criteria that determine this risk are drone weight, airspace classification (controlled or uncontrolled), operational scenario (presence of people or not), VLOS or BVLOS flight, flight height, etc.

Author Contributions: Individual contributions to this article are: "Section 1. Introduction", "Section 5. Application of the new European Drone regulation to mosquitoes-related operations" and "Section 6. Conclusions" by M.G.; "Section 2. New European Drone Regulation" by I.M., "Section 3. SORA Risk Analysis Methodology" by I.A., "Section 4. Integration into Airspace (U-Space)" by D.G., and A.O. and A.V. have contributed intensively by supporting and reviewing the article. In addition, M.G. has made the final compilation of the contents and I.M. has reviewed the document to give it coherence. All authors have read and agreed to the published version of the manuscript.

Funding: This work has been partially funded by the Spanish public organism CDTI through the iMOV3D R&D project (CER-20191007).

Institutional Review Board Statement: Not applicable.

Informed Consent Statement: Not applicable.

Acknowledgments: The authors especially thank Antonio Jiménez for his support in the definition of the scenarios which are of interest for this application.

Conflicts of Interest: The authors declare no conflict of interest, and the funders had no role in the design of the study; in the collection, analyses, or interpretation of data; in the writing of the manuscript, or in the decision to publish the results.

References

1. Tolle, M.A. Mosquito-borne Diseases. *Curr. Probl. Pediatr. Adolesc. Health Care* **2009**, *39*, 97–140. [CrossRef] [PubMed]
2. Ranson, H.; N'Guessan, R.; Lines, J.; Moiroux, N.; Nkuni, Z.; Corbel, V. Pyrethroid resistance in African anopheline mosquitoes: What are the implications for malaria control? *Trends Parasitol.* **2011**, *27*, 91–98. [CrossRef] [PubMed]
3. Hendrichs, J.; Vreysen, M.J.B.; Enkerlin, W.R.; Cayol, J.P. Strategic Options in Using Sterile Insects for Area-Wide Integrated Pest Management. In *Sterile Insect Technique*; Springer: Berlin/Heidelberg, Germany, 2005; pp. 563–600.
4. Dahmana, H.; Mediannikov, O. Mosquito-Borne Diseases Emergence/Resurgence and How to Effectively Control It Biologically. *Pathogens* **2020**, *9*, 310. [CrossRef] [PubMed]
5. Lowe, R.E.; Bailey, D.L.; Dame, D.A.; Savage, K.E.; Kaiser, P.E. Efficiency of Techniques for the Mass Release of Sterile Male *Anopheles albimanus* Wiedemann in El Salvador. *Am. J. Trop. Med. Hyg.* **1980**, *29*, 695–703. [CrossRef] [PubMed]

6. Bouyer, J.; Culbert, N.J.; Dicko, A.H.; Pacheco, M.G.; Virginio, J.; Pedrosa, M.C.; Garziera, L.; Pinto, A.T.M.; Klaptocz, A.; Germann, J.; et al. Field performance of sterile male mosquitoes released from an uncrewed aerial vehicle. *Sci. Robot.* **2020**, *5*, eaba6251. [CrossRef] [PubMed]

7. Ackerman, E. Drones make a special delivery-Mosquitoes [News]. *IEEE Spectr.* **2017**, *54*, 9–11. [CrossRef]

8. Chung, H.-N.; Rodriguez, S.D.; Gonzales, K.K.; Vulcan, J.; Cordova, J.J.; Mitra, S.; Adams, C.G.; Moses-Gonzales, N.; Tam, N.; Cluck, J.W.; et al. Toward Implementation of Mosquito Sterile Insect Technique: The Effect of Storage Conditions on Survival of Male *Aedes aegypti* Mosquitoes (Diptera: Culicidae) during Transport. *J. Insect Sci.* **2018**, *18*, 2. [CrossRef] [PubMed]

9. Zhang, D.; Xi, Z.; Li, Y.; Wang, X.; Yamada, H.; Qiu, J.; Liang, Y.; Zhang, M.; Wu, Y.; Zheng, X. Toward implementation of combined incompatible and sterile insect techniques for mosquito control: Optimized chilling conditions for handling *Aedes albopictus* male adults prior to release. *PLoS Negl. Trop. Dis.* **2020**, *14*, e0008561. [CrossRef] [PubMed]

10. Deaconu, A.M.; Udroiu, R.; Nanau, C.-Ş. Algorithms for Delivery of Data by Drones in an Isolated Area Divided into Squares. *Sensors* **2021**, *21*, 5472. [CrossRef] [PubMed]

11. Udroiu, R.; Deaconu, A.M.; Nanau, C.-Ş. Data Delivery in a Disaster or Quarantined Area Divided into Triangles Using DTN-Based Algorithms for Unmanned Aerial Vehicles. *Sensors* **2021**, *21*, 3572. [CrossRef] [PubMed]

12. Esch, E.D.; Horner, R.M.; Krompetz, D.C.; Moses-Gonzales, N.; Tesche, M.R.; Suckling, D.M. Operational parameters for the aerial release of sterile codling moths using an uncrewed aircraft system. *Insects* **2021**, *12*, 159. [CrossRef] [PubMed]

13. Lo, P.L.; Rogers, D.J.; Walker, J.T.S.; Abbott, B.H.; Vandervoet, T.F.; Kokeny, A.; Horner, R.M.; Suckling, D.M. Comparing Deliveries of Sterile Codling Moth (Lepidoptera: Tortricidae) by Two Types of Unmanned Aerial Systems and from the Ground. *J. Econ. Entomol.* **2021**, *114*, 1917–1926. [CrossRef] [PubMed]

14. OJEU. Regulation (EU) 2018/1139 of the European Parliament and of the Council of 4 July 2018 on common rules in the field of civil aviation and establishing a European Union Aviation Safety Agency, and amending Regulations (EC) No 2111/2005, (EC) No 1008/2008. *Off. J. Eur. Union L* **2018**, *212*, 1–122.

15. EASA. Acceptable Means of Compliance (AMC) and Guidance Material (GM) to Part-UAS. UAS Operations in the "Open" and "Specific" Categories. 2019. Available online: https://www.easa.europa.eu/sites/default/files/dfu/AMC%20%26%20GM%20to%20Part-UAS%20%E2%80%94%20Issue%201.pdf (accessed on 1 December 2021).

16. Martinez, C.; Sanchez-Cuevas, P.J.; Gerasimou, S.; Bera, A.; Olivares-Mendez, M.A. SORA Methodology for Multi-UAS Airframe Inspections in an Airport. *Drones* **2021**, *5*, 141. [CrossRef]

17. Janik, P.; Zawistowski, M.; Fellner, R.; Zawistowski, G. Unmanned Aircraft Systems Risk Assessment Based on SORA for First Responders and Disaster Management. *Appl. Sci.* **2021**, *11*, 5364. [CrossRef]

18. Capitán, C.; Capitán, J.; Castaño, Á.R.; Ollero, A. Risk Assessment based on SORA Methodology for a UAS Media Production Application. In Proceedings of the 2019 International Conference on Unmanned Aircraft Systems (ICUAS), Atlanta, GA, USA, 11–14 June 2019; pp. 451–459.

19. Miles, T.; Suarez, B.; Kunzi, F.; Jackson, R. SORA Application to Large RPAS Flight Plans. In Proceedings of the 2019 IEEE/AIAA 38th Digital Avionics Systems Conference (DASC), San Diego, CA, USA, 8–12 September 2019.

20. OJEU. Regulation (EC) No 216/2008 of the European Parliament and of the Council of 20 February 2008 on common rules in the field of civil aviation and establishing a European Aviation Safety Agency, and repealing Council Directive 91/670/EEC, Regulation (EC) No. *Off. J. Eur. Union L* **2008**, *79*, 1–49.

21. OJEU. Commission Delegated Regulation (EU) 2019/945 of 12 March 2019 on unmanned aircraft systems and on third-country operators of unmanned aircraft systems. *Off. J. Eur. Union L* **2019**, *152*, 1–40.

22. OJEU. Commission Implementing Regulation (EU) 2019/947 of 24 May 2019 on the rules and procedures for the operation of unmanned aircraft. *Off. J. Eur. Union L* **2019**, *152*, 45–71.

23. Agency EASA. Opinion No 01/2018: Introduction of a Regulatory Framework for the Operation of Unmanned Aircraft Systems in the 'Open' and 'Specific' Categories. 2018. Available online: https://www.easa.europa.eu/sites/default/files/dfu/Opinion%20No%2001-2018.pdf (accessed on 1 December 2021).

24. JARUS. JARUS Guidelines on Specific Operations Risk Assessment (SORA). 2019. Available online: http://jarus-rpas.org/sites/jarus-rpas.org/files/jar_doc_06_jarus_sora_v2.0.pdf (accessed on 1 December 2021).

25. SESAR JU. European Drones Outlook Study Unlocking the Value for Europe. 2016. Available online: https://www.sesarju.eu/sites/default/files/documents/reports/European_Drones_Outlook_Study_2016.pdf (accessed on 1 December 2021).

26. SESAR JU. U-Space Concept of Operations. 2019. Available online: https://www.sesarju.eu/sites/default/files/documents/u-space/CORUS%20ConOps%20vol2.pdf (accessed on 1 December 2021).

27. FAA. Unmanned Aircraft System (UAS) Traffic Management (UTM) Concept of Operations. 2020. Available online: https://www.faa.gov/uas/research_development/traffic_management/media/UTM_ConOps_v2.pdf (accessed on 1 December 2021).

28. GUTMA. UAS Traffic Management Architecture. 2017. Available online: https://www.gutma.org/docs/Global_UTM_Architecture_V1.pdf (accessed on 1 December 2021).

29. SESAR JU. U-Space Blueprint. 2017. Available online: https://www.sesarju.eu/sites/default/files/documents/reports/U-space%20Blueprint%20brochure%20final.PDF (accessed on 1 December 2021).

30. EASA. Opinion 01/2020: High-Level Regulatory Framework for the U-Space. 2020. Available online: https://www.easa.europa.eu/sites/default/files/dfu/Opinion%20No%2001-2020.pdf (accessed on 1 December 2021).

MDPI

St. Alban-Anlage 66

4052 Basel

Switzerland

Tel. +41 61 683 77 34

Fax +41 61 302 89 18

www.mdpi.com

Applied Sciences Editorial Office

E-mail: applsci@mdpi.com

www.mdpi.com/journal/applsci